PRINCIPLES AND PROOFS

Richard D. McKirahan, Jr.

PRINCIPLES AND PROOFS

Aristotle's Theory of Demonstrative Science

Princeton University Press

Princeton, New Jersey

Published by Princeton University Press, 41 William Street,
Princeton, New Jersey 08540
In the United Kingdom: Princeton University Press, Oxford

Library of Congress Cataloging-in-Publication Data

McKirahan, Richard D.
Principles and proofs : Aristotle's theory of demonstrative science / Richard D. McKirahan, Jr.
p. cm.
Includes bibliographical references and index.
ISBN 0-691-07363-5
1. Aristotle. Posterior analytics. 2. Aristotle—Contributions in concept of demonstrative science. 3. Science—Philosophy. 4. Knowledge, Theory of. 5. Logic. 6. Metaphysics. I. Title.
B441.M525 1992
160—dc20 91-36774
CIP

This book has been composed in Linotron Times Roman
Princeton University Press books are printed on acid-free paper, and meet the guidelines for permanence and durability of the Committee on Production Guidelines for Book Longevity of the Council on Library Resources
Printed in the United States of America
10 9 8 7 6 5 4 3 2 1

TO MY PARENTS

CONTENTS

ACKNOWLEDGMENTS

THE PRESENT BOOK has had a lengthy gestation. It grew out of my 1973 doctoral dissertation, entitled *Aristotle's Theory of Demonstrative Science*, which I wrote under the supervision of the late Professor G. E. L. Owen. The research and writing of the dissertation were supported by graduate fellowships at Harvard University and a position as tutor at Kirkland House. Revision for publication began in summer 1975 when I was awarded a Summer Stipend by the National Endowment for the Humanities. The next lengthy stretch of work was carried out at Cambridge University in spring and summer 1979. On this occasion I was supported by the generous sabbatical policy of Pomona College and had the honor of being associated with St. John's College, Cambridge. Spring 1983, the occasion of my next sabbatical leave, I spent at the American School of Classical Studies at Athens, and during that spring and summer I was able to complete a draft of the entire book. During summer 1985 at a National Endowment for the Humanities Summer Seminar at Yale University, I was able to spend much of my time revising the manuscript. Professor G. E. R. Lloyd, then of King's College, Cambridge, and Professor Ian Mueller, of the University of Chicago, had the kindness to read and make extensive comments on this draft; their criticisms were the basis of further revisions. The manuscript was submitted to Princeton University Press in 1988 and accepted in early 1989. On the basis of anonymous readers' reports I spent the year 1990, especially the summer, again at the American School of Classical Studies, supported by a second N.E.H. Summer Stipend, making extensive final revisions.

The people for whose assistance I am grateful are too many to recount here. My deepest debt is to Professor G. E. L. Owen, who was the model thesis-supervisor both in the advice he offered and in the scholarly ideal he embodied. The book has benefited from the careful discussions of the *Posterior Analytics* by the West Coast Aristotelian Society in which I participated, and from the opportunity afforded by the Department of Classics at the University of California at Berkeley to teach a graduate seminar on my work. I also wish to thank Stanford University, the Society for Ancient Greek Philosophy, the "B" Club, the University of California at Berkeley, and Pomona College for inviting me to read papers on this subject. Finally I want to express my gratitude to my wife, Voula, whose support as well as scholarly advice has enabled me to bring the project to completion.

ABBREVIATIONS

THE FOLLOWING ABBREVIATIONS are used for works of Aristotle:

APo = *Posterior Analytics*
APr = *Prior Analytics*
Cat. = *Categories*
De An. = *De Anima*
De Int. = *De Interpretatione*
De MXG = *On Menexenus, Xenophanes, and Gorgias*
EN = *Nicomachean Ethics*
G&C = *De Generatione et Corruptione*
Gen An. = *De Generatione Animalium*
HA = *Historia Animalium*
Met. = *Metaphysics*
Meteor. = *Meteorologica*
PA = *De Partibus Animalium*
Phys. = *Physics*
Probl. = *Problems*
SE = *Sophistici Elenchi*
Top. = *Topics*

The following abbreviations are used for other works:

Barnes = Barnes 1975a
Bonitz = Bonitz 1870
DK = Diels and Kranz 1956. W. K. C. Guthrie's translations of DK fragments are so identified, and are taken from Guthrie 1971–81; all other DK translations are mine.

D.L. = Diogenes Laertius, *Vitae Philosophorum*
LSJ = Liddell, Scott, and Jones 1968
Mure = Mure 1928
Philoponus, *In An. Post.* = Philoponus 1909
Proclus, *Comm.* = Friedlein 1873
Ross = Ross 1949
Simplicius, *In Phys.* = Simplicius 1882–95
Themistius, *In Post. An.* = Themistius 1900

PRINCIPLES AND PROOFS

INTRODUCTION

THE INTEREST and importance of the *Posterior Analytics* lie concealed behind a thicket of difficulties—not only those found in most of Aristotle's works but also some unique to this one. As usual, we find obscure arguments, unclear transitions and cross-references, and inadequate examples, all of which place stringent demands on the reader. But in addition the work seems rough and unfinished, a series of jottings on different aspects of its subject matter. Regardless of the view we adopt on the time and manner of its composition, it does not read like a finished treatise, making definitive statements, but like an inchoate collection of thoughts, some worked out more thoroughly than others. The first task for interpreters is therefore reconstructive—even constructive. They must begin from the general sketch of the theory in the first chapters of the work and on that basis piece together Aristotle's disjointed remarks with the twin goals of making sense of each passage in its immediate context and fitting all the passages together to form a coherent whole. To take Aristotle's points in isolation, without regard for their place in the overall theory, risks distorting the author's intentions and can even sink to the level of taking cheap shots.

The subject of *APo* is *epistēmē*: science or scientific knowledge. As the first elaborated theory in the Western philosophical and scientific traditions of the nature and structure of science, it richly deserves study for its historical importance alone. In addition, the philosophical content of the theory is extremely interesting, since it embodies a fascinating vision of the structure of knowledge and objective reality, which is well worth investigating. Even if we do not share Aristotle's vision, the effort to understand his theory as a whole may lead to healthy critical reflection on our own outlook.

Aristotle's theory can be stated briefly. Each science (including branches of natural science and mathematics, and perhaps other fields as well) treats a limited range of things or phenomena, its subject genus, as arithmetic studies numbers and geometry spatial magnitudes. The subject genus can be regarded as a structured collection of subjects and attributes in certain relations. For example, points, lines, triangles, and figures are subjects in geometry; attributes include intersecting and having angles equal to two right angles; and these items are related in such ways as these: lines intersect in points, triangles have angles equal to two right angles, triangles are figures, and triangles are composed of lines. The task of a science is to explicate the network of relations in its subject genus.

A science proceeds by proving conclusions. Using unprovable principles as a basis, it forms proofs of their consequences. The relations in its subject genus are the body of unprovable principles and provable conclusions that constitute the science. Demonstrations are also explanations. To prove *that* a conclusion holds is simultaneously to show *why* it holds. Thus, proofs do more than indicate logical relations among propositions; they also reveal real relations among facts, and scientific knowledge involves knowing not only that but also why. Further, science deals primarily in the universal and necessary, not the particular and contingent. It treats individuals not in their own right but as falling under universals, and the facts (relations) it studies always hold, of necessity, and in all cases.

A demonstration of a fact is a syllogistic argument with that fact as conclusion and unprovable principles as premises. It is arguable that Aristotle developed his theory of the syllogism as a tool for science. Since the syllogistic requires premises and conclusions to be in one of the four forms—"All *S* is *P*," "No *S* is *P*," "Some *S* is *P*," and "Some *S* is not *P*"—the distinction between subjects and predicates (attributes) is important for the theory of *APo*. Another crucial distinction is that between principles and conclusions. Conclusions are logical (syllogistic) consequences of the principles. In fact, all such consequences are provable facts treated by the science in question. Thus, the principles determine the entire content of the science.

In a sense, then, demonstrative sciences consist of principles and proofs, so that in order to understand *APo* we must get clear on both. Regarding the former, we need to know why Aristotle believed there are indemonstrable principles, and what their role in sciences is. We also need to learn what types of principles there are, how each is related to the science's subject genus, what form each has, and what distinctive contribution each makes in demonstrations. In addition, we must consider whether the kinds of principles identified are (or whether Aristotle had reason to think they could be) an adequate basis for science; and finally, we must examine what he says about how we gain knowledge of principles and what sort of knowledge we have of them, since even if the theory is otherwise attractive, failure on these counts will render it useless. Regarding proofs, we must investigate how they work—how the principles work together to entail the conclusions and whether demonstrations yield scientific knowledge as Aristotle conceives it. We must also ascertain what types of proofs sciences contain and how each proceeds. We must explore how proofs constitute explanations of their conclusions, and what kind of explanation is intended. Finally, we must survey the information Aristotle provides on how to construct proofs, to see if there is a method for organizing a body of facts into a demonstrative science.

APo provides material on all these topics, but in many cases fails to discuss them clearly, definitively, or in a single place, so that the only hope of making coherent sense is to "massage" the text, gradually building up an overall picture by assembling material on one topic while constantly bearing others in mind. Discussion of some topics is facilitated by bringing passages from other Aristotelian works. To a certain extent this method is justified by occasional comments from works as early as *Topics* and as late as *De Anima*, which suggest that, whenever Aristotle composed *APo* (I follow the standard view that it is a fairly early work), the idea of a demonstrative science was with him for a very long time.

The general and abstract level of discourse in *APo* renders it extremely difficult to grasp what a demonstrative science looks like, and in particular how demonstrations proceed. If Aristotle had composed a single scientific treatise on the lines indicated in *APo*, many uncertainties would be resolved. Euclid's *Elements* offers a great deal of help. With appropriate qualifications the principles of the *Elements* correspond to the principles Aristotle recognizes for sciences, and Euclidean proofs to Aristotelian demonstrations. (Euclid was quite possibly influenced by *APo*.)

In general, the present treatment of Aristotle's theory is synthetic and sympathetic. Its goal is to recapture Aristotle's vision of the nature of science and scientific knowledge, not to dismantle a text. Although *APo* has shortcomings, I believe it is possible and worthwhile to reassemble it, with help from outside sources, as indicated above. The result might be more accurately called an Aristotelian theory than Aristotle's theory, since the fact that it must be based in part on texts over and above *APo* indicates that Aristotle himself never put together the theory exactly as presented here. In some places it also seems appropriate to point out improvements possible within the overall framework of the theory. This practice brings us yet further from *APo*, but it may help reveal the philosophical potential of Aristotle's insight.

The plan of the present work is as follows. The scientific background of *APo* is surveyed in Chapter 1, in order to provide an intellectual context for the work and to indicate a possible motive for Aristotle to develop a model for science. Chapter 2 examines the general statement of the nature of demonstrative science and the list of criteria for scientific principles in *APo* I.1–2. Chapter 3 establishes that *APo* recognizes three kinds of scientific principles: definitions, existence claims, and axioms. The nature of the subject genus and an apparent exception to the rule that distinct sciences have distinct subject genera occupy Chapters 4 and 5. Next comes a systematic treatment of the kinds of principles—axioms, in Chapter 6; definitions, in Chapters 7 through 9; and existence claims, in Chapter 10—followed by consideration of the question whether the kinds of principles indicated are an adequate basis for a science, in Chapter 11. This last issue is taken up in connection with Euclid's *Elements*, and leads to an investigation of the unproved principles on which Euclid's geometry is based and their relation to Aristotle's principles. The next chapters take up issues connected with demonstrations. Chapter 12 considers, in general terms, how demonstrations proceed—what kind of things can be proved and how the various kinds of principles function. Once again the *Elements* provides help. The three types of demonstrations recognized in *APo* are then treated in Chapters 13 through 15, and the syllogism of essence, sometimes thought to be another kind of proof, is analyzed in Chapter 16. The final issue addressed in connection with demonstrations is the nature of scientific explanation, covered in Chapter 17. The last chapter of the book brings together passages from several Aristotelian works to supplement the material in *APo* on the topics of how we come to know scientific principles and how we organize facts to form demonstrative sciences.

My strategy in this book is to base my interpretation of Aristotle's theory of demonstrative science on a close reading of relevant passages in *APo* and other works of Aristotle. The not infrequent obscurity of the Greek and the existence of conflicting translations and interpretations of the text have at times forced me to proceed at a slow pace. Except in a few places, I have made my own translations, intending them to be as literal as possible, taking account of the fact that they are meant to be translations into English. Aristotle's Greek is highly stenographic; where a literal English rendering would be unintelligible, I have supplied the missing words in square brackets. Where the text or translation is obscure or controversial, I have offered justifications, defenses, and apologies in a section labeled "Notes on Translation" immediately following the translation. Where a translated passage as a whole requires interpretation, I have offered one in a section labeled "Discussion." The result is a book that in places resembles a commentary. Readers who want to follow my reconstruction of Aristotle's theory without wading through this material are invited simply to skip over these sections.

Although I have tried as far as possible to make the book serviceable for the Greekless, I have had to include some Greek, either because of difficulties in finding satisfactory and uncontroversial English equivalents (as in the case of, for example, *noûs*, *epagogḗ*, and *gnṓrimos*, all discussed in Chapter 18 below) or because of controversies about what the Greek text actually is or how it should be translated. I have adopted the following conventions in transliteration. Epsilon appears as *e*, eta as *ē*, omicron as *o*, omega as *ō*, upsilon as *u*. Iota subscript is written adscript. The rough breathing appears as *h*. In compound words formed by prefixing a preposition to a word beginning with a rough breathing, I have retained the *h* (for example, *dihóti*). Accents are written as they appear in Greek. The resulting transliteration may seem cluttered, but it more accurately reflects the original. It also has the advantage of distinguishing between *ḗ* ("or") and *hêi* ("qua"), both of which words are crucially important for understanding certain aspects of Aristotle's theory.

The Motivation for the Theory of Demonstration

CHAPTER I

WHY DID Aristotle invent the theory of demonstrative science? What is the theory's purpose, and what features of Aristotle's intellectual heritage and of fourth century philosophy and science lie behind his enterprise? Although thorough treatment of these questions would require a long discussion of the methods and achievements of Aristotle's predecessors and contemporaries, which would be out of place here,[1] a brief discussion of some of the issues will help set the scene for Aristotle's theory.

Even a rapid reading of the remains of the Presocratic "philosophers" reveals that these men pursued what we would call science at least as much as what we would call philosophy, though since no distinction had been made between the two areas of inquiry, it is unfaithful to the Presocratics to isolate parts of their work as scientific, however necessary this practice may be for modern discussions, including the present one. It is also apparent that the Presocratics discuss an extremely wide range of topics—topics that from a modern point of view would be said to fall within the fields of astronomy, biology, chemistry, geology, meteorology, physics, physical anthropology, and other sciences as well. No attempt was made to distinguish these fields by subject matter or methods of inquiry, and there was no conception that such differentiation was possible, let alone beneficial.

A third striking feature of early Greek science is the multitude of competing systems, reflecting a wide divergence in the principles on which the theories are based. The lack of agreement amounts to something close to anarchy. This situation too contrasts markedly with today's practice, where in all but exceptional cases, scientists in a given field are agreed on the basic entities and the range of phenomena they discuss and the kinds of explanations they accept.

Fourth, a typical Presocratic system accounts for a very wide field of phenomena by means of a very small number of principles, which are declared to apply through all of nature. The resulting economy in admissible types of explanation is purchased at a price. Theories are vague and imprecise in their application and of little or no use in predicting phenomena. As a consequence they are difficult or impossible to confirm or refute.

Fifth, there is no consensus on the phenomena to be explained. To be sure, some phenomena (especially astronomical and meteorological events) are treated by most Presocratics if not all, but different theories are better suited to explaining different kinds of phenomena, and each unsurprisingly concentrates on the phenomena it explains best.

Sixth, the Presocratics rely on familiar facts of ordinary experience and only rarely extend the range of phenomena by close observation or experiment. They use evidence to corroborate rather than test their theories or decide among rival theories.

These are only some characteristics of Presocratic science. They suggest a conclusion borne out by the surviving writings of the Presocratics themselves: that after its promising start in sixth-century Ionia, Presocratic science floundered. The century between the Milesians and the post-Parmenidean systems of the late fifth century brings important differences. But these are clearest in the area of foundations, since the late-fifth-century natural philosophers constructed their systems with eyes on issues raised by the Eleatics. In the range of phenomena discussed, the detail in which they are treated, and the inherent plausibility of the explanations offered, this period achieved little. Anaximander's theory that the phenomena of the world are due to interplay of opposites as they "make just recompense to one another for their injustice according to the ordinance of time"[2] is no different in spirit from Empedocles' "They never cease thus to alternate continually, now all coming together into one through love, and now again drawing apart by strife's hatred."[3] Likewise Xenophanes' statement that god "always remains in the same place, not moving at all, nor indeed does it befit him to go here and there at different times, but without toil he makes all things shiver by the impulse of his mind"[4] does not sound very different from Anaxagoras's pronouncement that Mind "is the finest and purest of all things, and has all judgment of everything and greatest power; and everything that has life, both greater and smaller, all these Mind controls; and it controlled the whole revolution, to make it revolve in the beginning."[5]

On the other hand the anarchy is not absolute. Despite the disagreement, there is a fundamental unanimity of purpose. To judge from what is known about the early theories, a principal goal of the Presocratics from Anaximander (if not Thales) through the atomists was to understand how the universe is structured, how it functions, its history and its future, and in particular, with what kinds of things it is populated, how they behave, and what their basic constituents are.[6]

Further, their approaches are similar. They present their views in an account or story, which sets out from an appropriately chosen starting point (such as the Beginning) and proceeds to relate how the world came to be as it is—simultaneously telling *how* it is. We may infer that the criteria for a good account include economy, consistency, and plausibility of principles and of the details of development, and a reasonable fit between what their theory says is the case and what is observed to be the case.

After the earliest theories, the Eleatic attack on the world of appearances forced the later natural philosophers to attend to the philosophical foundations of their subject and to wrestle with such issues as the nature of motion, change, plurality, coming-to-be and passing-away, and the conditions that account for them. The work of the later Presocratics in these areas, which constitutes the most significant scientific advance of the later fifth century, was evidently undertaken because accounts of the world unable to cope with the problems raised by the Eleatics were no longer plausible. Standards of plausibility became higher as people became more philosophically sophisticated.

There is also agreement about types and sources of evidence: mainly facts familiar from everyday experience; only occasionally material drawn from specially conducted observation, and even here specialized equipment was unknown and all observations could easily be made by anyone in the right place. Empedocles' use of the clepsydra[7] and Xenophanes' observation of fossils on Paros, Malta, and in the quarries of Syracuse[8] are typical of the "specialized observation" of the Presocratics. Also explanations are based on familiar principles: substances such as earth, water, air, and fire, and familiar occurrences, such as the conflict of opposites, condensation and rarefaction, separation and mixture, love and strife.

This agreement marks Presocratic philosophy as different from anything Greece and earlier cultures of the Eastern Mediterranean and Middle East, including Egypt and Babylonia, had produced. The Presocratics have much in common, but their common ground leaves considerable room for differences. The persistence of the diversity needs to be understood, both its causes and its effects.

A multiplicity of competing theories is characteristic of an immature science whose practitioners have yet to agree on a basic framework. The history of later science would lead us to expect the initial diversity of theories to be followed by a period of consolidation when agreement on principles and on modes of explanation and the range of explananda would be reached. Then would come rapid progress within that framework. Instead, Greek science shows increasing sensitivity to philosophical issues, a concern for foundations largely due to Eleatic influence. But even if the later Presocratics are agreed on the need to account for plurality, change, and the other phenomena challenged by Parmenides and his followers, so that there is a consensus on subject matter and on what counts as a successful theory, still, the differences among the three great post-Parmenidean systems, those of Empedocles, Anaxagoras, and the Atomists, show that the tolerance of diversity is still very wide.

The reasons for the failure to reach this level of maturity in the fifth century are of two kinds, which can be labeled conceptual and cultural. The conceptual reasons reflect science's infant state. Scientists did not agree on a single theory because no one theory proved indisputably superior to the others. They were content with familiar explanations and data because they were unaware of other approaches. They did not specialize in their work and gain the advantages of specializing because they had no reason to suppose that their inquiries should be differentiated.

These conceptual reasons for science's slow progress are also its effects. Since none of the competing theories had advanced far it was unclear which, if any, to follow, or even if knowledge of the world could advance beyond the level it had reached. Also, since no branch of natural science had progressed significantly beyond any other, there was no model of success to guide the rest. Since specialization had not begun, there was no notion of the advantages it brings. Nor was there a notion of technical vocabulary or distinctively scientific procedures of reasoning and fact gathering, and no reason to suspect that familiar data and kinds of explanations are inadequate. In their ignorance of a better way, the Presocratics pursued the way they knew.

Moreover, certain features of the cultural milieu of fifth-century Greece forced scientific discourse to remain intelligible to the general public. This is a highly relevant consideration since only when a science passes that point, when it develops specialized methods, language, and problems, can real scientific progress be made.[9] Paradoxically, the same features of Greek life that inhibited science also promoted moral and political thought and contributed to the fifth-century enlightenment.

Many of the philosophers and scientists of the sixth and fifth centuries were prominent citizens of their states, or known for nonphilosophical and nonscientific achievements. Thales was a respected political adviser and a "practical genius and man of affairs,"[10] Pythagoras became the leading citizen of Croton, Parmenides was a lawgiver at Elea, Melissus a Samian politician and admiral, Empedocles a champion of democracy in Acragas, and Anaxagoras the associate of Pericles. Unlike modern scientists the Presocratics (with the exception of Democritus) probably did not devote much of their lives either to thinking or to writing about nature.

Further, there is reason to believe that most of these thinkers deliberately made their views accessible to the public. Too little survives of the Milesians to tell one way or the other, though the striking language of Anaximander's surviving fragment may be intended to stick in people's minds. Xenophanes, however, is a good case in point since he may have made his living as a bard,[11] giving recitals that included, we may suppose, poems on scientific and philosophical subjects. As a public performer he had every incentive to keep his ideas easily comprehensible.

Xenophanes is our only bard, but Parmenides and Empedocles probably wrote in verse at least partly from the desire to make their thoughts memorable. Admittedly, Parmenides' subject matter and approach are not the stuff of great poetry.[12] But a comparison of Parmenides' style with that of Gorgias's prose treatment of a closely related subject in his "On What Is Not, or On Nature"[13] shows the advantages of verse even here.[14]

In addition, these philosophers may be imagined to share Lucretius's stated reasons for writing philosophy in verse:

> Since this philosophy full often seems too bitter to those who have not tasted it, and the multitude shrinks back away from it, I have desired to set forth to you my reasoning in the sweet-tongued song of the Muses, and as though to touch it with the pleasant honey of poetry, if perchance I might avail by such means to keep your mind set upon my verses, while you come to see the whole nature of things, what is its shape and figure. (Lucretius I. 943–50, trans. Bailey 1947)

Only a few Presocratics expounded their theories in verse, but all needed to keep their ideas accessible to the public. For in the sixth and fifth centuries, the specialized scientific community was virtually unknown. Two apparent exceptions are the Pythagoreans and the Atomists. But in their earliest years the Pythagoreans were not engaged in scientific work.[15] Their main contributions are to mathematics, not natural science, and so are irrelevant to the present discussion.[16] Their physical theories mark no improvement on other contemporary systems. The Atomists, especially Democritus, are more likely to be a real exception. To judge from the catalog of his books,[17] his output exceeded that of the other Presocratics by far. He seems to have been the first to expound in detail the fundamentals of his system and its systematic application to a wide variety of phenomena. However, two things should be remembered. First, Democritus lived thirty or forty years into the fourth century. Second, though Diogenes Laertius[18] asserts that Democritus had many followers, including Nausiphanes, the teacher of Epicurus, there are no signs of a specialized atomistic "school" before Epicurus.

Thus, any audience for scientific works was of necessity a public audience, and we find indications that the Presocratics presented their views publicly.[19] For Parmenides[20] the evidence is indirect but significant. Not only did he write in verse, but we also have testimony that his views were ridiculed for their absurd and contradictory consequences.[21] Even though we do not know who the hecklers were, it is clear that Parmenides made his views known outside the circle of his followers.

With Zeno the picture is more definite. Whatever we make of the story that his book was published against his wishes in a pirated edition,[22] when he and Parmenides came to Athens for the Great Panathenaea, he read his treatise to Socrates and many others. He may have presented it more than once, since his host Pythodorus had previously heard it.[23] Pericles also heard it through,[24] perhaps on the same visit. Moreover, Zeno claims his book attacked Parmenides' hecklers by showing that even greater absurdities follow from their belief in plurality than attended Parmenides' monism. Also, presumably because of his arguments against plurality, motion, and the validity of the senses, Zeno was called the father of dialectic[25] and gained a reputation for developing a way to refute people by involving them in self-contradiction and thus reducing them to perplexity.[26] A final piece of evidence that Zeno aimed his book at a general audience is the statement that he produced no fewer than forty arguments for monism.[27] This is the sort of arsenal a public speaker might assemble to drive home his point.

What were the occasions for such performances, and how public were they? We do not know if Parmenides and Zeno were Speakers' Corner orators or presented their work to relatively small and select groups. But some philosophers and scientists delivered their views at the Olympic festival, where they reached the largest possible audience. Empedocles delivered his *Katharmoi* there,[28] and Hippias, whose polymathy included astronomy and mathematics,[29] regularly attended the Olympics and offered to answer any question on any subject.[30] A passage from Lucian reveals that the practice of going to the Olympic festival to make known one's work was common in the fifth century:

> As soon as he [Herodotus] had sailed from his Carian home for Greece, he concentrated his thoughts on the quickest and easiest method of winning a brilliant reputation for himself and his works. He might have gone the round, and read them successively at Athens, Corinth, Argos, and Sparta; but that would be a long toilsome business, he thought, with no end to it; . . . his idea was, if possible, to catch all Greece together. The great Olympic Games were at hand, and Herodotus bethought him that here was the very occasion on which his heart was set. He seized the moment when the gathering was at its fullest, and every city had sent the flower of its citizens; then he appeared in the temple hall, bent not on sight-seeing, but on bidding for an Olympic victory of his own; he recited his *Histories*, and bewitched his hearers. . . . That was what he made out of his *Histories*; a single meeting sufficed, and he had the general unanimous acclamation of all Greece; his name was proclaimed, not by a single herald; every spectator did that for him, each in his own way. The royal road to fame was now discovered; it was the regular practice of many afterwards to deliver their discourses at the festival; Hippias the rhetorician was on his own ground there; but Prodicus came from Ceos, Anaximenes from Chios, Polus from Agrigentum; and a rapid fame it brought, to them and many others. (Herodotus and Aëtion, 1–3, trans. Fowler and Fowler 1905)[31]

A common feature of major festivals may have been scientists and philosophers propounding their latest theories in *epideixeis*, display pieces, and receiving either applause or, as Zeno's words in the *Parmenides* suggest, mockery. It may be no coincidence that Parmenides and Zeno visited Athens during the quadrennial Great Panathenaea. A different kind of display at the Olympics gained fame for Anaxagoras, who was known for his predictions of natural phenomena such as the fall of a meteorite at Aegospotami in 467 B.C.[32] He wore a sheepskin into the stadium after predicting rain, even though the Olympic festival fell during the dryest part of the year, and sure enough, rain began to fall.[33] Also the astronomer and mathematician Oenopides dedicated at the Olympic festival a bronze tablet inscribed with his astronomical calculations.[34] Public presentations were not reserved for special occasions. Hippias boasted in the Athenian agora of the range of his wisdom[35] and Anaxagoras spoke frequently of his scientific views.[36]

The best evidence for the public nature of science is Gorgias's display piece, *Encomium of Helen*, which argues that persuasive speech can affect the mind however it wishes (sec. 13). Two pieces of evidence are these statements: "the words of the natural scientists who replace one belief with another and make the unbelievable and unclear apparent to the eyes of opinion" and "philosophical debates that show how easy it is for the quickness of our judgment to alter our confidence in our opinions." Gorgias seems to be speaking of phenomena known to his readers, who we may assume were familiar with philosophical debates and presentations of scientific theories. Moreover, since in Gorgias's time there was no clear distinction between science and philosophy, the range of subjects covered by "philosophical debates" would have included scientific topics. Whether or not there were competitions with prizes,[37] Hippias's boast that he had never been bested at anything by anyone at the Olympics or elsewhere suggests a public competitive setting in which to declaim or debate.

It was virtually inevitable that fifth-century Greek science would be presented in a public forum, given the nature of Greek society and the prominence social and political developments of the fifth century gave to public speaking and debate, and particularly given the absence of a scientific community to serve as a specialized audience. Moreover, remaining in the public domain might bring some advantages. No doubt scientists wished to present their theories in public to promote the truth they claimed to have found. The desire to disseminate new knowledge may have arisen from disinterested motives of benevolence and love of wisdom, but concerns of ego and self-interest were often present too.

Thus Xenophanes may have earned his living by singing his poems. Hippias, who made money by teaching, would have been eager to appear in public at least partly in hopes of attracting new students. No doubt many spoke in public for the reason Herodotus did. The desire for fame was a strong incentive for the egotistic Hippias to speak in public. It may have been one of Empedocles' motives for reciting his *Katharmoi* at the Olympic festival and may also have prompted Pythagoras, the other charismatic figure among the Presocratics. Some may have hoped to acquire followers by public presentations: Pythagoras again, Parmenides, and perhaps the early Milesians and early Atomists. A final benefit that might come from the fame gained by public speaking was proximity to the politically powerful. Anaxagoras, we are told by Plato,[38] became a close associate of Pericles because the latter was impressed with his scientific views, "on which Anaxagoras was continually discoursing."[39] In some cases the goal may have been political power itself: Parmenides the lawgiver and Pythagoras.[40]

Virtually the only major Presocratic not mentioned above is Heraclitus. Even that isolated figure may have attempted to teach his difficult doctrines to his fellow citizens,[41] but his surviving fragments show that his teaching was not readily intelligible. His isolated position among the Presocratics and the little influence his ideas exerted, especially his scientific ideas, in effect make him the exception that proves the rule.

Fifth-century science was in an important sense publicly accessible, and the powerful reasons why it was so contributed to its failure to advance more rapidly. The price of remaining intelligible to the general public was heavy. It meant not creating a technical vocabulary, not developing unfamiliar techniques of argument and evidence, not working on abstruse problems. To a large degree it meant not specializing, since the ability to treat (however superficially) a broad range of problems would seem more impressive than the ability to solve (however thoroughly and perhaps tediously) a narrow range. More generally, the time and intellectual effort expended in preparing and delivering public presentations would have distracted scientists from scientific endeavors. And the nature of the public debating situation encouraged participants to attack their opponents rather than look critically at their own theories.[42] These circumstances impede progress in any area of inquiry. And the situation was far worse in the fifth century, before the possibility of such progress had been conceived.

Contrast today's practice, where scientists present their contributions to their professional peers, normally only those in their own specialty, or even those concerned with the same range of specific problems.[43] Scientific papers are normally unintelligible to the layperson, and certainly unclear, if not wholly unintelligible, to scientists not in the particular field. When one's audience is expert there is no need to start from the beginning each time. An agreed-upon body of facts and theories, a common vocabulary, and shared ideas of the standards and criteria of success mark off specialist from layperson. And since they can take these for granted, scientists have no need to go over elementary material and debate about the foundations of their subject each time.

Return now to Greece. With no body of experts to appeal to and no common vocabulary, theories, or standards and criteria of success, each time a scientist presented his theories, not only his latest results but also his earlier results and the entire basis of his science were subject to public scrutiny. The most acceptable theories would have been the most understandable; hence the need for familiar material. Further, with no common framework of concepts, methods, and standards of explanation and accuracy, that is to say, no agreed-upon basis, so that scientists were free to work from different principles, discussion can hardly be expected to have advanced beyond foundations and elementary results.

Kuhn calls this state of science the "pre-paradigm" stage. Although he does not discuss our period, some of his comments on physical optics before Newton could have been written about fifth-century Greek science. For example:

> Though the field's practitioners were scientists, the net result of their activity was something less than science. Being able to take no common body of belief for granted, each writer on physical optics felt forced to build his field anew from its foundations. In doing so, his choice of supporting observation and experiment was relatively

> free, for there was no standard set of methods or of phenomena that every optical writer felt forced to employ and explain. Under these circumstances, the dialogue of the resulting books was often directed as much to the members of other schools as it was to nature.[44]

He makes the following observations about his main example, the history of electrical research in the early eighteenth century:

> History suggests that the road to a firm research consensus is extraordinarily arduous. History also suggests, however, some reasons for the difficulties encountered on that road. In the absence of a paradigm or some candidate for a paradigm, all of the facts that could possibly pertain to the development of a given science are likely to seem equally relevant. As a result, early fact-gathering is a far more nearly random activity than the one that subsequent scientific development makes familiar. Furthermore, in the absence of a reason for seeking some particular form of more recondite information, early fact-gathering is usually restricted to the wealth of data that lie at hand. . . . Only very occasionally . . . do facts collected with so little guidance from pre-established theory speak with sufficient clarity to permit the emergence of a first paradigm. This is the situation that creates the schools characteristic of the early stages of a science's development. No natural history can be interpreted in the absence of at least some implicit body of intertwined theoretical and methodological belief that permits selection, evaluation, and criticism. . . . No wonder, then, that, in the early stages of the development of any science, different men in confronting the same range of phenomena, but not usually all the same particular phenomena, describe and interpret them in different ways.

How much wider still the realm of facts and interpretations when sciences are not differentiated and the whole universe, all of *phusis*, can supply one's materials! With so vast a field one could not know where to begin, how closely to look at one phenomenon before examining the next, or even how to look at a phenomenon "closely." Even supposing the technology adequate, there was little point in constructing specialized apparatus to wring new data from what was at hand, since it was not clear what data would be useful; and besides, the problem was not a lack of data but of consensus on how to treat the data at hand.

Adoption of a paradigm produces a radical transformation in scientific activity. For one thing, progress is quicker. The paradigm becomes a criterion for identifying problems that scientists have every reason to suppose they will be able to solve. They become more successful at solving them and their progress is more noticeable. They no longer need to choose problems that appear important to laypersons. Since the paradigm involves technical vocabulary, techniques for solving problems, and standards of success, scientists become specialists, isolated (as regards their work) from the rest of humanity in that they are free to choose esoteric problems and have their own criteria for successful solutions. Moreover, they can address their results to a specialist audience without concern for challenges from rival schools on fundamentals or the need to defend their results and methods before a lay audience.[45]

My suggestion is that fifth-century Greek natural science failed to make significant progress because it was in the pre-paradigm stage. Since *no* natural science had a paradigm, scientists did not know that anything was missing. Moreover, the public setting of scientific discussion with the lay public as arbiter presented further obstacles to securing the widespread agreement needed for a paradigm to be adopted. As long as science was a competitive, not a cooperative, pursuit, there could be little temptation for anyone to undertake the self-criticism needed to set science on a firmer footing.

Thus the techniques of rational argument that were crucial to the emergence of the Greek scientific tradition also contributed to the relative slowness of its progress. To advance, science needed a paradigm. Rather, it needed to be divided into a number of *sciences* and a paradigm needed to be found for each. For this to happen, reform had to come from outside, since natural science was in a rut from which it was not likely to free itself.

In fact, the remedy came from outside the realm of natural science, from mathematics, whose early history was quite different. The scarce source material for pre-Euclidean mathematics makes it certain that there was remarkable progress by the end of the fifth century, especially in geometry. It is sufficient to refer to two advances made by Hippocrates of Chios, whose activity is placed in the third quarter of the fifth century. His achievement of the quadrature of lunes,[46] is significant for present purposes not only for its sophistication in treating difficult problems, but particularly for the broad knowledge of geometry it represents. It presupposes knowledge of the geometry of angles, parallels, triangles (including the extension of the Pythagorean theorem to non-right triangles), polygons, circles (including polygons inscribed in circles), and similar figures. This material includes subjects treated in books I, II, III, IV, and VI of Euclid's *Elements*. Most strikingly it also requires the proposition that appears as *Elements*, book XII, theorem 2, whose proof in Euclid depends on the method of exhaustion, a sophisticated technique attributed to Eudoxus but undoubtedly anticipated by Hippocrates in his proof of the proposition.[47] In addition, by this time Pythagorean geometry knew material Euclid would present in books I, II, IV, and VI.[48]

It is clear that by the late fifth century the sum total of geometrical knowledge was considerable. Moreover, the first steps had been taken to organize it systematically. Hippocrates is called the first to compose a book of elements.[49] Although his book has disappeared along with all other pre-Euclidean books of elements, it is agreed that it had at least roughly the same organization as Euclid's work; that is to say, it depended on "a distinction between more and less fundamental propositions, and . . . the term 'elements' (*stoicheîon*) itself was used of primary principles or starting points of *some* kind—even though their nature may not have been clearly specified."[50] Hippocrates' work may have included most or all of the material mentioned above, although the theorems were not necessarily placed in the same order as in Euclid, the proofs may have been different and less rigorous, and the basic principles may not have had the same form as in Euclid. Also in number theory there is evidence that "implies the existence at least as early as the date of Archytas (about 430–365 B.C.), of an *Elements of Arithmetic* in the form which we call Euclidean; and no doubt text-books of the sort existed even before Archytas, which probably Archytas himself and others after him improved and developed in their turn."[51]

Whether mathematicians took their notion of proof from the dialectical practice of philosophy, especially that of the Eleatics,[52] or themselves first discovered that some geometrical and number-theoretical truths follow from others, and so invented the notion of proof, the idea of proving theorems was familiar before the end of the fifth century.

In the fourth century, books of elements multiplied, and the sort of arrangement we find in Euclid doubtless became standard. After remarking on Hippocrates, Eudemus mentions Leodamas, Archytas, and Theaetetus, contemporaries of Plato, who all added to the number of known theorems and to a more scientific arrangement of the subject. In addition, Leon (called a pupil of Neoclides, who was younger than Leodamas) "composed elements more carefully with regard to both the number and the usefulness of the propositions proved," Theudius of Magnesia "did a fine job of arranging the elements," and Hermotimus of Colophon "discovered many of the elements." Thus at least three men had composed books of elements by the mid–fourth century. The others Eudemus names may also have written books of elements, but even if they did not, their efforts were directed at least partly toward improving existing systematizations of geometry.[53]

Simultaneously with these advances in organization geometry was also experiencing unprecedented growth in new and more general and difficult results. In addition to Leodamas, Archytas, and Theaetetus, whose achievements were referred to above, Eudemus mentions that Neoclides, Leon, Eudoxus, Amyclas (or Amyntas) of Heraclea, Menaechmus, Dinostratus, Theudius, and Hermotimus as having advanced the subject and notes others who contributed as well. Still other names could be added, but the point is clear already. To judge by the number of mathematicians active at this time and their impressive results (one thinks above all of Eudoxus),[54] geometry was the outstanding example of a successful discipline.

Not the least reason for its success was its organization into systems of elements, from which several advantages accrued. First, interest in precise definitions and development of a technical vocabulary, both of which Aristotle attests,[55] minimized fruitless quibbling over meanings of statements, as well as enabling mathematicians to state theorems precisely and think more clearly and easily. Second, agreement on the form of elements would have fostered unanimity about the problems to tackle, the methods to employ in solving them, and what to count as successful solutions.

Third, the books of elements presented material mathematicians could take for granted, a common stock of knowledge on which to base further work. They may be regarded as textbooks, which serve the joint purposes of introducing the field to novices, familiarizing them with the language and methods of the subject, covering basic material that a person must master before going on to advanced work, displaying the range of the field by presenting a variety of topics, and presenting outstanding earlier work as models for students to emulate.[56] Euclid's *Elements* does not contain all the geometry known in his time and Theudius's book is called the "text-book of the time just preceding Euclid."[57] The distinction between textbook and research treatise may therefore go back to Hippocrates, and the subsequent editions of *Elements* presumably updated their predecessors at least partly by incorporating some new results.

Fourth, since mastery of the material covered in books of elements would be a minimum prerequisite for a person wishing to be a geometer, for the first time in a theoretical subject we can distinguish experts from laymen.[58] To be sure, many excellent geometers did not specialize in that field to the exclusion of all else; mastering the subject did not require so much. But once geometers could take a common background for granted they no longer had to worry about hecklers saying that a straightedge does not touch a circle at a point,[59] and no longer felt pressed to keep their work within the reach of the uneducated public. In effect, adopting such styles of elements as *the* proper form for mathematical textbooks and the type of proof they contained as *the* proper form of mathematical argument brought the advantages that result from adopting a paradigm. Mathematics matured more rapidly than natural science and reaped the benefits of its maturity.

An important reason why mathematical subjects flourished early is their abstract nature, which makes it relatively easy to identify items and properties to treat and which permits precise and significant results to be reached without meeting the conceptual obstacles that hindered natural science.

At an early date it was found that some natural phenomena can be described mathematically. The most famous early case is the discovery attributed to Pythagoras of the numerical ratios corresponding to the musical intervals of the octave (2:1), fifth (3:2), and fourth (4:3). This startling constellation of facts prompted further research and interest in a numerical theory of "harmonics" referred to by Plato[60] and Aristotle[61] and reflected in the *Sectio Canonis*.[62]

Before long mathematics was applied to other phenomena. In the fourth century the most famous examples are Eudoxus's and Callipus's attempts to account for the motions of the heavenly bodies, including sun, moon, and planets, by combining circular motions, the revolutions of concentric spheres. Mathematical astronomy also treated other problems, which Plato hints at[63] and Aristotle mentions,[64] and which can be seen in *On the Moving Sphere*, by Autolycus of Pitane. There was also a mathematical treatment of optics alluded to by Aristotle[65] and found in Euclid's *Optics*.[66]

For our purposes the point is not that mathematics applies to the natural world but that by the fourth century serious work was being done in proving results about the natural world by the kind of demonstration that had become standard in mathematics, *more Euclideo*. However, the extent to which mathematical reasoning penetrated the natural sciences must not be exaggerated. Euclid's *Optics* and the *Sectio Canonis* proceed by proving theorems in number theory or geometry and then applying the pure mathematical results to their subject matter according to previously stated rules of interpretation. For example, in *Optics*, proposition 4, Euclid proves that one thing appears larger than another by showing that the angle formed by the lines drawn from the observer's eye to the extremities of the one thing is larger than the angle formed in the same way for the other thing, and then invoking the principle that things seen under a larger angle appear larger.[67] Thus these works are applied mathematics more than natural science, but it remains that the fourth century had the conception of proving facts about certain areas of the natural world and arranging those proofs along the lines of a mathematical system of elements.

These are the scientific ingredients of the theory of demonstrative science. Aristotle, I submit, was struck by the rapid advances of mathematics and attributed its success at least partly to its arrangement. He knew some branches of knowledge could be mathematicized either because they obeyed mathematically expressed laws (for example, the "laws of motion" in *Physics* IV.8 and VII.5) or because they were amenable to mathematical reasoning and organization. He concluded that organization was the key to success and that any science organized like geometry might make quick progress. This belief was reinforced by a conviction that the world was so structured that disciplines presented in the form of "elements" are presented *correctly*, so that a "Euclidean" arrangement of a subject is not window dressing, but necessary to a full understanding of the subject. *APo* goes a significant step further, *defining* a science as a subject organized after the fashion of mathematics.

In taking this step Aristotle of necessity adapted and generalized, since geometry and other branches of mathematics present features that do not apply straightforwardly to natural science. His goal was a wholly general theory that can apply both to abstract mathematical branches of knowledge and also to concrete natural ones. These features are reflected in the kinds of principles Aristotle requires, the kinds of scientific demonstrations he recognizes, the role of different principles in them, and the notion of *aitía*, "cause," explanation, or grounds.

The previous discussion has said nothing of the philosophical as opposed to the scientific and mathematical background of the theory of demonstration. In fact, there was a philosophical background, found most clearly in Plato, for the idea of a single method for all branches of knowledge. To sketch Plato's thoughts on this subject and systematically set out *APo* as Aristotle's response is not part of my undertaking. Much work has already been devoted to the question,[68] although it would be worthwhile to reconsider the matter in the light of the results reached in the present study. In general, as frequently occurs, Aristotle is indebted to Plato for a brilliant and fertile idea, but one needing considerable development before it can be put to use.

In working out the conception of demonstrative science, Aristotle attempted to specify what was needed to set any intellectual discipline on the road to progress. In effect, he established a model for scientific paradigms—not a master science like Plato's dialectic, but a pattern according to which any discipline can be arranged. Given that natural science was in a "pre-paradigm" stage, Aristotle showed how it, or rather, each of its branches (for he recognized the need to differentiate science into different fields and suggested criteria for doing so), should be subjected to a paradigm. In fact, his model of science is definite enough that a precise notion of paradigm change (and therefore of scientific revolution) can be given in its terms, and is general enough that he could expect succeeding paradigms to fit the model. This subject too is beyond the scope of the present work, but it makes the conception of demonstrative science of even greater interest today.

The Approach to the Principles

CHAPTER II

THE PRINCIPAL SUBJECT of *APo* is the theory of demonstrative science. *APo* I.1–3 launches the discussion by introducing Aristotle's conception of demonstrative science, arguing that each such science is based on unprovable principles, and establishing a number of requirements that principles must meet. In this chapter I will summarize or translate the relevant parts of *APo* I.1–3 to show in detail how Aristotle presents this material and how he gradually and carefully develops the fundamental concepts of his theory. Following in his path, I will set out his ideas and arguments, discuss some of the difficulties they present, and prepare the way to showing that his conception of scientific principles makes for a theory that is consistent and economical.

I.1 71a1–11 The work begins with the sweepingly general assertion[1] "all intellectual teaching and learning arise from preexisting knowledge [*gnôsis*]" (a1–2). There follows a list of cases (a2–11) that corroborate this claim and also help to sharpen and limit it. Most significantly the claim is said to hold for both deductive and inductive arguments, for both kinds of arguments depend on prior knowledge—of the premises in the former case, and of the particulars in the latter. Moreover, other kinds of arguments—argument by example and by enthymeme—are parasitic on deduction and induction.[2]

I.1 71a11–17[3] Aristotle's second point is that two kinds of preliminary knowledge are required: (a) knowledge that something is and (b) knowledge of what the thing mentioned is. This time there is no justification, only examples: the example of (a) is the principle that everything may be truly affirmed or denied; the example of (b) that "triangle" signifies *this*; the unit is an example of both (a) and (b)—we must know both what it signifies and that it is.

This material presents difficulties. (1) The law of the excluded middle and the unit are of logically different types, yet we are said to know in advance that both of them *are*. (2) More generally, why should we believe that these two kinds of prior knowledge are needed? (3) Are these the *only* kinds of prior knowledge we need? (4) Do we need them just for scientific knowledge or for all types of "intellectual" teaching and learning? Nothing in the present context determines answers to these questions, and nothing elsewhere, as far as I know, is helpful for the fourth. The first three, however, reflect the theory of science as developed in *APo*. It will be necessary to return to them in Chapter 3 when additional material is at hand. For now, I shall merely observe that the passage anticipates some of the major concerns of *APo*, and therefore may have been intended to apply only (or in particular) to scientific teaching and learning. It is not justified in its immediate context, and cannot be fully understood until the theory of demonstrative science is presented in some detail.

The lines that follow (71a17–b8) go closely with a1–11 since they discuss a kind of learning that could be taken as an exception to the claim that all learning depends on previous knowledge.[4]

Chapter 2 begins by identifying two conditions for having scientific knowledge (*epístasthai*) of something:

> **I.2 71b9–12** "We think we have scientific knowledge [*epístasthai*][a] without qualification of each thing . . . when we think we know [*ginṓskein*] the grounds [*aitía*][b] of the thing [*prâgma*] as being its grounds, and that this cannot be otherwise."

NOTES ON TRANSLATION

a. "Have scientific knowledge of" is at the present stage of the proceedings an overtranslation of *epístasthai*, an ordinary Greek word for "know." The corresponding noun *epistḗmē* ordinarily means "knowledge." *APo* develops two special meanings for the noun: "scientific knowledge" (the cognitive state of the scientist) and "science" (the body of knowledge known by the scientist).[5] A geometer has scientific knowledge (*epistḗmē*) of the fact that all triangles have 2R (i.e., angles whose sum is equal to two right angles,) and geometry is a science (*epistḗmē*). Correspondingly, the verb *epístasthai*, "to have *epistḗmē*," will in either case mean "to have scientific knowledge." Since Aristotle has not yet specified the import *epistḗmē* will have in the work, the reader approaching the text for the first time will have no reason to take the word in this special way. On the other hand, anyone reading the work a second time will have every reason to do so. Also, the conditions for *epistḗmē* that Aristotle gives here are presumably meant to be individually necessary and jointly sufficient,[6] but if so, they do not identify all of what we would consider to be knowledge or all of what a Greek would consider to be *epistḗmē*[7]—an indication that Aristotle intends to use this ordinary word in an unordinarily restricted way.[8] It therefore makes for clarity to translate the words in their "loaded" senses throughout.[9]

b. I offer "grounds" as a translation of *aitía*, which is more usually rendered as "cause" (Mure, Ross)[10] or "explanation" (Barnes). The issue at stake will be discussed thoroughly in Chapter 17.

DISCUSSION

What are the objects of scientific knowledge: things, facts, or propositions? Aristotle, as a realist, believes that true propositions correspond to facts, and from this perspective he does not always distinguish carefully between knowing facts and knowing propositions. Moreover, facts are about things, and a true proposition that corresponds to a fact is in subject-predicate form, where the subject term names the "thing" that the fact is about. In some contexts, it is natural to talk of having scientific knowledge of things, as well as of facts and propositions. In general, knowledge of a thing will amount to knowledge of facts about the thing, and these facts will be expressed in propositions. Consequently I will not hesitate to refer to principles, conclusions, and so on as propositions, as facts, or both, as seems appropriate.

THIS PASSAGE IDENTIFIES two conditions for a person (*A*) to have scientific knowledge of *x*:

(a) *A* knows[11] the grounds of *x* (qua grounds of *x*)
(b) *A* knows that *x* cannot be other than it is,[12] that is, that *x* is necessarily true.[13]

Aristotle makes a brief attempt to justify this account of scientific knowledge in terms of general linguistic usage (71b12–16), but says no more at present about the significance of (a) and (b). Accordingly, our discussion of these requirements of scientific knowledge will be deferred.[14]

Scientific Knowledge and Scientific Demonstrations

I.2 71b16–19 Aristotle next declares (again without proof) that one way of having scientific knowledge (*epístasthai*) is knowing (*eidénai*) by means of demonstration (*apódeixis*) (b16–17).[15] He defines demonstration as a "scientific deduction" (*sullogismòs epistēmonikós*) (b17–18), and he defines this as a deduction our possession of which enables us to have scientific knowledge (*epístasthai*) (b18–19). The result of this section appears meager, but is significant: we have scientific knowledge by possessing a deduction that gives us scientific knowledge. Since Aristotle is committed to the possibility of scientific knowledge,[16] he asserts here that there is a kind of deduction that yields scientific knowledge; that is, by possessing it we satisfy the conditions of scientific knowledge. A demonstration, then, is a deduction that entails the necessary truth of its conclusion and that makes known the grounds for its conclusion.

The Premises of Scientific Demonstrations

Since we have to do with deductions, it is imperative that we know the nature of the premises. Since not every valid deduction is a scientific demonstration (71b23–24), the difference between proofs (I use this term synonymously with "demonstration") and other valid deductions will reside in the nature of the premises. Aristotle thus needs to identify requirements such that any valid deduction from premises meeting the requirements will be a demonstration.

Six Requirements for Scientific Principles

Aristotle acknowledges this task and identifies six requirements:

> **I.2 71b20–22** "Demonstrative scientific knowledge must depend on things that are (R1) true, (R2) primary, (R3) immediate, (R4) better known than,[17] (R5) prior to, and (R6) grounds of the conclusion."

DISCUSSION

Aristotle discusses these requirements shortly below, but first it is worth noting that R1–R3 are characteristics of the principles of a science in their own right, and R4–R6 refer to the relations of the principles to the conclusion. Also, only R2 and R3 apply to the principles and to nothing else. R1 applies to all the propositions of a science (cf. 71b25–26), and R4–R6 apply to any propositions used as premises in a demonstration, including previously proved propositions when used as premises for a further conclusion.[18]

The fact that conclusions can also be premises may be responsible for Aristotle's studied and neutral choice of words. He says not that R1–R6 are requirements for *premises* (if he did, R2 and R3 would entail that conclusions could not be used as premises),[19] but that demonstrative scientific knowledge (or demonstrative science) depends on, or proceeds from (Ross),[20] things that are true, and so on.[21] The point is made clear shortly below (71b26–29; cf. I.3 72b18ff.): even if the premise of one proof is the conclusion of another, every conclusion depends ultimately on unproved and unprovable principles. Thus a premise meeting requirements R1–R6 is a principle. But the notion of principle (*archē*) has yet to be introduced,[22] and it will depend on requirements R2 and R3. Aristotle therefore has reason to proceed cautiously.[23]

PRINCIPLES ARE PRIMARY AND IMMEDIATE

> **I.2 71b26–29** "[Scientific knowledge] depends on things that are primary and indemonstrable[a] because a person lacking a demonstra-

tion of them will not have scientific knowledge. For having scientific knowledge of things of which there is demonstration (and not having it incidentally) is a matter of having a demonstration [of them].''

NOTE ON TRANSLATION

a. ''Indemonstrable'' is better as a translation of *anapódeiktos* than ''undemonstrated.''[24] The term is used equivalently to ''immediate'' (b21). In fact, in syllogistic logic the two terms are equivalent, but ''indemonstrable'' is a term that would be appropriate no matter what kind of deduction were allowed, whereas ''immediate'' gets its sense from the syllogistic: a proposition that is *ámesos* is, literally, ''lacking a middle term [*méson*],'' a term that can serve as a middle term in a syllogistic proof of the proposition. (''Unmiddled'' could be used as a translation if literalness were worth the price of barbarism.) For example, in the syllogism ''*A* belongs to all *B*, *B* belongs to all *C*, therefore *A* belongs to all *C*,'' if the premises satisfy conditions R1–R6, the syllogism is a demonstration of the conclusion, via the middle term *B*. In any syllogistic demonstration, which means, as far as *APo* is concerned, in *any* demonstration, the conclusion is proved via a middle term appropriately linked in the premises to the terms occurring in the conclusion. Conversely, any syllogism whose middle term is appropriately related to the terms in the conclusion is a demonstration of that conclusion. Thus, any proposition that ''has a middle''[25] can be proved, and any that ''lacks a middle'' cannot.[26]

DISCUSSION

APo I.2 71b26–29 sketches an argument that can be filled in as follows. Scientific knowledge depends on indemonstrable principles. Suppose it does not. Then it depends on demonstrable principles. In that case, there is demonstrative scientific knowledge of the principles, so that to have appropriate, scientific knowledge (*epístasthai*) of them requires having proofs of them, because where something can be proved, scientific knowledge of it requires having the proof. Therefore, without the proof of the principles, the conclusion based on them cannot be known, since (by I.1) prior knowledge of the principles is required for knowledge of the conclusion. But if the principles can be proved, their proof depends on further principles, which are *ex hypothesi* not indemonstrable. (Otherwise we could simply say that the original conclusion depended on *these* indemonstrable principles.) Therefore the prior knowledge of the original principles depends on still prior knowledge of these further demonstrable principles, which in turn depends on still prior knowledge of yet further principles, and so *ad infinitum*: a vicious regress, which Aristotle states more clearly below and attributes to opponents of demonstrative science (I.3 72b7–15). The regress is avoided by denying the hypothesis that scientific knowledge does not depend on indemonstrable principles (I.3 72b18–24).

THE PRIMACY AND immediacy requirements are taken up at 72a5, where the term "principle" (*archḗ*) is formally introduced:

> **I.2 72a5–8** "To depend on things that are primary is to depend on appropriate principles [*archaì oikeîai*], for I call what is primary and a principle the same thing. A principle of a demonstration[a] is an immediate proposition,[b] and an immediate proposition is one to which no other proposition is prior."

NOTES ON TRANSLATION

a. The Greek could also mean "a principle is an immediate premise of a demonstration" (see note b). Nothing hangs on which translation we choose. The translation I have given is slightly preferable from the point of view of Greek style.

b. I follow Barnes in translating *prótasis* as "proposition," not "premise." What is conclusive for the translation here is Aristotle's speedy review just below (72a8–14) of definitions of certain logical terms, including *prótasis*, which is taken straightforwardly as "proposition" (a proposition affirming or denying one thing of one thing).[27]

DISCUSSION

A principle is what is primary, and principles are immediate propositions. I take the first of these claims as a definition of "principle," the second as a further assertion, whose proof requires the additional premise "all things that are primary are immediate propositions." Aristotle does not state or argue for this premise, but here and above (71b26–29) he treats "primary" and "immediate" as equivalent.[28] Interestingly, the definition of "immediate proposition" as a proposition to which no other proposition is prior might be thought more appropriate as a definition of "primary proposition" (*prôtos* as the superlative corresponding to the comparative *próteros*)—another sign that the terms are equivalent.

Aristotle thus introduces the concept of a principle by the equation

Principle = Primary Proposition = Immediate Proposition.

Since scientific knowledge depends on primary, immediate propositions, it depends on principles. And since, as I.1 implies, prior knowledge of principles is needed for scientific knowledge of the conclusions, all scientific knowledge depends on knowledge of the principles.

The present passage uses still another significant term, "appropriate" (*oikeîos*). At 71b23, premises meeting requirements R1–R6 are called appropriate principles of what is proved. Now premises meeting the primacy requirement, R2, are called appropriate principles. Since a principle is defined as a thing that is primary, it is redundant to call principles appropriate. Still, "appropriate" is not devoid of meaning. Rather, it alludes to the relative nature of principles: a principle is a principle *of something*. In a demonstrative science, the principles are the principles on which the demonstrations and demonstrated conclusions depend. Different sciences have different principles. Proofs and conclusions of geometry, not arithmetic, depend on the principles of geometry, which are thus "appropriate" to the proofs and conclusions of geometry, but not to those of arithmetic.[29]

The claim at 71b23, that premises satisfying conditions R1–R6 are appropriate principles of the conclusion, is therefore made with special reference to R4–R6. But 72a6–7 may seem to present an obstacle to this interpretation in saying that a premise satisfying R2 is an appropriate principle. Two ways out of this problem suggest themselves. (1) Since "primary" is the superlative of "prior," interpret R2 terms of R5: for a principle to be appropriate to a conclusion, it must be prior to it, and if it is to be a principle it must be primary, that is, must have no other premise prior to it (cf. 72a8). Therefore, to say that scientific knowledge depends on primary premises is equivalent to saying that it depends on premises that are appropriate principles. (2) If we bear in mind that a principle is a principle only for the demonstrations and conclusions based on it, it adds nothing to call it "appropriate." However, for some purposes it needs to be emphasized that a principle is *not* a principle for everything, *only* for the demonstrations and conclusions that depend on it. Aristotle does this by speaking of "appropriate" principles. Thus "appropriate principle" is equivalent to "principle of *x*," where *x* stands for one or more of the demonstrations and conclusions depending on the principle. And when Aristotle omits "appropriate" in 72a6–7, he immediately afterwards highlights the relative nature of principles by saying "a principle *of a demonstration* is an immediate proposition."[30]

I.3 In I.2 Aristotle commits himself to the existence of knowable unprovable (immediate, primary) principles on which all else in a science is based. *APo* I.3 treats two rival views, both maintaining that demonstration is the only way of acquiring knowledge (72b15–16; cf. b12–13). But where one says that scientific knowledge is impossible (since there are no knowable unprovable principles on which to base it), the other says that not only is it possible, but all the propositions in a science can be proved, and therefore known, because circular proof is possible (so that there is no need for unprovable principles).[31] After sketching these arguments (b7–18), Aristotle replies to the assumption shared by the two positions[32] that all scientific knowledge depends on demonstration (b18–25). Disappointingly, the meat of the reply consists in a reaffirmation of his own view, rather than an argument in rebuttal:

> **I.3 72b18–22** "*We* say that not all scientific knowledge is based on demonstration[a] but that of the immediates is not,[b] and this is obviously necessary,[c] for if it is necessary to have scientific knowledge of the things that are prior and on which the demonstration depends, it[d] stops at some point,[e] and these immediates[f] must be indemonstrable."

NOTES ON TRANSLATION

a. "Based on demonstration" (*apodeiktikḗ*): sometimes translated "demonstrative."

b. "Not [based on demonstration]" (*anapódeikton*): "indemonstrable" or "undemonstrated"; the former is preferable.

c. It is necessary in that it follows necessarily from the following consideration.

d. "It" refers to the process of basing premises on still prior premises.

e. The comma is Barnes's punctuation, following Solmsen (1975, p. 104 n. 2), and is defended by Verdenius (1981, p. 346).

f. "These immediates": the *ultimate* premises.

DISCUSSION

The inadequacy of this response is obvious. Aristotle's opponents agree that scientific knowledge requires scientific knowledge of the premises on which it is based, and they simply deny that the process ever stops. In fact, this idea is the basis of their objections.

ARISTOTLE CONTINUES:

> **I.3 72b23–25** "This is our view on these things, and we say that there is not only scientific knowledge [*epistḗmē*], but also a principle of scientific knowledge [*archḕ epistḗmēs*)[a] by which we recognize the definitions.[b]"

NOTES ON TRANSLATION

a. Aristotle says first that there is indemonstrable *epistḗmē* of immediates (b19–20) and now that aside from *epistḗmē* there is also an *archḕ epistḗmēs*. He seems to dither on the question whether knowledge of the principles is *epistḗmē*, but the situation is a bit more complex. Since he has not yet introduced *noûs* and clarified its relation to *epistḗmē* (as he does at II.19 100b5–17), it is not clear how he could express himself better. According to his final view, *all epistḗmē* is demonstrative, but there is *noûs* of the indemonstrable immediates. But to say this at 72b18ff. would befuddle the discussion. Rather than dithering, Aristotle may be conscientiously refusing to use undefined terms in his discussion.

b. "Definitions" is my interpretation of *hóroi*. Other interpretations are possible: either "terms" or "limiting propositions" (Ross). *Horismós* is the word most frequently used for "definition" in *APo* (e.g., at I.2 72a21).[33] Clearly the sense must be "by which we know the immediate principles." This sense can be achieved either by taking *hóroi* to cover all kinds of principles,[34] or by taking *hóroi* to mean "definitions" and supposing that Aristotle is speaking as though definitions were the only kind of principle,[35] or by taking it to mean "definitions" and supposing that he is using definitions as a specimen type of principle.

DISCUSSION

The rest of I.3 raises objections to the possibility of circular proof, which need not be considered here.[36] We are left, then, with an undefended assertion of the existence of unprovable but knowable principles. Aristotle's position is coherent, but he needs to show that such principles do exist. He takes up this task in book II (explicitly in II.19, but in effect elsewhere too). His strategy is to show how we come to have knowledge of indemonstrable principles. If his account succeeds, it follows that there are knowable indemonstrable principles.

PRINCIPLES ARE TRUE

> **I.2 71b25–26** "They [principles] must be true because it is not possible to (a) have scientific knowledge of what is not the case, e.g., that the diagonal [of a square] is commensurate [with its side]."

DISCUSSION

This elliptical argument is Aristotle's entire defense of R1. I explicate it as follows. The nature of deduction guarantees that true premises yield true conclusions.[37] One or more false premises may occur in a deduction with a true conclusion, but false premises may also lead to false conclusions (e.g., that the diagonal is commensurate with the side).[38] Since scientific knowledge is based on deductions, the only way to ensure that a person with scientific knowledge of a conclusion knows it to be true is to require that the premises of the deductions be true.[39]

PRINCIPLES ARE BETTER KNOWN, PRIOR, AND GROUNDS

> **I.2 71b29–33** "[The principles] must be grounds, better known, and prior: they must be grounds because we have scientific knowledge precisely in cases when we know the grounds; they must be prior if they are grounds; and they must be previously known not only in that we understand them, but also in that we know that they are the case."

DISCUSSION

Aristotle's argument for R6 comes straight from 71b9–12, where knowing the grounds of x is one of two requirements for having scientific knowledge of x. Further support is needed. Granted that having scientific knowledge of x consists in possessing a scientific demonstration of x, that this is a deduction based on indemonstrable principles, and that scientific knowledge of x requires knowing the grounds of x: even so, why must the grounds be the principles rather than an intermediate conclusion used as a premise in proving x, or rather than some wholly different kind of consideration?[40] Aristotle takes up the former question in II.17–18. The latter question challenges the Aristotelian conception of demonstration, since it asks in effect why we should believe that a demonstration *explains* anything. This issue will be taken up in detail in Chapter 17 below. For the present I shall anticipate some of the results of that discussion by saying that the conception of grounds (*aitía*) in *APo* is tailored to the needs of demonstrative science. A demonstration is an explanation, and the principles are the (ultimate) grounds of the conclusion. Aristotle is using the word *aitía* acceptably, even if not in its only sense, or its most common one. He could defend himself by saying that he has sharpened the ordinarily fuzzy conception of *aitía* into a philosophically serviceable tool, according to his normal procedures.[41] At most he has shown at 71b30–31 that the principles of a scientific demonstration must *in some sense* be the grounds of its conclusion. He returns later (especially at I.13 and II.8–18) to the task of specifying that sense.

R4 AND R5 ARE TREATED together, apparently as equivalent:

> **I.2 71b33–72a5** "Things may be prior and better known in two ways: for what is prior in nature [*têi phúsei*] and what is prior in relation to us (*pròs hēmâs*) are not the same, nor what is better known and what is better known to us [*hēmîn*]. I call prior and better known in relation to us the things that are nearer to perception, and prior and better known without qualification [*haplôs*] the things that are farther. The things that are farthest [from perception] are those that are most universal [*kathólou málista*] and those that are nearest are the particulars; and these are opposed to one another."

DISCUSSION

Again, it will be best to put off full discussion of this distinction until Chapter 17,[42] but a few preliminary points may be made. First, Aristotle does not use fixed formulas. On the one hand, things may be prior or better known either "in relation to us" (*pròs hēmâs*) or simply "to us" or "for us" (*hēmîn*). On the other, they may be prior or better known "in nature" (*têi phúsei*) or "without qualification" (*haplôs*), or they may be simply "prior" or "better known" (as at a1).

Second, although "better known" is the best translation of *gnōrimṓteron*[43] and it seems important to keep a single English word for both ways of being *gnōrimṓteron*, the difference between the two points to "more intelligible" as a translation of the word when we have *gnōrimṓteron* "in nature" and "better known" when we have *gnōrimṓteron* "to us."

Third, at a1–4 Aristotle seems to *define* what is prior and better known to us as what is "nearer to perception." However, as Barnes points out (p. 101), it is difficult to reconcile this definition with *Topics*. VI.4 141b36–142a4, which indicates that there is no single answer to the question "What things are most familiar to us?" Aristotle is probably not defining the term, but giving a general indication of the kinds of things that ordinary people who are not experts at a science are more readily able to grasp, think about, and understand (in a limited unscientific way).[44]

Fourth, the contrast between what is "better known to us" and what is "better known in nature" or "better known without qualification" does not imply an unbridgeable gulf between what "we mere humans" can know and what is knowable in "the highest sense." Plato in a bleak mood might have affirmed such a view, but Aristotle places no such limits on our ability to know reality. That the gulf can be crossed is made clear by the statement that we must make what is better known by nature better known to us (*Met.* Z.3 1029b1–12). Our situation can change; we can become experts from laymen. When he contrasts the terms as extremes, he has in mind "us ordinary laymen" with no scientific knowledge of the subject, at the opposite extreme from the expert to whom what is best known is what is most intelligible, the principles of the science.[45]

Fifth, Aristotle seems to treat R4 and R5 as if they were identical—most clearly in the "definitions" (a1–4). But if they are not definitions but what was indicated above, R4 and R5 need not be identical. They may be notionally distinct, while applying to the same items.

Sixth, the "definitions" are followed by assertions that particulars are nearest (to perception) and "the most universal things" are farthest (from perception). Aristotle changes from the comparatives ("nearer," "farther") found in the "definitions" to superlatives and then associates the superlative terms with other terms that we see as being of different logical types. The contrast between particular and universal is not one that can occur in degrees: something is a particular or a universal, but nothing is *more* universal than another, and therefore nothing is *most* universal. On the other hand, there *are* degrees of generality and specificity. To understand Aristotle here, we must see that he does not distinguish between universal and general, so that when he speaks of things that are most universal, he means the most general terms (which of course are universals), and at the opposite end of the scale he puts particulars, which he thinks of as the least general, most specific terms.[46]

In any case, the purpose of making the distinctions is to clarify in what way the principles of a science are prior and better known. Clearly they must be prior to and better known *in nature* than the conclusions.

WE MAY NOW return to 71b31–33. Principles are prior since they are grounds. Two current explanations of this point are that the principles are "objectively prior to, i.e., more fundamental than, the fact stated in the conclusion" (Ross, p. 509; cf. ibid., p. 54) and that Aristotle "has in mind here . . . priority '*in knowledge*' (*P* is 'prior in knowledge' to *Q* if knowledge that *Q* requires knowledge that *P*, but not *vice versa*)" (Barnes, p. 100). The passage surely favors Ross, since the priority of the principles is explained in terms of their being grounds, not in terms of their greater intelligibility. But is this all? As Barnes points out (ibid.), it depends on how we understand "and they must be previously known" (b31) to be connected to the passage. Is it a restatement of the "better known" requirement ("The principles must be prior since they are grounds.

They must also be previously known''), or a further reason for their priority (''The principles must be prior since they are grounds and also since they are previously known'')? To begin with, if ''previously known'' is understood in a temporal sense, it does not mean the same as the relevant sense of *gnōrimōteron* (''more intelligible''). In fact, the discussion of ''better known to us'' and ''better known in nature'' shows that in general the principles are *not* known before the conclusion[47] and also that Aristotle is inclined to think that the principles are the *last* things to be known.[48] A further objection to Ross's interpretation is that it makes the discussion of R4 anomalous. The stretch from 71b25 offers justification for the six conditions placed on scientific principles at b21–22. Each of the other five conditions is said to hold *because* (b25, b27, b30) or *since* (b31) something else is the case. What is wanted for R4, then, is not just a restatement, even in a somewhat more detailed form.

Barnes's interpretation avoids the second of these problems. The justification of R4, then, occurs at 71b33–72a5 in the discussion of ''better known in nature,'' and comes from the close association of R4 with R5 there, especially at a3. The principles are prior (without qualification), and the same things that are prior (without qualification) are also better known (without qualification). Therefore, the principles are better known without qualification. But now the problem with ''previously known'' arises again. How does it explain that principles are prior (in nature) to say that they are previously known, and why should we believe that they are previously known at all? The solution consists in noticing that although Aristotle *does* intend temporal priority,[49] he has in mind the context of demonstrative science. From this point of view, a person must first know the principles (in the appropriate way for principles to be known) and may then construct proofs based on them. We may know the truth of a derivative proposition before we know the principles, but we cannot know that proposition *scientifically* until we have the appropriate knowledge of the principles. This interpretation is confirmed by 72a25ff., especially a27–28, and a35, which says that anyone who is convinced (of a conclusion) by means of a demonstration must have previous knowledge (of the principles).

What, then, is the nature of the principles' priority? There is one way in which they are not temporally prior: they are not *true* earlier in time than the conclusions. Then is the priority (1) temporal as regards our knowledge (we must know them before we know the conclusions scientifically, (2) epistemological (knowledge of conclusions depends on knowledge of principles, but not vice versa), (3) ''causal'' (in some sense of that word),[50] or (4) logical (the truth of the conclusions follows from that of the principles)? The principles are prior in all four ways. *APo* I.2 71b31–33 points to (1) (''previously known'')[51] and (3) (''grounds''). (4) is implied by 71b28–29 (''having scientific knowledge of things of which there is demonstrative science is a matter of having a demonstration [of them]''), and (2) is implied by (4) together with the claim that having a demonstration depends on knowing the principles.[52] The fact that the principles are prior in a number of ways does not mean that these different sorts of priority come to the same thing, but that the theory of demonstrative science coordinates them. It reflects a commitment to a natural ordering of propositions and of the facts the propositions represent, and also to a natural ordering of intelligibility, so that the basic propositions/facts are the most intelligible and the dependence of one proposition/fact on others is reflected in the way its intelligibility (and our scientific knowledge of it) depends on the intelligibility of (and our knowledge of) the principles.

I.2 72a25–b4 Two more epistemological requirements are discussed in the final paragraph of the chapter: (1) we must know the principles more (*mâllon*) than the conclusions and believe more (*pisteúomen mâllon*) in the principles than in the conclusions, and likewise (2) we must know more and believe more in the principles than "the opposites to the principles, on which depends the deduction of the contrary mistake [*tês enantías apatês*]." The significance of the paragraph is suggested by the first sentence, which must not be misunderstood:

> **I.2 72a25–28** "Since it is necessary to believe in [*pisteúein*] and to know the thing by having the kind of syllogism we call demonstration, and this [kind of syllogism] results from the fact that the things on which the syllogism depends are *these*,[a] it is necessary not only to know the primary things[b] previously, either some or all of them, but also to do so more [*mâllon*]."

NOTES ON TRANSLATION

a. "These": the kinds of principles that have been discussed.
b. "Primary things": the principles.

DISCUSSION

The "since" clause is obviously false if understood to say that the only way to believe in something is by having a demonstration of it. Even the present chapter of *APo* contains two reasons against this interpretation. First, there are degrees of believing in something (as Aristotle knows—cf. *pisteúomen mâllon* at a32). Having a demonstration of something is grounds for believing very strongly in it, and so weaker belief will presumably come in other ways. Second, we must have some belief in the things Aristotle calls "better known to us." For those things that are "nearer perception," seeing is believing. And we can have genuine belief, genuinely felt confidence in such things, even without a scientific proof.

This problem too is solved by recognizing that Aristotle is speaking in the context of demonstrative science. We should understand the clause to mean "since if we are to have scientific knowledge of x we must believe in and know x by having a demonstration of x."

Aristotle's summary of (1) makes it clear that he intends this context:

> **I.2 72a37–39** "A person who is going to have the scientific knowledge that comes through demonstration must not only come to know [*gnōrízein*] the principles more and believe more in them than in what is proved . . . "

DISCUSSION

Does this mean that Aristotle is introducing a special kind of belief as well as a specialized kind of knowledge? He gives no special definition or account of belief here, so we have no reason to think that the word is not being used with its ordinary meaning. On the other hand, the context is a discussion of a kind of knowledge for which there are unusually strong requirements. When we have this kind of knowledge of *x*, we know that *x* is true, and also that *x* is necessary, and why *x* is necessarily true, and how it depends on the relevant principles. The kind of belief in *x* that is appropriate in this context will correspond: we must not only believe in *x* (believe that *x* is true), but also believe that *x* is necessary, and so on. This is what belief in something *as* a scientific fact consists in. In this way it is clear that we can have genuine (nonscientific) belief in the truth of *x* that is not based on belief in the principles, and also that when we have scientific knowledge of *x*, our belief that *x* is necessary and that *x* is necessarily true for such and such reasons is based on our belief in the principles, specifically our belief that the principles on which *x* depends satisfy conditions R1–R6. What is at stake, then, is not the *meaning* of the word "belief," but the conditions for using it in this particular context. The paragraph asserts that in the case of scientific knowledge, the grounds of our belief are identical with the grounds of our knowledge. An expert at a science believes in the conclusions of his proofs because of his belief in the principles. Faith and knowledge coincide.

The main difficulty is how to understand the force of the comparative in the claim that we know and believe in the principles *more* (*mâllon*) than the conclusions. Belief can vary in strength, we may suppose, but knowledge does not: either you know something or you don't. Barnes has a promising approach to this problem. He says:

> Aristotle was familiar with this line of argument . . . , and the text provides, implicitly, his answer to it: part of the notion of *knowing that P* is being *convinced that P*; since I can be more convinced that *P* than I am that *Q*, I can be said, in a reasonably intelligible sense, to know that *P* "more" than I know that *Q*. And [the assertion that, if *P* is a conclusion and Π its principles, and if *a* has demonstrative knowledge that *P*, then Π is more known to *a* than *P* is] can then be defended on Lockean grounds: *P* is infected by any uncertainty that infects Π and also by any uncertainty that affects the inference from Π to *P*. If we allow the possibility of complete certainty, then [the

> assertion] needs a slight modification: *P* cannot be more known than Π is. (Pp. 104–5)

The objection to this reading is that the "slight modification" destroys whatever point Aristotle is trying to make. For Aristotle, the expert's certainty in the conclusions of demonstrations is complete ("a person who has unqualified scientific knowledge must be unable to be persuaded to change his mind," 72b3–4), so that on Barnes's interpretation principles and conclusions are equally well known. An interpretation that is false neither to the text nor to Aristotle's recognition that knowledge does not have degrees must find a different interpretation of *mâllon*.[53]

THE MOST LIKELY approach is given by another statement that speaks of knowing and believing "more":

> **I.2 72a30–32** "Since we know and believe in [the conclusions] through [or on account of] the primary [principles], we also know and believe in them [i.e., the principles] more, because through [i.e., on account of] them also [we know and believe in] the subsequent [propositions]."

DISCUSSION

Experts know the principles "more" because it is through them that they know the conclusions. "More," then, seems to mean something like "more fundamentally," or "more basically," in the sense that knowledge of principles is primary, that of conclusions derivative. Experts are more committed to their knowledge of principles than to their knowledge of what depends on them. This greater commitment to principles is expressed by saying that we believe in principles more than the conclusions.[54] It is not necessary at present to say more about this paragraph.[55] The passage complements the other texts we have examined and helps us understand the nature of scientific principles and the relation the specialist has to them.

WE HAVE SEEN that the theory of demonstrative science is committed to the existence of indemonstrable principles that satisfy R1–R6. We have also seen, in a preliminary way, how R1–R6 fit together to make the theory consistent and economical. We have not seen what kinds of propositions serve as principles, how the requirement that the objects of scientific knowledge (the conclusions of demonstrations) be necessary (71b12) is accounted for by R1–R6, or how demonstrations make use of principles in proving their conclusions. These issues need to be discussed for us to understand the workings of demonstrative sciences. I shall begin with the first.

The Kinds of Principles

CHAPTER III

ARISTOTLE'S GENERAL VIEW on the kinds of principles found in demonstrative sciences is reasonably clear. There are three sorts of principles: axioms, which are principles occurring in more than one science and for that reason frequently called "common" (*koiná*); definitions of the subjects and attributes of the science; and assertions that the subject or subjects of the science exist. The definitions and existence claims are called proper principles (*ídia*) in contradistinction to the common axioms.

How such principles function in scientific demonstrations, what the precise nature of such kinds of principles is, and how reasonable it is to base a science on such principles will be taken up later. In the present chapter I examine passages bearing on the topic of the kinds of principles and establish that Aristotle's view is what I have asserted it to be. Because these passages can be read in many different ways, it is necessary to work through them with care. We shall see that in addition to confirming this interpretation, the passages reflect different points of view, depending on Aristotle's immediate purposes, and that there is a certain amount of fluctuation in his terminology for the various kinds of principles.

First, a methodological note. At several places *APo* contains lists. In I.1, there is a list of ways in which we must have preliminary knowledge (71a11–16). In I.2, a list of kinds of immediate deductive principles (72a14–24). In I.7, a list of "the things in demonstrations" (75a39–b2). In I.10 a list of "the principles in each genus" (76a31–36), an account of "the things they use in demonstrative sciences" (76a37–b10), and a list of three things every demonstrative science is concerned with (76b11–15, cf. b21–22). In I.32, a list of "two kinds of principles" (88b27–29). The lists do not agree in all details, but rather than simply charge Aristotle with inconsistency, I will try to account for the differences in terms of his varying points of view in the contexts of the passages.

I begin with I.10,[1] which contains the longest discussion of the types of principles:

> **I.10 76a37–41** "Some of the things they use in demonstrative sciences[a] are proper [*ídia*] to each science and some are common [*koiná*]. . . . Examples of proper ones are that a line is such and such, and that the straight is such and such. An example of a common one is that if equals are subtracted from equals, the remainders are equal."

NOTE ON TRANSLATION

a. "The things they use in demonstrative sciences": a vague phrase, which the following lines show is used of the principles of a science.

DISCUSSION

This passage distinguishes between proper and common principles, and gives examples of each. Aristotle's examples of proper principles are definitions. There is no need to mention the other kind of proper principles, existence claims, since Aristotle needs only to indicate that there *are* proper principles. Existence claims are identified as proper principles just below (76b3).

THE NEXT PARAGRAPH continues this discussion:

> **I.10 76b3–10** "Proper things[a] are (1) those things that are assumed to exist,[b] concerning which[c] the science investigates the attributes that belong to them per se.[d] For example, arithmetic [is concerned with] units, geometry with points and lines. For [arithmeticians] assume that these exist and that they are this.[e] (2) [Concerning] the per se attributes of these they assume what each signifies[f]—for example, arithmetic assumes what odd, even, square, and cube signify, and geometry what irrational or to inflect or verge signify. But that they exist, they prove through the common principles[g] and from the things that have been proved."[h]

NOTES ON TRANSLATION

a. "Proper things": in the previous paragraph, proper things were propositions; here they are items (subjects, attributes). The proper principles of a science are the principles that concern the proper things in the ways specified.

b. "Those things that are assumed to exist": one of Aristotle's ways of characterizing the subject genus of a science. Compare 76b12, translated and discussed in Chapter 4 below. The examples at 76b4–5 make it clear that these are the subjects (as opposed to the attributes) of a science.

c. "Concerning which": another characterization of the subject genus. Compare 76b22 and I.32 88b28, and also Chapter 4 below.

d. The signifcance of the technical expression "per se" (*kath' hautá*) is given in I.4 and elaborated in I.6. It is discussed below in Chapter 7.

e. The most economical interpretation of "that these exist and that they are this" is that both the existence and the definitions of the subjects are assumed. This accords with 76a31–36 (translated and discussed shortly below). It is not clear what else could be meant. But why "that they are this" instead of "what they signify" (cf. 76a32, b7)? The answer is provided at II.7 92b4–8: only if we know that *x* exists is it possible to know *tí esti x* ("what *x* is"). Otherwise, all we can know is *tí sēmaínei x* ("what *x* signifies"). This is precisely what is being said of the subjects of a science, as opposed to the attributes. The existence of the subjects is assumed, whereas that of the attributes must be proved, and at the initial stage at which we have only the unprovable principles, only the definitions of subjects whose existence is assumed can be characterized as *tí esti*; those of the attributes qualify only as indicating *tí sēmaínei*. (Aristotle does not make the further point that after their existence is proved the same formula that indicates *tí sēmaínei* will then qualify as indicating *tí esti*, but there is no need for him to explain this subtlety here.) The passage says "that they are this," not "what they are," but the difference is not significant. It presumably means "that it is precisely this sort of thing," which will be only an emphatic way of saying "what it is."

f. Regarding "what each signifies," see note e.

g. For discussion of the sense of "through" as it is used in "through the common principles," see the section "The Alleged Regulative Role of Axioms" in Chapter 6 below.

h. "From the things that have been proved": Aristotle fails to mention the proper principles as well. Zabarella (1966, pp. 791e–792a) correctly supplies this omission.

DISCUSSION

This passage distinguishes two kinds of proper things in a science: subjects and per se attributes of those subjects. Definitions of both are assumed as scientific principles. The subjects are assumed to exist, while the per se attributes are proved to exist, that is, proved to belong to an existing subject.

APo II.1–2 allows that the existence of some subjects can be proved,[2] thus prompting a distinction between primitive and derivative subjects. Can this distinction be found in *APo* I, or is Aristotle actually inconsistent on this important matter? The present passage does not seem to know such a distinction, since when it talks of "the attributes that belong to them per se," "them" should mean all subjects, not just primitive ones. On the other hand, the examples given of subjects (units, points, lines) are probably intended as primitive subjects in their sciences.

ARISTOTLE NEXT EXPANDS the distinction at 76a37–41 in the light of 76b3–10:

> **I.10 76b11–15** "For every demonstrative science is concerned with three things, (1) as many things as are posited to exist (these

are the genus whose per se attributes the science investigates), (2) the common principles, called axioms,[a] from which as primary [the science] forms its demonstrations,[b] and (3) the attributes, concerning which [the science] assumes what each signifies."[c]

NOTES ON TRANSLATION

a. In "the common principles, called axioms" (*tà koinà legómena axiômata*), "axioms" should be understood as a gloss on "common principles," the term that has been used until now in this chapter (76a38, a41, b10). Other versions are "what are called the common axioms" (Barnes), "the common axioms so-called" (Heath 1949, p. 51), and "the axioms which the mathematicians call common" (Ross, p. 540).[3]

b. "From which as primary [the science] forms its demonstrations" is only one possible translation. Others are: "which are the primary source of demonstration" (Heath 1949, p. 51), "the primitives from which it demonstrates" (Barnes), "which are the primary premises of its demonstration" (Mure). The occurrence of "primary" is puzzling: are the common principles being marked as prior to the other principles? But I.2 71b21 states that *all* principles are primary. I suggest that "from which as primary" marks axioms as principles as described in I.2, and that Aristotle points this out because in the present list they are the only item that *is* a kind of principle. The others are the subjects and attributes that are not themselves principles, although their definitions and existence claims (for some or all subjects) are.

c. "Of which [the science] assumes what each signifies" corresponds exactly to the description of the per se attributes at 76b6–7.

DISCUSSION

This passage summarizes the foregoing. A new feature is the term "genus" (*génos*), here identified with "as many things as are posited to exist": the subjects (or primitive subjects). At I.10 76b12–13 and I.7 75a42ff. the genus is conceived as all its subjects. In the present passage, the per se attributes are attributes of the genus, while at 76b6–7 they are attributes of the subjects. If the genus is identical with all the subjects, the two come to the same thing.

However, this passage is not decisive on the question whether *APo* distinguishes between primitive and derivative subjects. Aristotle has so many ways of characterizing the subject genus[4] that it is quite conceivable that he could describe it in terms of the primitive subjects (since everything else follows from them).

I.10 76b16–22 The next lines defend the doctrine of kinds of principles against cases that seem to violate the rules just laid down. A science may neglect to make some of its principles explicit if they are obvious. But that does not affect matters. Even if the assumptions are tacit they are still needed. This passage concludes:

I.10 76b21–22 "Nonetheless, these things are by nature[a] three: (1) that concerning which[b] [the science] forms its demonstrations,

> (2) what it demonstrates,[c] and (3) the things from which [it demonstrates]."[d]

NOTES ON TRANSLATION

a. "By nature" means that the nature of demonstrative science requires that there be these three kinds of things. The distinction between "by nature" and "with respect to us" of I.2 71b34ff. is relevant. Even if someone overlooks the need to assume some of the principles of the science, according to the real, fully intelligible state of affairs they are there: they are assumed implicitly even if not explicitly.

b. The singular of the pronoun shows that this is the genus. For the identity of the genus with the subjects, see 76b12–13 and 76b3–4. (Both passages are translated earlier in the present chapter.)

c. Comparison with 76b6–11 shows that these are the per se attributes, which it proves to exist.

d. These are the common principles, described similarly at 76b14.

DISCUSSION

This passage repeats familiar doctrine: in each science there are subjects, per se attributes, and axioms. Again, although this is not a list of types of principles (the only principles mentioned are the axioms), the definitions of the subjects and of the per se attributes, and the existence claims for some or all of the subjects, constitute the remaining principles.

THE PREVIOUS DISCUSSION makes it easier to understand the difficult first paragraph of I.10:

> **I.10 76a31–36** "I call principles in each genus those [regarding] which it is not possible to prove that they exist. What both the primary and the derivative things signify is assumed, but it is necessary to assume that the principles exist and prove that the others do. For example, it is necessary to assume what unit signifies, and what straight and triangle do, and that the unit and magnitude exist, but it is necessary to prove the rest."

DISCUSSION

The key to understanding this passage is the examples of "principles" at 76a35f.: "the unit and magnitude." Principles (*archaí*) here are subjects (or primitive subjects)—an exceptional (and unnecessarily confusing) use of the term.[5] Once this is recognized, the passage falls into line with the passages of I.10 already discussed. Again we have the distinction between things (here called primary [*prôta*] as well as principles) whose existence as well as whose definition is assumed, and derivative things (*tà ek toútōn*) whose definition alone is assumed.

This passage provides evidence that *APo* I distinguishes between primitive and derivative subjects. *APo* I.10 76a34–36 groups the subject triangle along with the attribute straight as an item whose definition is assumed but whose existence needs to be proved. Thus, unit is taken to be a primitive subject of arithmetic, other numbers being defined in terms of units, but triangle is a derivative subject (cf. II.7 92b15–16) whose existence depends on more primitive kinds of magnitude. In fact, the appearance of magnitude instead of triangle as an instance of a subject whose existence must be assumed looks like a deliberate move to point to the distinction between primitive and derivative subjects.[6]

ANOTHER BASIC PASSAGE on the kinds of principles is found in I.2:

> **I.2 72a14–24** "Among immediate deductive principles,[a] I call a thesis[b] one that it is not possible to prove, but that it is not necessary for a person to have if he is going to learn anything,[c] and an axiom one that it is necessary for a person to have if he is going to learn anything at all. For there are such things. For we are accustomed to apply the name especially to such things. Among theses,[d] one that assumes either of the members of the contradiction[e] (i.e., I mean, that something exists or does not exist)[f] is a hypothesis,[g] and one without this is a definition. For a definition is a thesis, for the arithmetician posits[h] that a unit is that which is indivisible in quantity, but it is not a hypothesis, for what a unit is and that a unit exists are not the same."

NOTES ON TRANSLATION

a. "Among immediate deductive principles" (*amésou d' archês sullogistikês*): that is, among principles of demonstrations. Demonstrations are a kind of deduction (71b18), and mention of the immediacy condition implies the context of demonstrative science, where the only deductions are demonstrations. The singular in the Greek can be explained by the structure of the sentence. Aristotle sets out to define "thesis" (singular) and does so in terms of "one that it is not possible to prove . . . " (singular). The Greek naturally uses no word to express "one," and the Greek for "that" (singular) refers back to "immediate deductive principle," which must be singular to preserve agreement. This is possible Greek, but English requires a plural.

b. The translation of *thésin* as "thesis" accords with translating *hupóthesis* as "hypothesis" (72a20, a23), but does not exhibit the etymological force of the explanation in terms of *títhetai* ("posits") at 72a21–22. Barnes's "supposition" for *hupóthesis*, "posit" for *thésis*, and "posits" for *títhetai* preserve the links among the words, but I prefer to keep the traditional translation "hypothesis," at the small cost of a brief explanatory note.

c. "If he is going to learn anything" has the same meaning as "if he is going to learn anything at all" just below.

d. Though English requires the plural in "among theses," the Greek (théseōs) is singular, not plural. The construction is somewhat like the one explained in note a.

e. "Contradiction" and "member of a contradiction" are defined above (72a12–14): "A contradiction is an opposition in which per se there is nothing in between. A member of a contradiction asserting something of something is an affirmation, and one denying something of something is a negation." For example, an affirmation is "Socrates is mortal," a negation is "Socrates is not mortal" and these two propositions together form a contradiction. Since there is a negation corresponding to any affirmative proposition and vice versa,[7] the expression "either of the members of a contradiction" is equivalent to "any affirmative or negative proposition." Note also that "*the* contradiction" is a more literal translation of *tês antipháseōs* than "*a* contradiction," which is the translation usually given (Mure, Barnes). Aristotle specifies which contradiction he intends in the following words.

f. "I.e., I mean, that something exists or does not exist" translates *hoîon légō tò eînaí ti ḕ tò mḕ eînaí ti*. We must make choices in translating two words in this clause. The term *hoîon* can mean either "for example" ("e.g.") or "that is to say" ("i.e."). Compare Bonitz, 502a1–6 for the former and 502a7–20 for the latter.[8] Also *eînai* can be taken as either "to exist" or "to be the case." The decision among these possibilities depends on our interpretation of the present passage and of Aristotle's doctrine of the kinds of principles. On the view that there are precisely three kinds of principles, and of these, hypotheses are existence claims, *hoîon* is taken as "i.e." and *eînai* as "to exist," and *tês antipháseōs* is taken as referring to *the* contradiction "*x* exists; *x* does not exist." If hypotheses are taken to include, but not to be restricted to, existence claims, either *hoîon* is taken as "e.g.," *eînai* as "to exist," and *tês antipháseōs* as "*a* contradiction," or *hoîon* is taken as "i.e.," *eînai* as "to be the case," and *tês antipháseōs* again as "*a* contradiction." Since the present passage identifies axioms, definitions, and hypotheses as principles and I.10 identifies axioms, definitions, and existence claims for subjects as principles, consistency with I.10 is achieved by translating *hoîon* as "i.e." and *eînai* as "to exist" so that the *only* hypotheses are existence claims. This choice also permits us to translate *tês antipháseōs* more naturally as "*the* contradiction" since it commits us to the view that the only relevant contradiction is the one formed by the propositions "*x* exists" and "*x* does not exist." Anyone who believes that the three kinds of principles identified in I.10 are not the only kinds intended will translate *hoîon* and *eînai* differently. But since nothing in this passage implies a doctrine inconsistent with that of I.10, it is best to assume consistency without a compelling reason to do otherwise.[9]

g. The clause ending with "hypothesis" (*hupóthesis*) is Aristotle's strict definition of hypothesis. He introduces other meanings of the term at I.10 76b23ff.

h. "Posits" (*títhetai*) is the verb from which "thesis" (*thésis*) derives. See note b.

DISCUSSION

This passage identifies three kinds of principles. It uses the procedure of division by dichotomy, first distinguishing theses from axioms and later subdividing theses into hypotheses and definitions. In I.10 axioms are called common principles as opposed to proper principles. If the two chapters tell the same story, then the theses of I.2 will be the proper principles of I.10.

Axioms here are said to be prerequisite for all learning—a strong claim, which may be softened somewhat. First, the first sentence of *APo* makes it clear that *APo* is concerned only with "intellectual" learning (71a1). Second, we may go farther and suppose that here the only learning that counts is the kind associated with demonstrative sciences (cf. *manthánontos*, I.10 76b33). Barnes interprets the definition of axiom as "*P* is an axiom for *S* $=_{\mathrm{df}}$ if anyone knows any proposition in *S*, then he knows that *P*," where *P* is a proposition and *S* is a science (p. 103), but this account fails to capture the idea that axioms apply in more than one science.

In saying "For there are such things. For we are accustomed to apply the name especially to such things," Aristotle defends his definition of axiom by insisting that there are things that satisfy his definition of axioms. He refers to current usage of the word and asserts that the things to which it is applied do satisfy his description. We conclude that the word "axiom" was in current use and that Aristotle thought that the things called axioms were very general; also that (since he found it necessary to point this out) the characterization of axioms Aristotle offers here was unfamiliar—which means either that axioms had previously been described differently (so that Aristotle is redefining the term) or that the term was in use without a definition, or without a generally accepted definition (so that Aristotle is filling a gap). I will discuss the nature of axioms below in Chapter 6.

Regarding hypotheses, notes e and f justify interpreting them as existence claims. This is not the only possible interpretation of the text. Its advantage is its agreement with I.10. A potential objection to the view is the thought that a science could not get very far starting from only axioms (which it shares with other sciences), definitions, and existence claims. One of the main theses of this book is that Aristotle believed that the three kinds of principles *are* an adequate basis for a science, so that the interpretation I have given of this passage is what Aristotle meant.

A further question arises out of the claim that a hypothesis assumes *either* of the members of the contradiction in question. How can a claim of *non*existence be a principle of a science? Surely sciences treat things that exist and not things that do not. *Physics* IV.6–9 may have some bearing on this question. These chapters contain dialectical (nondemonstrative) arguments to show that there is no void, and it is possible that the negative hypothesis "there is no void" would have been one of the principles of physics if Aristotle had organized the subject as a demonstrative science. Such a proposition could serve as a premise in those proper demonstrations that proceed by *reductio*.

The description of a definition as "one without this," by which is meant a thesis that does not assert existence or nonexistence, is odd, but in accordance with the procedure of division by dichotomy. In fact, a definition does *not* assert existence (or nonexistence).[10] This interpretation is more plausible than Barnes's view that a definition "does not suppose that anything is the case" (pp. 103–4). In a perfectly ordinary sense of the expression, we would say that, for instance, the definition of triangle supposes that it is the case that a triangle is a three-sided rectilinear plane figure.

Aristotle envisages the same principles in I.2 as he does in I.10. The differences are differences of emphasis. I.10 is concerned to distinguish between the kinds of indemonstrable assumptions relating to subjects (or possibly primary subjects) and those relating to attributes (and possibly derivative subjects too) and to distinguish these groups from axioms. I.2 sets out to identify the types of principles without reference to the kinds of things they apply to. The difference is only a difference of viewpoint, not of doctrine, and even the example used in the I.2 passage (viz., "unit") is used to make the same point in I.10 (72a22–24, 76a34–36, b4).

THE PRECEDING PASSAGES contain Aristotle's views on the kinds of scientific principles. A number of other passages bearing on this subject must be discussed too, since they can be taken in ways that raise problems for the interpretation being advanced. First is a paragraph that introduces among other terms a nonstandard use of "hypothesis":

> **I.10 76b23–34** "What must be[a] because of itself and must seem [to be because of itself][b] is not a hypothesis or a postulate.[c]. . . Whatever provable things a person assumes, if he has not proved them himself and if he is assuming things which seem [provable] to the learner,[d] he is hypothesizing[e] them—and it is not a hypothesis without qualification, but only in relation to *him*—and if he is assuming the same thing either when there is no opinion or even the opposite opinion, he is postulating.[f] And a hypothesis and a postulate differ in this, for what is contrary to the learner's opinion is a postulate—anything that can be demonstrated and that a person assumes and employs without having proved.[g]"

NOTES ON TRANSLATION

a. In "what must be," "be" means "be the case" since the present passage defines hypothesis and postulate in terms of propositions ("things that can be proved").

b. What immediately precedes ("What must be because of itself") shows "[to be because of itself]" to be the correct supplement. Translators usually supply simply "to be the case" or "to be true" (Ross, p. 540; Barnes), but this is not the sense that is wanted. Aristotle is offering a second characterization of scientific principles. The relevant point is not that they seem to be true, but that they seem to be *principles*. This epistemic characterization makes the description of principles correspond to the descriptions of hypothesis and postulate below.

c. This is the first occurrence of the term "postulate" (*aítēma*) in *APo*. It is Euclid's word in the *Elements* for one kind of principle of geometry, but Aristotle's use is different (cf. below, Chapter 11).

d. How are we to complete the thought implicit in "things which seem"? It is wrong to supply "to be because of itself" as at b24 because here we are dealing with provable propositions, not principles. The standard interpretation (as at b23) is "things that seem to be the case." I prefer to take it as "things that seem to be *provable*" (going back to *deiktà ónta* in b27)—a stronger condition than the standard reading since it excludes not only false propositions but also many true ones. It has the advantage of making the point more relevant in its context. In a demonstration all premises must be either principles or proved conclusions. The premises must not merely be true; they must be proved (or anyway provable).

e. The verb in "he is hypothesizing" (*hupotíthetai*) corresponds to the noun "hypothesis."

f. The verb in "he is postulating" (*aiteîtai*) corresponds to the noun "postulate." The verb is an ordinary word for "ask" or "demand" and in the middle voice also means "ask for one's own use." The noun translated as "postulate" derives from this verb and has the sense of "something asked, demanded, or asked for one's own use." Aristotle is imagining a teacher having to request or demand to be allowed to use a proposition, whereas if the pupil believed the proposition to be true (or provable) such a request would not be necessary.

g. I follow Hayduck (cf. Ross, p. 540) and Barnes in emending the text so that it merely repeats the definition of "postulate" (first the differentia, then the genus). Without emendation, the passage reads: "for a postulate is (1) what is contrary to the learner's opinion, or (2) anything that can be proved and that a person assumes and employs without having proved" (*ésti gàr aítēma tò hupenantíon toû manthánontos têi dóxēi, è hò án tis apodeiktòn òn lambánēi kaì chrêtai mè deíxas*). (The emendation consists in removing *é*, "or.") Unemended, the passage first repeats the specific difference of postulates given above and then defines postulates as the genus of which it was just said to be a species, and we must see Aristotle as (abruptly!) introducing a different definition of "postulate," thus showing "that it, like *hypóthesis*, has not yet hardened into a technical term" (Ross, p. 540). I prefer to emend, not because it is hard to believe that there were competing definitions of such terms, but because the present passage reads much more smoothly.

DISCUSSION

This passage introduces hypotheses (in a noncanonical sense) and postulates as kinds of assumptions made in connection with demonstrative sciences. The first sentence shows that they are not principles. Recall that principles are grounds of the other propositions in a science (I.2 71b22). Being unprovable, they are also ungrounded (cf. *ek mè aitiatôn . . . aitíōn* at I.9 76a20; also I.6

75a35–37, I.24 85b24–25), and hence they "must be the case because of themselves and must seem to be the case because of themselves."[11] Hypotheses and postulates are provable propositions a person assumes without himself having proved them. This makes it clear that the use of "hypothesis" here is different from that in I.2 72a14ff., where hypotheses are "immediate deductive principles."

What person would omit to prove provable theorems? Mention of "the learner" (b28, b33) suggests a teaching situation, so the "person" will be the teacher. But what Aristotle says does not apply to all ways of teaching. The method of teaching envisaged is demonstration, and the circumstances seem to be the sort that frequently occurs in teaching mathematics, where a teacher in proving one theorem uses another theorem whose proof is too difficult or time-consuming for presentation to the class.

Hypotheses and postulates differ according to the pupil's attitude. A proposition is a hypothesis if the pupil is inclined to believe it. (I argue in note d that the pupil believes not only that the proposition is true but also that it is provable.) Otherwise it is a postulate. (Parallel reasoning calls for the pupil to have no opinion on or positively to disbelieve that the proposition is provable.) Aristotle stresses the relative nature of hypotheses and postulates: "and it is not a hypothesis without qualification, but only in relation to *him*" (b29–30); "he is assuming the same thing" (b31)—the same thing can be both a hypothesis and a postulate, depending on the learner's attitude.

From the point of view of the structure of demonstrative science, hypotheses (in this sense) and postulates are totally unimportant. They are provable propositions, and it is only in a special context that their proofs are not given. Since this context is the teaching situation, it follows that there can be differences between the intelligible order revealed by the structure of a science and the order in which a science is taught, a fact that is a difficulty for those who believe that *APo* "offers a formal model of how teachers should present and impart knowledge."[12] The passage is important for showing that Aristotle envisages some relation between demonstrative science and teaching, and also for the insight it gives us into the unsettled state of scientific terminology. Nevertheless it does not teach us anything about the kinds of scientific principles.

ANOTHER PASSAGE that uses the terminology of scientific principles is the following:

> **I.10 76b35–77a4** "Definitions[a] are not hypotheses for they are not said to be or not be anything,[b] whereas hypotheses are among the propositions.[c] It is necessary only to understand the terms, but this is not a hypothesis (unless someone is even going to say that hearing is a hypothesis). Hypotheses are, rather, all those things[d] that are [true], such that by their being [true] the conclusion follows. Nor does the geometer hypothesize falsehoods, as some declare, declaring that we must not make use of what is false, but the geometer speaks falsely when he says that the line he has drawn is a foot long

when it is not a foot long or straight when it is not straight. But the geometer does not base any conclusion on the grounds that the line that he himself has spoken of is *this one* [which he drew], but on the things indicated by these [figures]. Further, every postulate and hypothesis is either general or specific,[e] whereas definitions are neither of these."

NOTES ON TRANSLATION

a. I accept the normal translation of *hóroi* (e.g., Mure; Ross; Heath 1949, p. 52) as opposed to "terms" (Fritz 1955; Barnes). Landor 1981 is helpful, and I follow his interpretation, though not his translation of the word as "definientia." If man $=_{df}$ rational animal, "rational animal," though technically the definiens, is frequently called the definition of man.

b. "Are not said to be or not be anything" follows the manuscripts, with Barnes. By itself, a definiens is not a proposition, though a proposition is formed simply by attaching the definiendum and the copula. "To be" *(eînai)* here is the copula (unlike the superficially parallel *eînai* at 72a20, which is existential).

c. "Propositions" is only one possible translation of *protásesin*. Another possible translation is "premises." In any case the only propositions they could be are premises. (They are contrasted with conclusions at b38–39.)

d. "All those things": "all" brings out the generality implicit in *hósōn*.

e. "Either general or specific" translates *è hōs hólon è hōs en mérei*, which is literally "either as [a] whole or as in [a] part." There is wide agreement (Mure, Ross, Heath, Barnes) that the intended contrast is between universal and particular, but since no scientific proposition is singular, "particular" would have to be taken as "specific" or "less general" and "universal" as "general" or "more general."[13]

DISCUSSION

The present passage contrasts *hóroi* and hypotheses. *Hóroi* usually means "definitions," but cannot here, since definitions are, strictly speaking, principles and therefore propositions, and b35–36 says that *hóroi* are not propositions.[14] The other translation, "terms," is ruled out by a4: *hóroi* are neither general nor specific (or, neither universal nor particular); terms *are*. Hence the attraction of interpreting *hóroi* as definientia ("rational animal") as opposed to full-blooded definitions ("man is a rational animal"). *Hóroi* do not constitute yet another kind of principle; not being propositions, they cannot be principles. A *hóros* is an element of a full-blooded definition, and so is related to one kind of principle.

Hypotheses here are not the hypotheses of I.2 72a19–20. There they are existence claims; here they are defined more broadly as any premises used in deductions (a38–39). Therefore "they are not said to be or not to be anything," cannot have existential force (as in Ross's paraphrase "not being assumptions of existence or non-existence," p. 538, which requires emending *légontai* to *légetai*). This third use of "hypothesis" (different also from I.10 76b23–34)[15] may be brought in from outside the theory of demonstrative science, since the definition is not framed in terms dependent on the theory.[16]

A point about mathematical practice is made at 76b39–77a2. The geometer draws a line *AB* and says "Let *AB* be straight." This is not a (false) assertion that *AB* is straight, but means "Let *AB* indicate or represent a straight line." This kind of assertion is important in the "ekthetic" proof procedure of geometry, and is a kind of premise used in geometrical proofs. It is not a definition and is not obviously an existence claim. Aristotle appears to have inadvertently brought in a kind of proposition that is a fourth kind of unprovable principle, though I shall argue later[17] that this is not the case. In the end the passage adds nothing to Aristotle's doctrine of the kinds of principles.

ANOTHER RELEVANT PASSAGE occurs at the beginning of *APo*:

> **I.1 71a11–17** "It is necessary to have two kinds of previous knowledge. (1) For some things it is necessary to suppose in advance that they are; (2) for others it is necessary to understand what the thing being said is;[a] and for yet others, both. For example, [regarding the principle] that everything may be truly affirmed or denied [it is necessary to suppose in advance] that it is, for the triangle [it is necessary to understand] that it signifies this, and for the unit *both* things—both what it signifies and that it is. For each of these is not similarly clear to us."

NOTE ON TRANSLATION

a. "What the thing being said is": equivalent to "what it signifies." Aristotle overlooks the distinction of II.7, discussed in note e, on I.10 76b3–11, earlier in the present chapter.

DISCUSSION

The examples are the key to the passage. The law of the excluded middle occurs elsewhere (I.11 77a22, a30, I.32 88b1) as an example of a common principle. In I.10 the unit appears prominently as a subject whose significance and existence both need to be assumed among the proper principles of arithmetic (76a32–35). Also triangle is a subject whose significance must be assumed (76a34–35). Therefore I.1 contains the standard doctrine of the three kinds of principles, despite its explicit statement that there are only two kinds of previous knowledge needed. In fact, one of the two is based on an ambiguity. Our preliminary knowledge of the law of the excluded middle, knowledge "that it is," is knowledge that *it is the case*, and our knowledge of the unit is knowledge that *it exists*. The present passage therefore anticipates the standard doctrine of I.10 and I.2, so the types of prior knowledge it treats are those needed for demonstrative science, or rather (cf. 71a1) for the teaching and learning that produce demonstrative knowledge. (Whether they are needed for other kinds of teaching and learning is not clear.) Also, they are apparently the *only* kinds Aristotle believed necessary for this teaching and learning.

This passage gives more evidence[18] that *APo* recognizes a distinction between primitive and derivative subjects. For the primitive subject unit but not for the derivative subject triangle we must have previous knowledge that it is (i.e., exists).

A FINAL PASSAGE presents a different picture of the principles:

> **I.32 88b27–29** "The principles are of two kinds: 'from which' and 'concerning which'; those 'from which' are common, those 'concerning which' are proper, for example, number, magnitude."

DISCUSSION

This passage contradicts Aristotle's standard doctrine that there are three kinds of principles. Those "from which" are the axioms (cf. I.10 76b14, b22), and those "concerning which" are subjects (as the examples indicate). What has become of the per se attributes? There is another sign that Aristotle wrote these lines without paying full attention to the details of his theory. The proper *principles* (*ídiai archaí*) are definitions and existence claims, not subjects. Subjects are proper things (*ídia*). In fact, it is wrong to say that *principles* are that "concerning which." Since the purpose of the passage is not to go into detail about the types of principles, but to establish that *there are* principles that are restricted to a single subject genus,[19] there is no call to see it as presenting a serious rival to the standard doctrine. All Aristotle is doing is recalling (casually) that there are proper principles in addition to the common ones.

Summary

Aristotle's views on the kinds of principles emerge mainly from I.10 76a31–b15 and I.2 72a14–24. There are three kinds of principles: axioms, definitions, and hypotheses. Hypotheses are assertions of the existence of subjects of the science. The axioms differ from the other kinds by the fact that they are common, relevant to more than one science,[20] whereas the definitions and hypotheses are proper, that is, restricted to one science.[21] Each science treats a body of subjects and their per se attributes. The definitions of all these subjects and attributes form one class of principles. Existence claims are needed for all or some of the subjects: the passages differ on this point. *APo* I.10 76b3–15 favors the former view; I.10 76a31–41, I.1 71a11–17 and I.28 87a38–39 favor the latter.

Although identifying the kinds of principles is an important step, it does not get us very far. In order to learn how demonstrative sciences work, we need to find out in detail how Aristotle conceived of the different principles. Only then will we be in a position to see what roles the different principles play in proofs and to appreciate Aristotle's conception of the nature of science and the structure of reality.

The Subject Genus

CHAPTER IV

I SHALL BEGIN my examination of scientific principles by looking at the nature of the genus or subject genus (*génos*, *hupokeímenon génos*), which is the basis for the proper principles (existence claims and definitions) of a science, and so is a central concept in the theory.

To put it in a rough and preliminary way, the subject genus is what a science is about,[1] the subject matter it investigates, as arithmetic investigates number and geometry spatial magnitude. Each science treats one genus (I.28 87a38). A genus consists of subjects and attributes, of which the proper principles state the definitions and (in some cases) posit the existence, and whose per se relations the demonstrations prove. The subjects and attributes are "proper to each science" (I.10 76a38) because they are parts of the science's subject genus.

Subject genera are not arbitrary but natural, natural kinds or groupings of related things.[2] Different types of things belong to different genera (I.32 88b1–2) and different genera are treated by different sciences ("The subject genus is different" is offered at I.9 76a12 as the explanation of why harmonics is a different science from arithmetic. Cf. also I.7 75b3–6).

Subject genera thus serve as principles of identity and individuation for sciences. A science always deals with its own genus (I.7 75b7–8), and no science can prove the conclusions of another (I.7 75b12–14). Two sciences are different if and only if they treat different subject genera. Two scientific propositions are treated by the same science if and only if their subjects and attributes belong to the same genus. Attributes can be proven to belong to subjects only if they belong to the same genus. Principles having to do with the subjects and attributes of one genus do not have to do with the subjects and attributes of another: different sciences have different principles.

This rule has two exceptions. First are the axioms or common principles,[3] which apply to a plurality of subject genera and are principles of otherwise unrelated sciences. Second are some of the principles of the "subalternate" sciences,[4] which sciences are described as being "one beneath the other" (I.7 75b15); in these hierarchies, Aristotle notes, the principles of one science are used to prove results in another. Even here the genera are the same "in a way" (I.7 75b9).

The doctrine of subject genera implies that the sciences are radically separated from one another. Several consequences of the doctrine indicate the nature of the separation.

First, there is no transference (*metábasis*) of proofs from one science to another. The argument for this consequence occupies most of I.7:

> **I.7 75a38–b20** "It is not possible to demonstrate by crossing from another genus; for example, [it is not possible to prove] what is geometrical by means of arithmetic. . . . In cases where the genus is different, as in arithmetic and geometry, it is not possible to apply an arithmetical proof to the accidents of magnitudes, unless the magnitudes are numbers. . . . *But arithmetical proof always has the genus with which the demonstration is concerned, and similarly for the others. And so the genus must be either the same without qualification or somehow the same if the demonstration is going to cross. Otherwise it is clearly impossible, for the extremes and the middles must be from the same genus. For if they are not per se, they will be accidents.* For this reason it is not possible to prove by geometry that the knowledge of opposites is single or that two cubes are a cube, nor [is it possible to prove] by one the matter of another, except for all that are so related to one another that one is under the other, as the facts of optics are related to geometry and the facts of harmonics to arithmetic. Nor [is it possible to prove by geometry] anything not belonging to lines qua lines and qua following from the proper principles [of geometry], for example, whether the most beautiful of lines is the straight line, or whether it is opposite to the circular line, for this does not belong qua the proper genus of them, but qua something common."

DISCUSSION

The heart of the argument (75b7–12) is italicized. A proof in one science cannot prove a conclusion in another because it "has the genus with which the demonstration is concerned": all the terms appearing in its premises and conclusion belong to the same genus. Since different sciences have different genera, a demonstration whose terms all belong to the genus of one science cannot have a conclusion whose terms are from the genus of another science, and therefore its conclusion cannot be a proposition relevant to that science.

Aristotle argues elliptically that all the terms in a proof belong to a single genus: "for if they are not per se, they will be accidents." This depends on two results of I.6: that the premises and conclusions of demonstrations assert per se relations among the terms of the science's subject genus and that a relation is accidental if and only if it is not per se.[5] The argument therefore is that unless all the terms in a proof are from the same genus, it will contain terms related accidentally, not per se, and so will not be a demonstration. The argument against genus crossing is thus based on a claim that terms related per se are in the same genus. This doctrine is important for understanding both the nature of the subject genus and that of per se relations. But why should we accept it? Aristotle does not argue for it. Later in this chapter I shall consider reasons that may have led him to adopt it.

A few examples make the point clear. Geometry cannot prove that the knowledge of opposites is single (b12–13). Presumably this is the job of metaphysics. Geometry cannot prove that two cubes are a cube (b13–14), that is, the *arithmetical* result that the product of two cubes is a cube: if x and y are whole numbers and $a=x^3$ and $b=y^3$ then $ab=(xy)^3$.[6] Moreover, "cube" is a term in geometry as well as in arithmetic, and its difference in the two sciences is reflected in the different definitions of the term and consequently in the different per se relations it has. The example thus calls attention to the importance of basing proofs on the per se relations appropriate to the genus of the science in question. Geometry cannot prove whether the straight line is most beautiful (b18–19) because in this proposition the subject belongs to the genus of geometry but the predicate does not. From the point of view of geometry, aesthetic qualities are irrelevant, incidental (*sumbebēkóta*). Beauty does not belong to lines qua (geometrical) lines and is not proved to belong to lines from the proper principles of geometry. The same holds in the final example: the predicate "is contrary" applies to geometrical objects not qua geometrical objects, but qua things that are (*ónta*).[7]

THE SECOND CONSEQUENCE is that there is no master science that proves the principles of all the other sciences.

> **I.9 76a16–23** "It is not possible to prove the proper principles of each thing,[a] for they[b] will be the principles of all,[c] and the scientific knowledge [or, science] of them[d] will govern all things.[e] For indeed the person whose knowledge comes from the higher[f] "causes" knows better,[g] since he knows from what is prior when he knows from "uncaused causes." And so, if he knows better and best, that science [or scientific knowledge] will also be better and best. But demonstration does not apply to another genus."

NOTES ON TRANSLATION

a. "Each thing" refers to each conclusion proved in a science (cf. 76a14).

b. "They" refers to the (ultimate) principles on which are based the proofs of the proper principles of each thing.

c. "Of all" (*hapántōn*) refers either to all the conclusions or to all the conclusions plus the (first-level) proper principles on which the conclusions are based. It cannot refer to all the principles or all the sciences, which would require a feminine, not a neuter, form.

d. "The scientific knowledge [or, science] of them" (*epistḗmē hē ekeínōn*) means either "the [indemonstrable] scientific knowledge of the ultimate principles" (admitting indemonstrable *epistḗmē* as in I.3 72b18–20) or "the scientific knowledge based on the ultimate principles" (although *ex ekeínōn* would be expected; cf. 75b38, 76a5, a7, a14, a19, etc.) or "the science to which they belonged" (Mure).

e. An alternative to "govern all things" (*kuría pántōn*) is "be authoritative over all things." The reference of "all things" is the same as that of "all" (a17). See note c. The idea that knowledge of the principles of *x* governs *x* extends the ideas that conclusions depend on principles and that knowledge of conclusions depends on knowledge of principles (I.2).

f. "Higher" is equivalent to "prior." "Causes" are prior to conclusions dependent on them (I.2 71b31). Ultimate principles are higher causes than first-level proper principles.

g. "Knows better" means "knows more fundamentally or more basically."[8]

DISCUSSION

The nature of a master science is revealed at a17–22: (1) its principles will be principles for the conclusions as well as the principles of all other sciences, and (2) the conclusions of all other sciences and our knowledge of them will be founded on the principles of the master science and our knowledge of them. Aristotle argues for (1) by saying that the (ultimate) principles that prove the proper principles of each thing will be principles of all. The argument can be read as invalid: "the ultimate principles for proving one thing are the ultimate principles for proving all things." But it can also be quite naturally interpreted as valid: "for each conclusion *x* in each science, there is a set of proper principles $p_{x1} \ldots p_{xn}$ on which *x* is based; let *P* be the union of the sets of all proper principles of all conclusions of all sciences. Then if it is possible to prove all members of *P*, each member *y* of *P* will be a conclusion of a proof based on a set of ultimate principles $q_{y1} \ldots q_{ym}$. Let *Q* be the union of the sets of all ultimate principles. The members of *Q* will be the (ultimate) principles of all things." All this would be difficult to express precisely in Greek, which perhaps accounts for the imprecision of Aristotle's language.

Aristotle argues for (2) as follows: the ultimate principles are "uncaused causes,"[9] so my knowledge based on them comes from prior or higher "causes," and so I know better. Indeed, I know best, and the science with these ultimate principles as its principles is also better and best. The superlative is justified by the ultimate nature of the principles in question. There can be no better knowledge because nothing is more ultimate. Apparently, being based on ultimate principles and therefore being better and best makes the master science authoritative over the principles and conclusions of the other sciences.

The possibility of a master science is rejected at 76a22–23 because "demonstration does not apply to another genus." If one science proved proper principles of different sciences it would prove results concerning different genera, which is impossible.

A THIRD CONSEQUENCE is that not all proofs are based on a single set of principles:

> **I.32 88a31–36** "And there are not even the same principles for all true deductions, since the principles of many are different in genus and do not apply. For example, units do not apply to points, for the former do not have position and the latter do. But it is necessary for [the principles] to fit either into the middle or above or below, or for [demonstrations] to have some of their terms inside and others outside."

DISCUSSION

Despite the obscure final sentence,[10] the argument clearly depends on the doctrines that different sciences have different genera and that there is no crossing from one genus to another in proofs.

FOURTH, FOR THE same reason it is impossible to prove all conclusions by means of the axioms:

> **I.32 88a36–b3** "There can be no common principles from which all things will be proved. I mean common ones like the law of the excluded middle. For the genera of things are different and some belong to quantities, others to qualities alone, with which they are proved through the common principles."

DISCUSSION

The interpretation of this argument is uncertain because of multiple possibilities of interpreting several key terms and phrases. Does "all things" (a37) refer to proofs? conclusions? attributes? "The genera of things" (b1) to subject genera of sciences? categories?[11] Does "some . . . others" (b2) refer to "all things"? "genera"? "things"? Or should it be interpreted as "principles" (Ross)? Does "with" (b3) mean "by means of"? "together with"? Does "which" (b3) refer to "some . . . others"? "quantities, qualities"? "the genera of things"? In "they are proved" (b3) does the subject pronoun refer to "some . . . others"? Or to something to be supplied without reference to the previous clause? Or is the verb used impersonally (= "proofs are constructed")?

A possible paraphrase is: "It is impossible to base the proofs of all conclusions on any common principles (like the law of the excluded middle) because the objects (subjects and attributes) of different sciences belong to different subject genera. The objects of some sciences are strictly in the realm of quantities, those of other sciences strictly qualities, and proofs having to do with these objects, even if they make use of common principles, must also make use of facts about the quantities or qualities in question—about the subject genus of the science.''

In any case, the point of the argument is clear. Common principles are insufficient to prove conclusions in particular genera. A conclusion in arithmetic, say, may depend on common principles, but it depends on arithmetical facts *too*, and the reason for this is the division of things into different genera.

OTHER PASSAGES REJECT interpreting the thesis that all conclusions have the same principles as meaning that they can all be proved from the totality of principles of all conclusions. One such passage is this:

> **I.32 88b17–21** ''This is really stupid.[a] For neither in the well-known branches of mathematics[b] does this happen,[c] nor is it possible in analysis.[d] For the immediate propositions[e] are principles, and a different conclusion occurs when an immediate proposition is added. But if someone should say that the *primary* immediate propositions are principles,[f] there is one in each genus.''

NOTES ON TRANSLATION

a. What is ''really stupid'' is the notion that this is a reasonable interpretation of the thesis.

b. ''The well-known branches of mathematics'' probably refers to arithmetic and geometry, which are so prominent in *APo* I.

c. ''Does this happen'' means ''does it happen that conclusions are based on all the principles.'' In the context this means that the proof of a given conclusion, say, that all triangles have 2R, does not employ all principles of arithmetic and geometry. It is also true that it does not employ all the principles of geometry.

d. ''Nor is it possible in the analysis'' means that the wrongness of this interpretation is also shown by the procedure of analysis, where we find the proof of a proposition *P* by working back from *P* to unprovable propositions (principles) that entail *P*.[12] This procedure shows that *P* depends on some principles, but not on *all*.

e. ''For the immediate propositions'' refers to the immediate propositions discovered by the analysis.

f. ''The *primary* immediate propositions are principles'' follows Ross's emphasis and punctuation (p. 604). Aristotle is not repeating the view just expressed (b18f.) but imagining his opponent as making a new point, and the addition of *prôtas* is the only difference.

DISCUSSION

The first two sentences of the passage contain three reasons against the interpretation. (1) Geometry, for example, does not depend on the totality of all principles of all sciences, only on those of geometry. (2) In analysis of particular geometrical propositions we always end up at geometrical principles, never with any other principles. In fact, (3) analysis will end up at a limited set of geometrical principles. Barnes (p. 188) interprets as follows. If the analysis reveals that *P* depends on certain principles, the fact that *P* does not depend on *all* principles is seen simply by tacking another immediate proposition (i.e., another principle) onto the end of the proof. This produces *another* conclusion, different from *P*, which depends on an additional premise, and so *P* does not depend on *all* premises. But this interpretation begs the question: it needs to be proved that there *is* another principle that can be tacked on. I prefer the following interpretation. The analysis of two different propositions will reveal that they depend on different sets of principles,[13] although these sets may overlap. The minimal difference in principle-sets of different propositions will be that one contains the other and one more principle besides. It follows that at least one of the propositions does not depend on *all* principles.

The final sentence is not a separate argument,[14] but a defense of the thesis Aristotle has just refuted, followed by Aristotle's refutation of the defense. The summary at b21–23 shows that only two views have been dealt with (b11–15, b15–21). The opponent suggests that there are primary immediate principles that are the principles he meant—the principles that entail the first-order principles and all the conclusions derived from them. Aristotle's reply is emphatic.[15] He effectively refutes the suggestion by noting laconically that however these primary immediate principles are selected, there will still need to be one (anyway) in each genus. The doctrine of no crossing from one genus to another still applies: even if you reduce the principles to the smallest possible number, there must still be at least one in each genus, and a conclusion in one genus is independent of proofs, conclusions, and principles (no matter how chosen) in other genera.[16]

A MODIFIED VERSION of the "really stupid" interpretation is refuted shortly below:

> **I.32 88b23–29** "The principles of all things are related,[a] but different conclusions are proved from different groups of principles. But it is clear that this too is impossible, for it has been shown that the principles of things that are different in genus are different in genus. For the principles are of two kinds: 'from which' and 'concerning which'; those 'from which' are common, those 'concerning which' are proper, for example, number, magnitude."

NOTE ON TRANSLATION

a. "Related" (*sungeneîs*) means "co-generic," in the same genus. The suggestion is that there is a single super-genus, which includes all the genera of the different sciences.

DISCUSSION

Aristotle's opponent retorts cleverly to the argument at 88b17–21. Aristotle has proved too much. Mathematical analysis shows not only that a given conclusion does not depend on all principles of all sciences, but also that it does not even depend on all the principles of any single science. But that is no reason to carve a science up into as many sciences as there are conclusions, each with its own set of principles. Principles are principles of the conclusions that depend on them, but are also principles of a science, and a given principle does not need to be used in all the proofs of the science. Aristotle would agree so far, but the opponent presses the point farther. Since a proposition may be a principle of a science without being a principle for all proofs in the science, all demonstrative knowledge can be considered as a single science, including among its principles the principles Aristotle would call proper to geometry, arithmetic, and so on. The proposition that all triangles have 2R will not depend on principles having to do with numbers, but that is no different from its not depending on principles having to do with (say) similar figures. Aristotle refutes this view (b26–27) by asserting "the principles of things that are different in genus are different in genus"—a consequence of his view that all the terms in a demonstration belong to the same genus (I.7 75b3–12, discussed earlier in the present chapter). Therefore all the *principles* in a proof belong to[17] the same genus, and so if conclusions belong to different genera, so do the terms in their proofs, and hence so do the principles of the proofs.

The last sentence contains some difficulties,[18] but clearly its purpose is to identify principles that *do* differ in genus. Aristotle invokes his views on different kinds of principles, granting that some principles are common but insisting that there are also proper principles concerned with the subject genus of a single science. He mentions number and magnitude as examples of different subject genera, and we are to understand that the definitions and existence claims for things belonging to those genera are the proper principles he means.

The Composition of the Subject Genus

Although Aristotle's general line on the subject genus is clear, a number of problematic passages must be examined in order to sharpen our understanding of it. First I shall consider a group of *prima facie* conflicting statements about the nature of the subject genus. The genus of a science is said to be (1) a single generic nature (e.g., number, spatial magnitude [*mégethos*]); (2) the individuals or types that embody the generic nature (numbers, magnitudes); (3) the subjects (or primitive subjects); (4) the per se attributes, perhaps together with the derivative subjects or all the subjects. Moreover, we are told that (5) the genus has per se attributes; (6) the conclusion of a demonstration belongs per se to some genus; (7) all terms in a demonstration must come from the same genus; (8) principles and conclusions are in the same genus; (9) a science "brings its proofs concerning a genus."

Evidence for (1): (a) I.10 76b16–19: "Nothing prevents overlooking some of these, for example, not hypothesizing that the genus exists, if it is clear that it exists. For it is not equally clear that number exists and that cold and hot do." (b) I.10 76b21–22: "Nonetheless, these things are by nature three: that concerning which the science forms its demonstrations, what it demonstrates, and the things from which it demonstrates." (c) I.32 88b26–29: "The principles of things that are different in genus are different in genus. For the principles are of two kinds: 'from which' and 'concerning which'; those 'from which' are common, those 'concerning which' are proper, for example, number, magnitude."

DISCUSSION: (a) indicates that number is a genus. (b) refers to the genus as "that concerning which,"[19] and (c) identifies number and magnitude as "that concerning which." These passages present problems too. Surely hot and cold are *attributes* of physics, whose genus is elsewhere described as "substance in which the principle of motion and rest is in itself" (*Met.* E.1 1025b19–21). At best, then, (a) shows that *one* way of thinking of the genus is as a single generic nature. The construction of passage (c) is not clear (see discussion of that passage in Chapter 3 above).

EVIDENCE FOR (2): (a) I.10 76a42–b2: "Each of these is sufficient so far as it is in the genus, for it will do the same even if [the geometer] does not assume it [as holding] of all things, but only for magnitudes, and for the arithmetician [it is assumed as holding] for numbers." (b) I.7 75b3–6: "In cases where the genus is different, as in arithmetic and geometry, it is not possible to apply an arithmetical proof to the accidents of magnitudes (unless the magnitudes are numbers)."

EVIDENCE FOR (3): (a) I.10 76b12–13: "As many things as are posited to exist (these are the genus . . .)." (b) I.7 75a42–b2: "the subject genus whose attributes, that is, per se accidents, the demonstration reveals."

DISCUSSION: (a) can be taken either as referring to the primitive subjects or to all the subjects of the science.[20] If we take (b) as shorthand for the view that proofs show that attributes belong per se to subjects, then Aristotle is identifying the genus with all the subjects, not a privileged subset of them.

EVIDENCE FOR (4): (a) I.28 87a38–39: "A single science is the science of a single genus—all the things that are composed of the primary things and are parts or per se attributes of these."

DISCUSSION: "All the things . . . " identifies the genus. It includes the per se attributes "of these." It is not clear what "of these" refers to ("all the things" or "the primary things" or both), what "the primary things" are, and how things are composed of the primary things. I interpret "the primary things" as primitive subjects, "all the things composed of the primary things" as derivative subjects, and "of these" as referring to both primitive and derivative subjects; so the genus consists of the derivative subjects and the parts and per se attributes of all the subjects. Primitive subjects are presumably in the genus too.[21] Whether the "parts" are the per se attributes[22] or are component parts or species is not clear. In any of these ways they will be subjects or attributes of the science.

EVIDENCE FOR (5): (a) I.10 76b13: "The genus whose per se attributes the science investigates." (b) I.7 75a39–b2: "There are three things in demonstrations: (1) what is demonstrated, the conclusion (this is what belongs per se to some genus) . . . (3) the subject genus whose attributes, that is, per se accidents, the demonstration reveals.''

EVIDENCE FOR (6): (a) I.7 75a40–41: ''The conclusion (this is what belongs per se to some genus).'' (b) I.6 75a28–29: ''All things that belong per se and as such belong of necessity concerning each genus.''

DISCUSSION: (b) can be taken as saying ''concerning each genus, things that belong per se and as such *to some subject* belong *to that subject* of necessity,'' so that ''concerning each genus'' is used generally and not closely with ''belong of necessity.'' But in view of (6a) and (5), it can also be taken as saying ''all things that belong to each genus per se and as such belong necessarily to it.''[23]

EVIDENCE FOR (7): (a) I.7 75b10: ''For the extremes and the middles must be from the same genus.''

EVIDENCE FOR (8): (a) I.28 87b2–4: ''It is necessary for them [the indemonstrable premises] to be in the same genus as the things that have been demonstrated. An indication of this is when the things proved through them are in the same genus and co-generic.'' (b) I.9 76a8–9: ''It is necessary for the middle to be in the same genus.''

EVIDENCE FOR (9): (a) I.11 77a24–25: ''By 'for the genus' I mean the genus concerning which [the science] brings its proofs.''

DISCUSSION: By this odd phrase Aristotle seems to express some or all of the points covered in (6)–(8).

THERE IS NO need to think that Aristotle spoke at random in characterizing the subject genus. Many of these statements are closely related. Statements (5), (6), (8), and (9) should be understood as consequences of (7). Also the tendency not to distinguish carefully between, on the one hand, things (subjects, attributes) in (1)–(5) and, on the other hand, terms or propositions in (6)–(9) is characteristic of Aristotle.

The chief remaining differences can be played down as well: viewing the genus as a single nature, as the things embodying that concept, as the subject (or primitive subjects), as the subjects and their per se attributes, and as including principles and conclusions come to much the same thing. For example, saying that geometry is the study of spatial magnitude is shorthand for saying that it is the study of spatial magnitude*s*, since geometry studies the kinds of magnitudes and the relations among them. Moreover, the magnitudes it studies can be regarded as the subjects of its proofs or as the subjects and their per se attributes that it proves to belong to them. Saying that geometry is the study of points, lines, triangles, squares, and so on (or of the point, the line, and so on) is no more correct than saying that it is the study of points, lines, and so on together with their attributes straight, isosceles, and so on. It is not even unreasonable to say that geometry is the study of points, lines, and circles (supposing that these are the primitive subjects from which all other subjects are derived),[24] or that it treats the principles and/or the conclusions concerning its subjects and attributes. All these are legitimate interpretations of the most general description of the subject genus, ''what the science is about.''

But what are we to make of assertions that attributes and conclusions belong per se to the *genus*? (See evidence for [5] and [6].) Possibly this is a loose way of saying that attributes belong per se to subjects in the genus[25] and that a conclusion states a per se relation holding between an attribute and a subject in the genus. But they can be given a reasonable interpretation as they stand, which points to a view of the subject genus that illuminates the general theory of science.[26]

When the genus is identified as the subjects, attributes, principles, conclusions, and so on, we tend to think of it as a collection of items. But we must keep in mind that it is a highly structured collection. Listing its contents represents its nature inadequately. The items are related to each other in various ways, which the proofs of the science reveal. The genus is better viewed as a network of subjects and attributes than as an unordered set. The links among the subjects and attributes are the immediate and nonimmediate per se relations that hold among them. The principles state which links are primary, and the proofs show that and how the derivative links depend on the primary ones.

A demonstrative science explicates the network that is its subject genus. In establishing facts about the subjects and attributes in a genus, the proofs also establish facts about the genus and so provide scientific knowledge about it. Knowledge that triangles have 2R is knowledge about spatial magnitude as well as about triangles (and the attribute 2R). Except perhaps for the definition of the generic nature (e.g., number) there are no other scientific facts aside from facts having to do with the subjects and attributes that belong to the genus. Moreover, it seems unlikely that the definition of the generic nature will appear as a principle of any demonstration. The same is true for a hypothesis that the generic nature exists. It is not even clear whether such existence claims are intended.[27]

On this interpretation, attributes and conclusions do belong per se to the genus. As we shall see,[28] a per se relation between a subject and an attribute is one that is due to their nature. But just as, for example, the fact that triangles have 2R is due the natures of the triangle and of the right angle, it is also due to the nature of spatial magnitude.[29] If the genus is a structured network of subjects and attributes, their per se relations are part of the nature of the network: if a subject had different per se relations it would be a different subject (Euclidean triangles are different from Riemannian triangles) and the network would be different. Thus to speak of attributes and conclusions as belonging per se to the genus, of demonstrations as revealing per se attributes of a genus, and of a science as investigating the per se attributes of its genus is only a slight extension of the more strict uses of "per se" discussed in I.4 and is a reasonable way to express the role of those facts and propositions in the science as a whole.

Interpretations of the Subject Genus

Finally, what philosophical justification might Aristotle offer for his view that sciences have distinct subject genera? The text contains no positive argument on this issue, so I will approach the question by considering whether the concept of subject genus is related to some firmly entrenched area of Aristotle's philosophy. If it is, we may conclude that the concept of subject genus is not ad hoc, but is of a piece with his larger metaphysical view. The parts of his philosophy I shall discuss are his doctrines on genus and species and his view that what is (*tò ón*) has genera.

First, did Aristotle think of subject genera as genera articulated according to a genus/species division, as commentators frequently think?[30] On this view, the subject genus (e.g., spatial magnitude) is the *summum genus* of the science, and its species are the subjects of the science (e.g., point, line, triangle). The definitions, which are the source of the per se relations of the science, will be in the form of genus plus differentia.

A number of considerations favor this suggestion. The subjects of demonstrative sciences are not particulars, but, like species, universals (see, e.g., I.31 87b28–39). Also, Aristotle's belief in natural kinds and natural subdivisions of those kinds into species may remind us of the natural groupings of things into subject genera of sciences and the natural constitution of those subject genera out of the subjects of the science. The subject genus also has the logical relations to the subjects of the science that genera have to their species: the subject genus and its definition are predicable of them but not vice versa, it applies to all particulars to which the different species apply, and so on.[31] Further, it separates its subjects from everything else (*Top*. VI.3 140a27–28) and is not an accident of its subjects (*Top*. VI.6 144a24–25).

Moreover, this interpretation explains how things related per se are in the same genus.[32] As I.4 and I.6 make clear,[33] things related per se are such that one is part of the essence of the other or is a consequence of the other's essence. Since the essence of something is stated in its definition and since Aristotle defines a species in terms of its genus and specific difference, every member of the subject genus (except the *summum genus*) is defined in terms of another member of the subject genus and so is related per se to it. Also, the only things related per se to members of a subject genus are other members of it, and every member (including the *summum genus*, which appears in some definitions) is related per se to others. The present interpretation of a subject genus thus entails that things related per se are in the same subject genus—as long as the only "things" under consideration are subjects.

One objection to this interpretation is that a single science may cover more than one genus, as biology covers all genera of animals. But Aristotle will call a single thing now a genus, now a species, depending on whether he is considering it as a class containing subdivisions or as itself a subdivision of a wider class.[34] This flexibility means that not everything that can be called a genus can be a subject genus of a science, but it does not mean that the subject genera of sciences are not genera.

This flexibility can be applied to the subalternate sciences. The subject genera of geometry and arithmetic are plausibly described as species of quantity, which could be the genus of universal mathematics, which Aristotle occasionally mentions.[35] We might also think (less plausibly) that the subject genus of optics (visible lines) is a species of the genus of geometry.[36]

A more serious objection stems from difficulties in arranging the subjects of a science into a taxonomic chart with the subject genus as *summum genus*. In geometry, figure might be divided into triangle, quadrilateral, and so on, but this would involve an infinite number of members of a single division, a situation that might well cause serious doubts about the legitimacy, as well as the intelligibility, of the division. In biology not only the species of animals but their parts will be subjects in the subject genus, but parts of animals have no place in a genus/species chart.

More seriously still, there is no room in a genus/species arrangement for the per se attributes, so this interpretation of the structure of the subject genus is incompatible with evidence that the genus contains attributes as well as subjects.

Also, the ordering that the proofs of a science reveal is not identical with genus/species ordering. The primitive subjects (e.g., units, or points, lines, and circles) are not the genus concept, and the existence proofs for derivative subjects do not in general argue from species to genus or vice versa. Also, many of the relations proved to hold among subjects (e.g., that any rectilinear figure can be broken into triangles, or that a square can be constructed equal in area to any given rectilinear figure) are not reflected in a genus/species arrangement.

Finally, since nothing in the theory of demonstrative science requires that subject genera have genus/species structure, and since there is clear evidence that a subject genus contains more than just its subjects, we must concede that Aristotle's conception of subject genera is different in origin, purpose, and nature from his conception of genus/species hierarchies.

The other possibility is to connect the view that different sciences have different subject genera with the doctrine that "being falls immediately into genera."[37] Here the answer is a qualified affirmative. Immediately after the statement just quoted, Aristotle says, "for which reason the sciences too will correspond to these genera."[38]

We must read these statements with care. Aristotle does not say there is a science for every genus of being, and although his words may be so interpreted, they are best taken as meaning that each science corresponds to *some* natural division of reality.[39] The genera into which being immediately falls are probably the categories,[40] but there is no hint of a doctrine that there is one science for each category.[41] Rather, there are sciences of some of the genera into which the categories fall,[42] as physics is the science of a large genus of substances: "substance in which the principle of motion and rest is in itself" (*Met.* E.1 1025b19–21). Similarly, geometry is the science of magnitude, which is defined as a species of quantity (*posòn metrētón*, *Met.* Δ.13 1020a9) and number, the genus of arithmetic, is defined as "limited plurality" (*plêthos tò peperasménon*, *Met.* Δ.13 1020a13), where plurality is a species of quantity (*posòn arithmētón*, *Met.*Δ.13 1020a8–9).[43] This view is supported by statements that sciences carve off or delimit some part of being and investigate it (*Met.* Γ.1 1003a24–25, E.1 1025b8–9).

The view that subjects of sciences are natural divisions of what there is also provides a way to ensure that all subjects and attributes related per se to one another are in the same genus (see discussion of I.7 75a38–b20 earlier in the present chapter). In investigating a subject genus, a science investigates the subjects of that genus qua belonging to that genus. The same thing may be treated by more than one science. For example, man is treated qua a thing-that-is by first philosophy, qua a substance whose principle of motion and rest is in itself by physics, and qua an animal by biology. The attributes of man treated differ from science to science,[44] depending on which are relevant to understanding man qua member of the subject genus in question. To consider a subject qua member of a given subject genus is to consider its (nonaccidental, per se) relations to certain attributes. Thus, just as the subjects of a science are naturally related by virtue of being in the same natural division of things-that-are, a subject of a science is naturally related to its per se attributes in that it has those attributes qua belonging to the subject genus of the science.

On this account we can have it both ways. The subject genus of a science considered as a division of things-that-are is the collection of subjects of the science, but the things-that-are considered as belonging to the subject genus are things-that-are considered in a certain way, as subjects concerning which it is relevant to investigate whether certain attributes belong per se to them, and irrelevant to investigate whether others do.[45] The attributes can count as parts of the subject genus since it is the nature of a thing's relation to the attributes that determines whether that thing belongs to the subject genus.

The importance of the subject genus as the principle of identity and individuation for sciences can hardly be overstated. It is what guarantees the independence of sciences from one another[46] and what confutes the possibility of a master science.[47] The doctrine of separate genera is normally and surely rightly interpreted as a rejection of Plato's view[48] that the science of Ideas can establish principles for geometry and other sciences[49] starting from the Idea of the Good.[50] But its significance is more than historical. It declares the unity of science and scientific knowledge to be a myth and proclaims specialization, not empty generalization, to be the key to scientific progress.[51] At the same time, the theory of demonstrative science is a theory of the unity of science at a second level. Despite their differences in subject matter, all sciences have the same types of principles and proofs. All the different genera of things are susceptible to the same treatment.

As the basis of the separation of sciences, the subject genus is also the basis of the proper principles of each science: the definitions of the subjects and attributes that make up the subject genus and existence claims for some or all of the subjects. Before taking up these types of principles, I will discuss the areas in which sciences are not wholly independent of one another: the subordinate sciences and the common principles.

The Subordinate Sciences

CHAPTER V

THE SUBORDINATE SCIENCES are the only exception to the general prohibition against genus crossing in proofs. They are first mentioned in connection with this rule, and they recur several times in *APo* I, usually as exceptions to or special cases of general claims about demonstrative sciences. *APo* I.7, I.9 76a4–15, and I.13 78b32–79a16 provide most of the information needed to understand the nature of the subordinate sciences, although other passages, especially I.27, are relevant too, together with surviving scientific treatises (such as Euclid's *Optics*).[1]

That the subordinate sciences are important for Aristotle is clear from his argument for the claim that there is no genus crossing in demonstrations (I.7), which is written with those sciences very much in mind.

I.7 75a38–b20. (This passage is translated in Chapter 4 above.) If the terms in a proof do not come from the same genus, they will not be related per se, and so the conclusion can only be accidental and therefore not scientific. But there are exceptions to this rule. Geometry somehow proves facts in optics, yet optics is a separate science. The subordinate sciences thus threaten the doctrine of the subject genus, according to which if geometry proves results in optics, geometry and optics have the same genus, which means that they are the same science, whereas if they are different sciences, then geometry cannot establish optical facts.

Aristotle needs to establish some intermediate status between identity and complete separation of the sciences, and he does so with what seems to be a waffle: the subject genera of the two sciences are not "the same without qualification," but are "somehow the same," so that optical proofs and geometrical proofs have terms from the same genus, and the conclusions of geometrical proofs can apply to optics. But this is less a solution than a promise of a solution, and Aristotle's hesitation on the matter is indicated by a straightforward assertion that the subject genus of harmonics is different from that of arithmetic (I.9 76a12).[2]

He needs to explain how genera can be the same in a qualified way, and although some of what he says about subordinate sciences helps one understand their partial identity, the picture does not come clear until we look beyond the Aristotelian material. Euclid's *Optics* shows that the science of optics was the general study not of vision but of the geometrical properties of visual sensation abstracted from all other aspects, that is, vision qua geometrical. Optics studies geometrical properties of *visual* lines, and optical properties, such as how large different objects appear, depend on geometrical properties, such as how large an angle they subtend in the visual field. Visual lines are not identical with geometrical lines. Lines as treated by geometry (lines qua geometrical, geometrical lines) are treated without reference to their material nature (I.13 79a9), but in the case of optical lines (lines qua visual, visual lines) the "material"[3] nature is not ignored. Thus the subject matter is different. But since optics studies the geometrical aspects of visual lines, the manner of treatment is geometrical.

What this means in practice is made clear by the proofs in the *Optics*. A proposition about visual properties is stated as a theorem ("*A* appears larger than *B*"). Then a strictly geometrical proof follows, leading to a strictly geometrical conclusion ("*A* subtends a larger angle than *B* does"). Finally, this result is applied to the subject matter of optics by means of "bridge principles" ("Things seen under a larger angle appear larger").[4]

Geometry can be invoked because the items in the subject genus of optics are in the genus of geometry *too*. Visual phenomena as well as physical objects are treated by geometry, when their geometrical properties are abstracted from the various kinds of matter in which they occur. Also, the properties abstracted for the sake of geometry are largely the same as those abstracted for geometrical optics (the only exceptions being the optical properties having to do with spatial appearance). Visual lines are susceptible to geometrical treatment because they *are* geometrical lines—not *all* geometrical lines, but certain ones, and those lines qua visible, that is to say qua having their geometrical properties and their visual ones *as well*. This compound nature is allowed because the two types of properties are related. Geometrical ones are the grounds for visual ones. The reason why *A* appears larger than *B* is that *A* subtends (and so is seen under) a larger angle than *B*. Hence the legitimacy of the bridge principles, which amount to rules for interpreting geometrical facts optically. These are the only principles used in optical proofs that are not also principles of geometry.

Thus Aristotle can say that conclusions in a superior science and in a subordinate one are proved from the same principles (I.12 77b1–2) and that their principles have the common nature needed for proofs to cross genera. Thus he can also say that conclusions in a subordinate science are proved in the same way as those in its superior science, but that there is a difference in that the fact (*tò hóti*) is in the province (subject genus) of the subordinate science, while the superior one provides the explanation (*tò dihóti*) in terms of relations that hold per se among *its* subjects and attributes (I.9 76a10–13). Euclid's *Optics* therefore enables us to understand the most important features of the relation between superior and subordinate sciences.

One other feature of subordination is mentioned in the main discussion of the way a superior science provides the explanation of the facts treated by a subordinate one:

> **I.13 79a4–6** "Often the practitioners of the superior science do not know the fact, just as people who investigate the universal frequently do not know some of the particulars through lack of observation."

DISCUSSION

Aristotle is not faulting the geometer for failing to notice facts of optics he should have noticed. The context shows that the point is that the geometer qua geometer does not investigate facts of the subordinate science. Geometry's brief is to discover and prove geometrical facts, not to apply them. Geometrical results apply elsewhere,[5] but other sciences make the applications.

THE PASSAGE SUGGESTS without entailing that the superior science is more general than the subordinate one, which means that its subject genus extends more widely. This can be true in two ways. First, some of the subjects and attributes treated by the superior science are irrelevant to the subordinate one (the geometry of the circle may be irrelevant to optics). Second, the subordinate science treats only one possible application of the superior science. Optics deals with visual lines, but mechanics may concern itself with lines qua having weight.[6] The second way is of special interest because it points toward a different kind of relation among sciences than Aristotle seems to have considered, and one that enables us to subsume the doctrine of common principles under the doctrine of subordinate sciences.[7]

This relation is found in modern abstract algebra, a number of whose branches can be arranged in order of generality: set theory, group theory, ring theory, and field theory.[8] Groups are defined as sets with certain additional properties, rings as groups with additional properties, and fields as rings with additional properties. Any property of sets holds for groups, rings, and fields too; any property of groups holds for rings and fields; and any property of rings also holds for fields. Proofs in one branch of algebra may make use of principles and conclusions of more general branches, but not vice versa.

Now Aristotle's account of the relation between superior and subordinate sciences describes some features of the relation between a more and a less general branch of abstract algebra. The genera are different, but the same "in a way." The principles of the more general branch are among those of the less general branch; also, the conclusions of a more general branch can be used in proofs of theorems falling under a less general one, although the more general science does not investigate the theorems of the less general one.

There are also differences between the two cases. Since proofs in less general branches of algebra frequently make substantial use of the principles that distinguish them from more general branches (unlike proofs in the *Optics*, which use the special principles of optics only to apply a geometrical result to the subject matter of optics), it is wrong to say that the more general branches provide the explanation of the facts of less general branches. Also, Aristotle's description of subordinate sciences as "perceptual" (*aisthētikôn*) as opposed to the "mathematical" (*mathēmatikôn*) superior sciences (I.13 79a2–3) does not hold.

Nevertheless, the relations among these areas of mathematics are closely related to those between superior and subordinate sciences, and the modifications in the theory of subordination needed to accommodate them are minor and may have been dimly perceived by Aristotle himself.[9]

These subtleties apart, the nature of subordination even as it stands is important to the theory of demonstrative science. It has the dual advantages of making the theory applicable to several existing disciplines and doing so in a way congenial to Aristotle's views on the nature of abstract entities. The doctrine of qualified identity of subject genera thus both solves the problem that evoked it and finds firm roots in Aristotle's metaphysics. It is a corollary, however, and not one of the central ingredients of the doctrine of *APo*.

Axioms or Common Principles

CHAPTER VI

SCIENCES ARE INDEPENDENT of each other since they have different subject genera. A science's subject genus provides the subjects and attributes it investigates, and these are its own concern. Definitions and existence claims for these things are proper principles and contain in latent form all the content of the science, which its proofs explicate. But just as the absolute separation of subject genera is violated by the subordinate sciences, so the rule that different sciences have different principles is broken by a certain class of principles.

The "common" principles are with us from the beginning of *APo*. The law of the excluded middle is cited in I.1 (71a14) as an example of one kind of prior knowledge needed for scientific knowledge. Similarly, the passage in I.2 (72a14–24 translated above in Chapter 3) that identifies the kinds of principles describes an axiom as "an immediate deductive principle . . . which it is necessary for a person to have if he is going to learn anything at all," and sets axioms apart from other kinds of principles. Definitions and hypotheses, called "theses" (*théseis*), are "not necessary for a person to have if he is going to learn anything." This distinction points to the contrast between axioms as "common" and other principles as "proper" that is made later on. In I.7 we learn that axioms can be the same in different sciences (75b2–3),[1] an indication of their special nature, which I.10 elaborates, again distinguishing them from proper principles and calling them "common."

> **I.10 76a38–b2** "common, but common by analogy, since a thing is useful insofar as it is in the genus that falls under the science. . . . For example, the principle that if equals are subtracted from equals the remainders are equal. Each of these is sufficient insofar as it is in the genus. For it will do the same even if [the ge-

> ometer] assumes it [as holding] not for all things, but only for magnitudes, and for the arithmetician [it is assumed as holding] for numbers."

Like I.7 (75a39ff.), I.10 identifies axioms as one of three ingredients needed for a demonstrative science (76b11ff.)[2] and stresses their "common" nature.[3] Finally, I.11 discusses their role in proofs.

> **I.11 77a10–13** "No demonstration assumes the law of noncontradiction except when it is necessary to prove the conclusion too in that way. It is proven by assuming that the first term is true of the middle and that it is not true to deny it [of the middle]."

The law of noncontradiction will not appear as a premise of a demonstration unless the conclusion is of the form "*A* is *B* and not not-*B*."[4]

> **I.11 77a22–25** "*Reductio ad impossibile* demonstration assumes the law of the excluded middle, and these [i.e., the law of the excluded middle and the law of noncontradiction] are not always [assumed] universally, but so far as is sufficient, sufficient for the genus. By 'for the genus' I mean the genus concerning which [the science] brings its proofs."

Other passages of *APo* I refer to the axioms, but do not add anything of importance.[5]

The data on which to base an interpretation of Aristotle's view of the nature and role of axioms is meager: three examples—the law of noncontradiction (LNC), the law of the excluded middle (LEM), and the principle that if equals are subtracted from equals the remainders are equal (EQUALS)—and a number of assertions that seem incompatible with other views on the structure of sciences and do not even fit all three examples.

Interpretations vary widely, commentators tending to take two tacks. Some focus on some but not all of the examples and statements, and either ignore the difficulties presented by the rest or attempt with limited success to squeeze the rest into the mold. Others supplement Aristotle's statements, and typically claim that there are two or more different kinds of axioms, but sometimes fail to distinguish between what Aristotle actually says and what they believe his doctrine demands.[6] My approach is to determine the distinctive role axioms play in demonstrative sciences and evaluate the evidence from that perspective. It will turn out that all the examples and almost all the statements can be accommodated by a single coherent interpretation.

The feature most frequently associated with axioms is that they are common (*koiná*); axioms are principles of more than one science.[7] LNC, LEM, and EQUALS apply to both numbers and spatial magnitudes, and so have a place in both arithmetic and geometry.

But beyond this, EQUALS seems to have little in common with LNC and LEM. LNC and LEM apply everywhere, whereas EQUALS applies only in the category of quantity,[8] a limitation that violates the statement that an axiom is something a person must have in order to learn anything at all (I.2 72a16–17).[9] Further, LNC and LEM are basic laws of thought, which are presupposed in reasoning and regulate it, but are not frequently used as premises of arguments,[10] whereas EQUALS, according to Aristotle, is standardly employed in demonstrations as a premise. If we want to infer syllogistically that $C = F$, given that $A - B = C$, $D - E = F$, $A = D$, and $B = E$, the principle EQUALS appears as the major premise:

Equal belongs to the remainders of equals subtracted from equals.
C and *F* are the remainders of equals subtracted from equals.
Therefore, equal belongs to *C* and *F*; that is, $C = F$.[11]

The regulative role of LNC and LEM is frequently thought to be distinguished from the role of axioms as premises by the preposition *diá*, "through," "in accordance with"—as opposed to *ek*, "from"—in the statement that proof proceeds "*through* or *in accordance with* the common principles" (I.10 76b10, I.32 88b3).[12]

These differences admitted, LNC and LEM still have important similarities to EQUALS. Aristotle calls them all (1) principles of demonstrations (I.2 72a14–17)[13] and (2) "common," used in more than one science (e.g., I.10 76a38); so they are (3) distinct from the "proper" (*ídia*) principles, which depend on the subject genus specific to the science (I.10 76a38; also see p. 50 above). Further, (4) knowledge of them is prerequisite for all learning (I.2 72a16–17), and (5) *all* sciences use them (I.11 77a26–27). So far we have seen that (1), (2), and (3) hold of "categorial" principles like EQUALS, and (2), (3), (4), and (5) hold of the "transcendental" principles like LNC and LEM.[14]

But (1) holds for LNC and LEM too. For Aristotle states that the transcendental principles can be premises in proofs.[15] Commentators are more struck by the statement that LNC and LEM occur as premises only in special circumstances (I.11 77a10–25) than by the admission that they can be premises at all. Given that they *can* be, they can only be common principles.

Moreover, categorial principles are (like the transcendental ones) prerequisite for learning. EQUALS is fundamental to thought respecting quantities, and understanding it is prerequisite for learning arithmetic or geometry. This point can be made in two ways. First, EQUALS has a "regulative" function in demonstrations concerned with quantities, in that it and the other quantitative axioms are principles in accordance with which demonstrations are organized, which guide the strategy of proofs, and which affect the choice of propositions to be proved. Second, they are an essential part of our understanding of the nature of quantity (and therefore of numbers, etc.). Without a grasp of them a person's knowledge of quantity will be fundamentally undermined. Just as a person who did not know how to use LNC could not make proper sense of propositions, so a practical understanding of the axioms of quantity is needed to make proper sense of propositions about quantities.

Admittedly EQUALS is less general than LEM and LNC, and fails to satisfy (4) and (5). However, I shall argue that the most important feature of axioms is that they are used in *more than one* genus, but not all, so that claims of their universality may be discounted. Moreover, after calling axioms "necessary for a person to have if he is going to learn anything at all," I.2 continues, "for there are such things. For we are accustomed to apply the name especially to such things" (72a17–18), as if other things called axioms were not so widely applicable[16]—a hint that (4) and (5) are not absolute requirements.

The Analogical Nature of Axioms

Principles applicable to more than one science raise an immediate problem, since all the terms in a demonstration must fall within the subject genus (I.7 75b10–11, discussed in Chapter 4 above). How can distinct sciences share a single principle?

Aristotle's solution is that the common principles are common "by analogy" or "proportionally" (*kat' analogían*, I.10 76a38–b2, quoted earlier in the present chapter): as the principle expressed in terms of spatial magnitudes stands to geometry, so the principle expressed in terms of numbers stands to arithmetic.[17] A science uses common principles only insofar as they apply to its own genus.[18] As far as the geometer is concerned, EQUALS could be stated specifically—thus: "if equal spatial magnitudes are subtracted from equal spatial magnitudes, the remainders are equal spatial magnitudes"—since this is the only use he makes of the principle even if he states it generally. *APo* I.10 thus includes the common principles within the genera of all the sciences to which they apply. (Cf. I.10 76a39, a42.)

Two questions remain about this "analogical" use of common principles. First, if a science uses them only qua restricted to its own subject genus, why classify them as a distinctive kind of principle? Why not state them in their restricted form and consider them as proper principles? The issue here is whether they are different from the proper principles just because other sciences use them (or analogous versions of them) or because they play a different role in demonstrations. If the latter, they are reasonably identified as a separate kind of principle.

The latter is in fact the case. The proper principles are definitions and existence claims, whereas LNC, LEM and EQUALS are neither of these.[19] Also they do not say anything *in particular* about any of the various subjects or attributes within the genera of the sciences to which they apply. They are principles that *apply to* genera, rather than principles that *are of* or *are in* or *are about* genera. They are regulative principles or inference licenses for going from propositions about a given genus to other propositions about it.

Take EQUALS for example. Although for geometry it could be stated in the specific way mentioned above ("if equal spatial magnitudes . . . "), by itself it says nothing concrete about circles, lines, and so on, but only provides a way to infer a true conclusion from true premises. To be sure, proper principles can be used in this way too. The definition of a circle as the locus of points equidistant from a given point can be used to infer from the fact that *A* and *B* are points on a circle with center *C* that $AC = BC$.[20] But this is not the only function of definitions (which make substantive claims about the nature of the terms in the subject genus), whereas it is the only function of EQUALS.

Since common principles apply to different subject genera, it is hard to see how they could say anything in particular about any single subject genus. Since a science is the study of a genus and is concerned with discovering per se relations among its subjects and attributes, any principles that also apply outside the genus will be unable to contain anything specifically about the subjects or attributes of the genus. Such information must come from within. The only other role left is for such principles to give ways of combining results derived about the subject genus to produce other results.

The second question concerns the analogical use different sciences make of common principles. When Aristotle says in effect that geometry uses the principle "if equal spatial magnitudes . . . " and arithmetic uses the principle "if equal numbers . . . ," does he mean that the two sciences use separate principles that are analogous, or instead that the two sciences use a single principle, EQUALS, in analogous ways?

This traditional problem is usually put by asking if common principles are univocal or equivocal. Zabarella[21] treats the matter extensively, starting from the view that accidents such as equal are equivocal if and only if quantity, the *summum genus* to which they belong, is also equivocal. There are difficulties either way since if EQUALS is not common univocally to arithmetic and geometry, then quantity is not a common thing univocally, whereas *Categories* treats it as a univocal genus.[22]

If EQUALS is common univocally, quantity is a univocal genus, so there can be only one science of quantity, and arithmetic and geometry will be two parts of the same science. Yet Aristotle repeatedly calls them separate sciences, each with its own genus. Zabarella stays with the *APo* view that arithmetic and geometry have different genera and concludes that their common attributes are equivocal. Therefore quantity is not a univocal genus, and the doctrine of the *Categories* must be discounted. For Zabarella, therefore, common principles are equivocal.[23]

However, there are reasons to deny that this is so. First, it is things, not words, that are equivocal. A geographical feature and an article of clothing are equivocal because each is called a cape, and the definition of "cape" is different in the two cases. Is this so for equal? Two given lengths and two given numbers may both be called equal, but does "equal" have a different definition in each case? No more than "animal" does because elephants and ants can both be called animals. There are different types of equality just as there are different types of animals, and the criteria for the different kinds of equality are different, just as the criteria for different kinds of animals are different. Moreover, "equal" has a single definition that applies to all quantities: "things are equal whose quantity is one" (*Met.* Δ.15 1021a12), so even if equality in arithmetic is equality in number and equality in geometry is equality in spatial magnitude, these are not different (and circular!) definitions of equality, but specifications of different kinds of equality, where "equality" has the same meaning throughout. The distinction between being univocal and being equivocal is unhelpful in the present matter. Common principles apply differently in different sciences, but not because of differences in the meanings of the terms they contain.[24] Rather, EQUALS is a single principle, which applies generally to quantities, and therefore specifically to numbers and spatial magnitudes.

This interpretation agrees with the text, which says not that geometry and arithmetic use different yet analogous principles, but that EQUALS is the same principle (I.7 75b2–3). It is "useful" in geometry only insofar as it applies to spatial magnitude; it is "sufficient" as long as it applies to that subject genus; and it would make no difference to the geometer if he assumed it qua applying to magnitudes alone and not generally to all quantities (I.10 76a39–b2, quoted earlier in the present chapter).

If EQUALS is a single principle that applies to all quantities and so to all kinds of quantities, then, according to Zabarella, it follows that (1) quantity is a univocal genus, so that (2) there is a single science of quantity, and therefore (3) arithmetic and geometry are only parts of this science. Are any of these consequences unwelcome? (1) agrees with Aristotle's tendency to call the categories genera[25] and his treatment of quantity in *Categories* 6. (2) does not follow from (1), since it is mistaken to assimilate the subject genera of sciences to the genus/species hierarchy of nature.[26] Even so, there is some evidence that Aristotle acknowledged such a science. His reference to the general proof of the law of alternating proportions (I.5 74a18–25) and his references to "universal mathematics"[27] may indicate that a start had been made on such a general science of quantity, or that the materials were at hand for doing so.[28]

Evaluation of Zabarella's final inference, that if there is a general science of quantity then (3) arithmetic and geometry are merely two parts of it, depends on what constitutes a "part" of a science. Zabarella assumes that if quantity is a single genus there is only one science for any and all quantities, so that arithmetic and geometry are parts of that single science, as the geometry of the circle is part of geometry. He objects to this view because arithmetic and geometry are different sciences, each with its own distinctive subject genus, whereas the circle is only part of the subject genus of geometry. Now arithmetic and geometry *are* part of knowledge about quantities, since they are knowledge of kinds of quantity. But they are not part of the science of *quantity in general*, because that science treats quantities generally and its principles and conclusions must apply to the subject genera of *both* arithmetic *and* geometry, whereas distinctively arithmetical facts do not apply to all kinds of quantity. Thus there can be distinct sciences whose genera are, respectively, quantity in general, number, and magnitude, and so the principle of the autonomy of the subject genus remains intact.

The Alleged Regulative Role of Axioms

It remains to examine the "regulative"[29] function of the axioms, which is supposed to be reflected in the use of different prepositions to describe the relations of axioms to the conclusions of proofs. The statement that demonstration proceeds "*through* the common principles" or "*in accordance with* the common principles" (*dià tôn koinôn*, I.10 76b10, I.32 88b3) seems to point to the regulative function of logical laws, though elsewhere the axioms are called the things "*from* which" (*ex hôn*) demonstration proceeds (I.7 75a42, b2, I.10 76b14, b22, I.11 77a27), which seems to describe their use as premises. At I.2 71b20 and I.10 76b10, for example, *ek* is used of the principles and prior conclusions that appear as premises of proofs. Still, it is wrong to place much weight on this varied use of prepositions. *Diá* and *ek* are used interchangeably in I.10 (76b10, b14, b22). *APr* I.25 speaks of conclusions as coming to be both *diá* (42a36) and *ek* (42a32) the premises. Also, *Metaphysics*, which treats LNC and LEM in their capacity as laws of thought, not premises of proofs, refers to them as demonstrative principles and calls them "the common opinions from which [*ex hôn*] all people make demonstrations" (*Met.* B.2 996b26–31). Since both *diá* and *ek* can express the relation between conclusion and premises, and *ek* can express the dependence of all thought on the logical laws, and *diá may* do so too, the relation the axioms have to other scientific propositions cannot be ascertained from the prepositions. Interpretation must depend on the immediate context of particular statements and the requirements of the theory of demonstrative science. The following four passages are the best evidence for the regulative function of axioms.

(1) I.10 76b14: One of the three things with which every demonstration is concerned is "the common principles, called axioms, from which as primary [*ex hôn prṓtōn*] [the science] forms its demonstrations." I have proposed[30] that this passage refers to the role of axioms as ultimate premises of scientific demonstrations, since they are the only principles mentioned in the list. "Primary" signifies not that axioms are prior to other principles, but simply that they *are* principles.

(2) The same explanation accounts for I.7 75a41–42. One of "three things in demonstrations" is "the axioms (axioms are 'that from which [*ex hôn*]')."[31]

(3) I.10 76b10: "they prove through [*diá*] the common principles and from [*ek*] the things that have been proved." Since Aristotle is talking of how proofs are constructed, it is the function of axioms as premises, not as regulative principles, that is relevant. The different prepositions are used for the sake of variety.[32]

(4) I.32 88a36–b3, translated and discussed in Chapter 4 above. The words of interest are "with which [*meth' hôn*] they are proved through [*diá*] the common principles" (b3). The point is that a conclusion in arithmetic, say, may depend on common principles, but it depends on arithmetical facts *too*. The prepositions underline that the axioms need to be accompanied in their role as premises by proper principles.

The evidence for the "regulative" role of axioms has collapsed. There is no reason to think that Aristotle had any other feature of EQUALS, LEM, and LNC in mind than their role as premises of demonstrations, even though the logical laws might be used this way only rarely. (Actually, it is only LNC he says will be used *in unusual* circumstances. Since LEM is used in *reductio* proofs [I.11 77a22–23] we may expect to see it used frequently.[33] It is probably significant that LEM is often mentioned as a common principle, LNC only once.) This result places the axioms on much the same level as the proper principles, while still reserving for them a distinctive role of their own. On the other hand, there is no reason to deny that they have regulative functions. *Metaphysics* Γ is not to be ignored. But neither is it to be thought of as containing Aristotle's only thoughts on axioms.

Summary

The axioms are therefore a distinct class of principles with a distinctive role as premises of demonstrations in the sciences to which they apply. An axiom is a single principle that is used in different sciences. Moreover, it is used in the same way in each of the sciences in which it is found, although criteria for application may be different in different sciences. (For example, the criteria for using the word "equal" and therefore the principle EQUALS are different in geometry and arithmetic: different kinds of comparisons are employed, and different ways of deciding whether things are equal.) The "analogical" use of axioms in different sciences means not that there are *different but analogous* versions of EQUALS or LEM used in different sciences, but that each science uses these principles only qua applying to its own subject genus.

Axioms and Subordinate Sciences

I will end by establishing a link between the doctrine of axioms and the theory of subordinate sciences that enables us to understand how axioms can be "common" to more than one science.

LEM and LNC hold everywhere because all sciences are concerned with things-that-are, and these principles apply to all things-that-are (*Met.* Γ.3 1005a22–28); EQUALS applies in geometry and arithmetic because the genera of these sciences are kinds of quantity (*Met.* K.4 1061b20–21). When distinct sciences share a common principle it is because their genera have a common character, a similar internal structure. The genera of sciences that share a common principle are subclasses of the class for which the principle holds.

One implication of the doctrine of axioms that Aristotle did not fully recognize is that any completely general consequence of a common principle will hold for the same class as the common principle and also for the subclasses that are the subject genera of the specific sciences. In this respect, common principles are like the law of alternating proportions, which is a theorem, not a principle (I.5 74a18–25): this law applies to numbers and spatial magnitudes, and so can be used in both arithmetic and geometry. The same holds true for the entire theory of proportion invented by Eudoxus and reworked by Euclid to form book V of the *Elements*. (The law of alternating proportions is book V, prop. 16.)

In fact, the law of alternating proportions creates an embarrassment for Aristotle, since it is "common" and the conclusion of a proof, yet the only "common" elements Aristotle recognizes are unprovable principles. However, my discussion of the subordinate sciences points to a solution.

If the Eudoxan theory of proportion is the science (or part of the science) of quantity, arithmetic and geometry are subordinate to it in the way one branch of modern algebra is subordinate to another.[34] The theory of proportion is more general than geometry and arithmetic, whose genera are defined as quantities with additional properties. It does not investigate the facts of arithmetic and geometry, but those sciences use its principles and conclusions together with their own proper principles and conclusions to prove further results.

This second kind of subordination not only covers the relation between arithmetic and geometry and the law of alternating proportions together with the rest of the Eudoxan theory of proportion, but also provides a framework for understanding the nature of the axioms. For they turn out to be facts occurring in superior sciences. For example, Euclid's *Elements*, book V, proposition 1, makes use of common notion 2, "if equals are added to equals, the wholes are equal," which Aristotle would place alongside EQUALS as a common principle. Like EQUALS, this axiom is found in "universal mathematics." Also LEM and LNC will be principles of the science of things-that-are. They, together with other principles and conclusions of that science, apply to all things-that-are and so can be used by any of the special sciences.

The doctrine of the qualified identity of subject genera of superior and subordinate sciences[35] also explains how axioms apply in more than one science. The relation between quantity in general and spatial magnitude is the same as the relation between spatial magnitude and the subject genus of optics. Just as visual lines are geometrical lines, but not *all* geometrical lines, and are treated qua having geometrical properties and visual ones *as well*, so spatial magnitudes are quantities, but not all quantities, and geometry treats them qua having properties of quantities in general and geometrical ones as well. Each axiom of a science is a proper principle or a conclusion of a superior science. This is guaranteed by the fact that axioms are common to more than one science and are common in virtue of a common character of their subject genera, and so have a place in the science whose genus is appropriately general.

Appendix: I.11 77a10–21 and the Law of Noncontradiction

The following difficult passage claims to show how LNC is used as a premise in demonstrations. I shall provide a translation and discussion of the passage and then indicate the role LNC plays.

I.11 77a10–12 "No demonstration assumes that it is impossible simultaneously to assert and deny [the same predicate of the same subject] unless it is necessary to prove the conclusion too in that way."

DISCUSSION

Since it would be vacuous to say that LNC is assumed only when it is needed, the passage must be asserting that LNC is used as a premise only when the conclusion is of the form "it is impossible for *C* simultaneously to be *A* and not-*A*." In the following discussion nothing is made of the concepts "impossible" and "simultaneous," and the conclusion is considered to be of the form "*C* is *A* and *C* is not not-*A*." The sense of this proposition is difficult to capture in the quantification of the syllogistic. Probably the best version is (III): (*A*a*C*) & (not-[{not-*A*}a*C*]), with the first "not" negating the proposition ([not-*A*]a*C*) and the second "not" negating the term *A*.

I.11 77a12–13 "It is shown by assuming that the first is true of the middle and not true to deny."

DISCUSSION

If *A* is the "first" (the major) term in the syllogism and *B* the middle, then Aristotle says that the major premise of the syllogism is "it is true that *A* belongs to *B* and not true that *A* does not belong to *B*," that is, (I): (*A*a*B*) & (not-[{not-*A*}a*B*]), where again the first "not" negates a proposition and the second "not" negates a term.

I.11 77a13–14 "It makes no difference to assume that the middle is and is not."

DISCUSSION

It makes no difference that the middle is and is not true of the minor. Aristotle says that the minor premise may be of the form: (*B*a*C*) & ([not-*B*]a*C*); that is, it may violate LNC without impeding the syllogism. III follows from I and the minor premise, II: *B*a*C*. But III also follows from I and II′: (*B*a*C*) & (not-[{not-*B*}a*C*]), though only the first member of II′ is relevant. The second member "makes no difference" to the conclusion because its predicate is (not-*B*); it would have to be *B* in order to form a syllogism with the major.

I.11 77a14–15 "And similarly for the third."

DISCUSSION

The third term is the minor term *C*. It makes no difference whether we assume the minor term only affirmatively or both affirmatively and negatively (i.e., with term negation, not-*C*) in the minor premise. III will follow regardless of whether the major premise I combines with the minor II or II′ (as was shown above) or even II″: (*B*a*C*) & (*B*a[not-*C*]), or II‴: (*B*a*C*) & (*B*a[not-*C*]) & ([not-

B]a*C*) & ([not-*B*]a[not-*C*]). As we have seen, only II is needed for the syllogism. I and II″ yield (*A*a*C*) & (not-[{not-*A*}a*C*]) & (*A*a[not-*C*]) & (not-[{not-*A*}a{not-*C*}]). The conjunction of the first two conjuncts is III, and the other two conjuncts are irrelevant to reaching III. These other two conjuncts are the consequences of the second member of II″: (*B*a[not-*C*]). Thus, adding this member "makes no difference" to reaching the desired conclusion, III. I and II‴ yield precisely the same conclusion as I and II″. Again, the additional conjuncts "make no difference."

> **I.11 77a15–18** "For if it was given that that of which it is true to say man [e.g., Callias] (even if it is also true to say not-man), but if only [it is given] that man is an animal and not a not-animal, it will be true to say that Callias (even if also not-Callias) is nevertheless an animal and not a not-animal."

DISCUSSION

This is offered as a specimen inference. The major is "man is an animal and not a not-animal"; that is, being animal belongs to (all) man and being not-animal does not belong to (all) man. This has the form: (*A*a*B*) & (not-[{not-*A*}a*B*]). The conclusion is "Callias is an animal and not a not-animal"; that is, being an animal belongs to Callias and being not-animal does not belong to Callias. Here as frequently Aristotle uses a singular term instead of a universal. For the sake of convenience I shall follow Barnes (p. 141) in treating "Callias" as a universal. The form of the conclusion is, then, (*A*a*C*) & (not-[{not-*A*}a*C*]). To reach this conclusion we only need the proposition "Callias is a man," that is, II: (*B*a*C*), for the minor premise. But even if the minor premise is "Callias is a man and also a not-man," that is, (*B*a*C*) & ([not-*B*]a*C*), the conclusion follows. Moreover, even if the minor premise is such that it follows that Callias and also not-Callias is an animal and not a not-animal, it makes no difference, since the desired conclusion is reached. The desired conclusion, (*A*a*C*) & (not-[{not-*A*}a*C*]), is part of and therefore entailed by the more complex conclusion (*A*a*C*) & (not-[{not-*A*}a*C*]) & (*A*a[not-*C*]) & (not-[{not-*A*}a{not-*C*}]), which results if the minor premise is either "Callias and not-Callias are men," that is, II″: (*B*a*C*) & (*B*a[not-*C*]), or "Callias and not-Callias are men and not-men," that is, II‴: (*B*a*C*) & (*B*a[not-*C*]) & ([not-*B*]a*C*) & ([not-*B*]a[not-*C*]).

> **I.11 77a18–21** "The reason for this is that the first is predicated not only of the middle but also of another, since it applies to many, and so even if the middle is both it and not it, it makes no difference to the conclusion."

DISCUSSION

The "first" is the major term as at a12. Aristotle explains how a conclusion like "Callias and also not-Callias is an animal and not a not-animal"—(*A*a*C*) & (not-[{not-*A*}a*C*]) & (*A*a[not-*C*]) & (not-[{not-*A*}a{not-*C*}])—is possible. Although it is false that everything that is not Callias is an animal, it is true that there are other kinds of animals than men. If not-Callias proves to be a tiger, then indeed not-Callias as well as Callias is an animal and not a not-animal. Thus if the middle term is "both it and not it," that is, *B* and not-*B*, and if not-*B* is understood to be not the simple contradictory of *B*, but something else in the range of *A* (in the example, another kind of animal), the same conclusion follows. Let *D* stand for tigers. The middle term is now "men and tigers," *B* and *D*. The major premise is (*A*a*B*) & (not-[{not-*A*}a*B*]) & (*A*a*D*) & (not-[{not-*A*}a*D*]). The minor premise is (*B*a*C*) & (*B*a[not-*C*]) & (*D*a*C*) & (*D*a[not-*C*]). The conclusion is (*A*a*C*) & (not-[{not-*A*}a*C*]) & (*A*a[not-*C*]) & (not-[{not-*A*}a{not-*C*}]), which is the same conclusion that follows if the minor premise is simply (*B*a*C*) & (*B*a[not-*C*]), and all reference to *D* is omitted. As Aristotle says, the addition of *D* "makes no difference to the conclusion."

THIS READING OF the text is close to Barnes's (pp. 140–42), but differs in two respects. First, it takes "even if also not-Callias" (a17–18) as referring to the conclusion—it is shorthand for "Callias and not-Callias are animals and not not-animals"—whereas Barnes calls the phrase "obscure" and suggests emending the text to read "Callias is also not a man." Second, at a18–21 (a passage Barnes finds "desperately obscure") Barnes takes the explanandum to be that II′ is superfluous; II is all that is needed in the minor premise. I suggest that Aristotle is explaining how to understand the proposition "not-Callias is an animal and not a not-animal" in a way that makes sense. As Barnes points out, Aristotle is not justified in holding that (*A*a*B*) & (*A*a[not-*B*]), but the reading I have proposed finds an acceptable way of taking that premise.

The final question is how LNC is used in the argument. It is not any of the stated premises. The closest approach to it is in the description of the major premise (a12–13): "the first is true of the middle and not true to deny": (*A*a*B*) & (not-[{not-*A*}a*B*]). This is not itself LNC, but it is possible to see LNC as a premise of an argument leading to this proposition. If we know *A*a*B* and want to have the conclusion of our argument be in the form (*A*a*C*) & (not-[{not-*A*}a*C*]), it is necessary to have the major premise in the form (*A*a*B*) & (not-[{not-*A*}a*B*]). And the major premise follows from (*A*a*B*) and LNC. The reasoning would be: LNC states that it is impossible simultaneously to assert and deny the same predicate of the same subject. It follows that (not-[{*A*a*B*} & {not-*A*}a*B*]). But (*A*a*B*). Therefore, (not-[{not-*A*}a*B*]). Therefore, (*A*a*B*) & (not-[{not-*A*}a*B*]).

Definitions, I: The Per Se

CHAPTER VII

DEFINITIONS OF SUBJECTS and attributes are one kind of proper principle (I.10 76a37ff.). Together with existence claims they contain the indemonstrable facts about the subject genus. Since the role of existence claims is to introduce certain items as legitimate subjects of proofs, but not to describe their natures,[1] the information contained in the definitions must somehow entail all the nonimmediate facts that the science demonstrates.

Although the official word for definition is *horismós*, this word occurs sparingly in *APo* I. More frequently we find the paraphrases "what it is" (*tí esti*) and "what it signifies" (*tí sēmaínei*)—which are the source of such expressions as "that it signifies *this*" (*hóti todì sēmaínei*, I.1 71a15) and "that they are *this*" (*tó . . . todì eînai*, I.10 76b6) both of which mean "that *this* is the definition"—and "what the thing being said is" (*tí tò legómenón esti*, I.1 71a13), a longer form of "what it is." The distinction between "what it is" and "what it signifies"[2] is a subtlety that this chapter can ignore, and although I shall follow Aristotle in using the former expression, most of the discussion below also applies to "what it signifies." However, another subtlety is crucial. The "what it is" is not exactly identical with a definition. The relation between them is encapsulated in the statement "the definition is called a verbal statement of the 'what it is' [*lógos toû tí esti*]" (II.10 93b29):[3] the definition is "an immediate deductive principle" (I.2 72a14–15), a proposition that can serve as a premise in a demonstration, whereas "what it is" is not a linguistic entity, but the real nature of the definiendum.

The connection between definitions and "what it is" forms the missing link between immediate principles and necessity. Since the object of scientific knowledge (the conclusion of a proof) "cannot be otherwise" (I.2 71b12, translated and discussed in Chapter 2 above),[4] it is necessary (I.4 73a21–22). Why, then, does the list of requirements for scientific principles (I.2 71b19ff., translated and discussed in Chapter 2 above) omit to say that they are necessary?[5] And if the principles do not have to be necessary, why should we accept that the conclusions are necessary? The answers to these questions need a good deal of justification, but can be stated briefly. Aristotle does not say that principles are necessary because not all principles are necessary.[6] The necessity relevant to conclusions of proofs stems from the necessity of the per se relations found in definitions. This is the topic of the present chapter.

The necessity of definitions stems from the fact that they state "what it is." For, as a crude approximation, things so related to each other that one is an ingredient of the "what it is" of the other are related per se, and per se relations are necessary. When these definitional relations are put into the appropriate deductive forms, their consequences are necessary too. What, then, is it for things to be related per se? Since this complex issue is a major concern of *APo* I.4, that chapter must be expounded in detail. *APo* I.4 begins with what appears to be an argument to prove that demonstrations depend on necessary principles:

> **I.4 73a21–24** "Since that of which there is unqualified scientific knowledge cannot be otherwise, that which is known by demonstrative knowledge is necessary. Demonstrative knowledge is the knowledge we have by having a demonstration. Therefore [a] demonstration is [a] deduction dependent on necessary [principles]."[a]

NOTE ON TRANSLATION

a. I say "dependent on" rather than "has as premises" in order to allow the use of previously proved conclusions as premises in proofs of subsequent conclusions,[7] though it is not clear that Aristotle has this practice in mind.

DISCUSSION

This passage begins by pointing out that the requirement for scientific knowledge stated at I.2 71b15–16, "cannot be otherwise," entails that the conclusions of demonstrations are necessary.[8] The train of thought from "that which is known by demonstrative knowledge is necessary" to "[a] demonstration is [a] deduction dependent on necessary [principles]" is incomplete. The argument does not "commit him to the view that 'if *P* is necessary and *P* is deducible from *Q*, then *Q* is necessary' "[9] since in the case in point we do not just know that *P* is necessary; since we have scientific knowledge of *P*, we know *P* on the basis of principles that are prior to, more intelligible than, and grounds of *P*.

ALTHOUGH 73a21–24 ASSERTS rather than proves that proofs depend on necessary principles, I.6 offers several arguments of varying quality to support this claim (74b13–39), the strongest of which runs as follows:

> **I.6 74b26–32** "That the deduction must depend on[a] necessary principles is obvious from these considerations too. For if a person

who lacks an account of why, when there is a demonstration,[b] does not have scientific knowledge (it might be in the circumstances that *A* belongs of necessity to *C*, and *B*, the middle through which it was 'proved,' is not of necessity),[c] he does not know why [*A* belongs necessarily to *C*].[d] For this is not through the middle.[e] For it is possible for it not to be,[f] but the conclusion is necessary."

NOTES ON TRANSLATION

a. For my use of "depend on" rather than "has as premises" see note a, on 73a21–24, just above.

b. "When there is a demonstration" can mean either "when the conclusion can be proved even though the person does not know the proof" or "when the person has what he *thinks* is (but is not really) a proof." The first is more likely (see below, note e), but the second is made possible by the talk of the "proof" through the middle term *B* (b29–30). Nothing depends on which we choose.

c. "Is not of necessity" may mean either "*B* does not belong of necessity to *C*" (cf. b29) or "*A* does not belong of necessity to *B* or *B* of necessity to *C*" (cf. 75a4–6). Since the point is that scientific knowledge requires all the premises to be necessary, either of these readings constitutes sufficient reason for saying that the conclusion is not known scientifically.

d. The supplement "[*A* belongs necessarily to *C*]" could also be "[*A* belongs to *C*]" (cf. I.2 71b9–12).

e. "For this is not through the middle" means that the faulty proof of *A*-*C* through *B* does not employ the middle term of the actual proof of *A*-*C*. That there is such a proof (and therefore such a middle) is guaranteed on the first reading of "when there *is* a demonstration" (b28) (see note b above).

f. "For it is possible for it not to be" can in view of the two interpretations of "is not of necessity" discussed in note c, mean either "it is possible for *B* not to belong to *C*" or "it is possible for *A* not to belong to *B* and for *B* not to belong to *C*." Another possibility is "it is possible [sc., as far as the 'proof' shows] for *A* not to belong to *C*."

DISCUSSION

Despite the multiple uncertainties of interpretation, the reasoning is clear. Demonstrations have necessary principles because otherwise we do not know *why* the conclusion holds and is necessary, and so lack scientific knowledge and demonstration of it (cf. I.2 71b10–12). If we have the argument "*A* belongs to *B*, *B* belongs to *C*, therefore *A* belongs to *C*," where either or both premises are not necessary connections, then even if *A* belongs necessarily to *C*, the argument is not a demonstration that *A* belongs to *C*. Supposing that there is a demonstration, it must proceed through a different middle term, since the present argument does not yield knowledge why *A* belongs to *C*. Aristotle's reasoning rests on the view that the grounds for a necessary fact and for scientific knowledge of that fact are themselves necessary: how can a necessary conclusion be accounted for by contingent facts, and how can we know that it is necessary if our knowledge is based on contingent premises?[10]

THE NECESSITY OF principles is grounded in their being grounds (not merely premises) of necessary conclusions. Further, the grounds of the conclusion are also the grounds for the conclusion's being necessary. The principles of a demonstration account not only for the fact that the conclusion holds but also for the fact that the conclusion holds of necessity.

This result is employed in the initial lines of I.6, which argue that proofs depend on premises that state per se relations:

> **I.6 74b5–12** "Now if demonstrative science (or knowledge)[a] depends on necessary principles[b] (for what he knows cannot be otherwise, and things that belong per se are necessary to things[c] . . . [d] it is obvious that the demonstrative deduction[e] will depend on such premises.[f] For everything belongs either thus or per accidens,[g] and accidents are not necessary."[h]

NOTES ON TRANSLATION

a. "Demonstrative science (or knowledge)" gives two possible renderings of *apodeiktikḕ epistḗmē*.

b. Regarding "depends on," see note a on 73a21–24 earlier in the present chapter.

c. In "things that belong per se are necessary to things," we may understand "to things" with "belong" and "necessary." If *A* belongs per se to *B*, *A* is necessary (belongs necessarily) to *B*.

d. I omit b7–10, which repeat the characterizations of two kinds of per se relations at I.4 73a34–b5. These relations will be discussed shortly below.

e. "The demonstrative deduction" means "the deduction which is a demonstration," "the demonstration."

f. "Will depend on such premises," namely on premises stating that one thing belongs per se to another.

g. "For everything belongs either thus or per *accidens*" means that if *A* belongs to *B* then either *A* belongs per se (in one of the two ways identified at b7–10; this proviso applies throughout the discussion of this passage) to *B* or *A* belongs per accidens to *B*. We may further understand "but not both."

h. In "accidents are not necessary," "accidents" (*sumbebēkóta*) is equivalent to "things belonging per accidens" (*tà katà sumbebēkós*).

DISCUSSION

This passage argues to the conclusion (1): Proofs depend on principles that state per se relations. The premises are:

2. Proofs depend on principles that state necessary relations (stated at I.4 73a24 and argued for at I.6 74b12ff.).

3. If *A* belongs per se to *B*, then *A* belongs necessarily to *B* (from I.4 73b16–18).
4. If *A* belongs to *B*, then either *A* belongs per se to *B* or *A* belongs per accidens to *B* (but not both) (loosely anticipated at I.4 73b4–5).
5. If *A* belongs per accidens to *B*, then *A* does not belong necessarily to *B*.

The proof is as follows:

6A. If *A* belongs to *B* and *A* does not belong per accidens to *B*, then *A* belongs per se to *B* (from 4).
6B. If *A* belongs to *B* and *A* does not belong per se to *B*, then *A* belongs per accidens to *B* (from 4).
7. If *A* belongs necessarily to *B*, then *A* does not belong per accidens to *B* (from 5).
8. If *A* belongs necessarily to *B*, then *A* belongs per se to *B*; that is, necessary relations are per se relations (from 6A and 7).
1. (From 2 and 8). Q.E.D.

The proof notably does not make use of (3). However, (3) is instrumental in two further consequences:

9. *A* belongs per se to *B* if and only if *A* belongs necessarily to *B*; that is, all and only necessary relations are per se (from 3 and 8).
10. *A* belongs per accidens to *B* if and only if *A* does not belong necessarily to *B*; that is, all and only nonnecessary relations are per accidens (from 9 and 6B).

Not only do proofs depend on per se relations, but also the necessary/contingent distinction coincides with the per se / per accidens distinction.

Since scientific necessity is grounded in per se relations, we must look closely at what Aristotle says about the nature of such relations. The principal discussion is found in *APo* I.4, a chapter that introduces and analyzes three relations of importance to the theory of demonstrative science: "in every case" (*katà pantós*), per se (*kath' hautá, kath' hautó*), and "universal" (*kathólou*).

In Every Case

I.4 73a28–34 "Whatever is not [such as to apply] to some cases but not to others, or at some times but not at others, this I say holds in every case. For example, if animal holds in every case of man, it follows that if it is true to say that this is a man, it is true to say that it is an animal too, and if it is now true to say the one, it is now true to say the other too. Likewise, if in every line [there is] a point. Evidence for this is that we raise objections to assertions either if something fails to apply to something or if it fails to apply at some time, as if we were asked [to believe that it holds] in every case."

DISCUSSION

A holds "in every case" of $B =_{df} A$ belongs to all *B*'s at all times. The first example (animal holds "in every case" of man) is clear. The account also fits "universal" relations (73b26): 2R belongs "in every case" to triangle. The second example, "in every line [there is] a point," raises a difficulty, since point is not predicated of line. It is said to be "in" (*en*) every line, not to *hold in* (*katá*) or *apply to* (*epí*) every line. Now nothing depends on the alternation between "hold in" and "apply to," which both refer to predication. But even though the *Categories* uses "in" of a kind of predication (*Cat.* 2), it deliberately distinguishes that kind of predication from the part/whole relation that holds between point and line.[11] To cover both the predication and the part/whole[12] relations by a single expression, we need something like "whenever there is a *B* there is also an *A*" as a paraphrase of "*A* holds in every case of *B*." But is this a solution or just a patch? As long as we take both of Aristotle's examples seriously we should accept it as a solution. Since the text declares that *A* holds in every case of *B* only when *A* and *B* are related in such different ways, it is best to avoid regimentation[13] and allow a variety of relations to obtain between *A* and *B* ("is a genus of," "is a part of," "is a necessary attribute of," etc.) such that if *A* is related in any of these ways to all *B* at all times, *A* holds in every case of *B*.

Per Se

The heart of I.4 is the treatment of ways in which things are related per se (73a34–b24). Since this part is most important for the purpose of understanding the nature of definitions and necessity, and since it presents several difficulties, it merits careful treatment.

Per Se 1

I.4 73a34–37 "All the things that belong [to something] in 'what it is,'[a] [I say hold] per se, for example, line [belongs per se] to triangle and point to line. For their being depends on these[b] and they are predicated[c] in the statement that states 'what they are.' "

NOTES ON TRANSLATION

a. For a defense of this translation of *hupárchei en tôi tí estin* (a34–35), see Zabarella, cited by Ross, p. 520.

b. In "For their being depends on these" (*hē gàr ousía autôn ek toútōn estí*), "being" can mean either "essence" or "existence." "Depends on" preserves the translation elsewhere adopted for *ek*. (Other interpretations are possible—see shortly below.) "Their" refers to "triangle," "line." "These" refers to "line," "point." In the next clause the first "they" refers to "line" and "point," the second to "triangle" and "line." The antecedents of the pronouns are clearer in Greek than in English.

c. "Are predicated" translates *enhupárchei*. The difference between *hupárchein* (translated as "belong") and *enhupárchein* is not significant. Even though *hupárchein* has a "rather general significance" and is "non-technical and ambiguous" (Ross, pp. 60–61, 520–21), *hupárchein tinì en tôi tí estin* is equivalent to *enhupárchein tinì en tôi tí estin* (Bonitz 1866, p. 367; cf. Jaeger 1911, pp. 31–32).

DISCUSSION

If *A* belongs to *B* in "what *B* is," then *A* belongs per se 1 to *B*. Taken most simply this says that all the things occurring as terms in the definition of something belong per se 1 to it. If square $=_{\text{df}}$ right-angled equilateral quadrilateral, then right-angled, equilateral, and quadrilateral all belong per se 1 to squares, and the definition yields three unprovable propositions that can be immediate premises of demonstrations: "all squares are equilateral," "all squares are right-angled," and "all squares are quadrilaterals." This interpretation fits the two examples. "Line" occurs as a term in the definition of triangle as "plane figure bounded by three lines," and "point" occurs in the definition of line as "flux of a point."[14]

The passage gives two reasons to justify the examples: (1) the being of triangle (line) depends on line (point), and (2) line (point) is predicated in the statement that states "what triangle (line) is." I take it that each of these reasons is meant as a necessary condition for the per se 1 relation.

(1) How is the being (*ousía*) of triangle (line) dependent on (*ek*) line (point)? The key word is *ek*, a word that covers a large number of relations. (a) Aristotle's standard view is that a definition consists of genus plus differentia. This interpretation points to the most natural interpretation of "what it is" as definition, but does not fit the examples, since line (point) is neither genus nor differentia of triangle (line). (b) *Ek* can mean "composed of," and Aristotle may be maintaining that triangle (line) is composed of lines (points). This reading applies easily to the case of triangle and line, but runs afoul of Aristotle's conviction that a line is *not* made up of points,[15] although we may not feel bound to abandon the reading because of this Aristotelian nicety. (c) *Ek* might be taken in the sense "determined by," in a way that fits both examples. I shall return to this issue below.

(2) This condition also presents difficulties. "Line" *occurs* in the definition of triangle but is not *predicated* of triangle. Also, line (point) and triangle (line) are both subjects, and though one subject can be mentioned in the definition of another when neither is a genus of the other, it is not clear how one can be *pred-*

icated of another except as a genus. On the other hand, since a triangle is a figure bounded by lines, "bounded by lines" is predicated of triangle. One suggestion is to take the examples with a grain of salt in just this way.[16] When Aristotle says that line belongs to triangle he means that the attribute "bounded by lines" belongs to triangle, and similarly for point and line. A parallel occurs in the way the attribute "having angles whose sum is equal to two right angles" is abbreviated "two right [angles]."[17] This suggestion must itself be taken with a grain of salt. Replacing "line" by "bounded by lines" yields an attribute predicated in the definition, but neglects the fact that *line* is per se 1 to triangle because the being of triangle depends on lines. Aristotle would not say that the being of triangle depends on bounded by lines or on being bounded by lines.

I propose to take the examples at face value (line, not bounded by lines, belongs per se to triangle), and to interpret the account of per se 1 as follows: if "*A*" appears as a term in the definition of *B*, *A* belongs per se 1 to *B*. Condition (1) amounts to this: either *A* is predicated of *B*, or *A* is a component of *B*, or the existence or essence of *B* depends on *A*, or *B* is bounded by *A*, or *A* and *B* stand in any of the other relevant relations that *ek* can express.[18] This allows things related in a multitude of ways to be related per se 1 to one another, and the different kinds of relations between things are expressed in their definitions. Condition (2) is satisfied as follows. If *A* belongs per se 1 to *B*, there is *some* expression containing "*A*" that is predicated in the definition of *B*. In the definition "a triangle is a figure bounded by three lines," "bounded by lines" is predicated of triangle. In the definition "man is a rational animal," "animal" and "rational" are both predicated of man. The subject-predicate propositions formed from these expressions ("animal belongs to man," "bounded by lines belongs to triangle") are the per se 1 premises on which demonstrations depend.

My interpretation distinguishes between *things* (triangle, line) related per se 1 to one another and per se 1 *propositions*, which are derived from definitions, express the relations between things related per se 1 to one another, and serve as principles of scientific proofs.

Per Se 2

> **I.4 73a37–b5** "Also [I say] all of the things belonging to them, to which they are predicated in the account indicating what it is, [belong per se].[a] For example, straight belongs to line, and so does circular, and odd and even to number, and prime and composite, and equilateral and oblong. And of all these there are predicated in the account stating what it is, in the one case line and in the other number. As also with the others [i.e., the things that belong per se 1], I call such things [as odd and even, as regards number] per se to each thing, and whatever does not belong in either of these two ways [I call] accidents,[b] as musical or white [belongs] to animal."

NOTES ON TRANSLATION

a. The first sentence is at least as obscure in Greek as it is in the translation given: *hósois tôn huparchóntōn autoîs autà en tôi lógōi enhupárchousi tôi tí esti dēloûnti*. But the examples and the justification ("And of all these . . . ") show that it refers to cases where *B* is a subject (number) and *A* (odd) belongs to *B*, and *B* is predicated in the definition of *A* ("number of the form 2n + 1 is odd"). "Them" (*autoîs*) is the subject, *B*. "All of the things belonging" (*hósois tôn huparchóntōn*) are the "belongers" like *A*. "The account indicating what it is" refers to the definition of the "belongers" (*A*). "They" in "they are predicated" (*autà . . . enhupárchousi*)[19] is the subjects (*B*).[20] The grammar of the sentence leaves obscure whether the "belonger" is related per se to the subject or vice versa, but since 73b22–24 (quoted shortly below) refers to even and odd as per se things (*tà kath' hautá*), things related per se to their subject, the "belongers" belong per se to the subjects.

b. Regarding "accidents" (*sumbebēkóta*), see note c on 73a34–37 earlier in the present chapter.

DISCUSSION

If *A* belongs to *B* and *B* is predicated in the definition of *A*, then *A* belongs per se 2 to *B*. It follows that all *A*'s are *B*'s and that *B*, individual *B*'s, and the species, subspecies, and genera of *B* are the only things to which *A* can belong.

The final sentence of the passage asserts that if *A* belongs to *B* and *A* does not belong per se 1 or per se 2 to *B*, then *A* is an accident of *B*.[21] Since Aristotle offers no justification for this claim, he seems to have regarded it as obvious. Its metaphysical underpinning will be discussed below in Chapter 8.

THE PER SE 1 AND PER SE 2 relations are explicitly linked with scientific necessity a few lines below in a passage that presents some difficulties:

> **I.4 73b16–24** "Therefore the things that, in the case of what is the object of unqualified scientific knowledge, are called per se in such a way as to belong to the things predicated or to be belonged in,[a] are on account of themselves and by necessity. For it is not possible [for them] not to belong either without qualification or the opposites,[b] as straight or bent belongs to line and odd or even to number. For the contrary is either a privation or a contradiction in the same kind,[c] as even is what is not odd *among numbers*, insofar as it follows.[d] And so if it is necessary to assert or deny, it is necessary also for the per se attributes to belong."[e]

NOTES ON TRANSLATION

a. The first clause of the first sentence identifies things related per se 1 and per se 2. In this extremely compressed expression, the work is done by active and passive forms of *enhupárchein*. I follow Barnes in taking *toîs katēgorouménois* to mean predicates and seeing the passage as referring first to per se 2, where the subject "belongs in" the definition of its predicate, and then to per se 1, where the subject "is belonged in" by the predicates, in that the per se 1 predicates occur in the definition.[22] The sense of the first sentence is "In the case of the objects of scientific knowledge, what are called per se in such a way that the subjects belong in the definitions of the predicates (i.e., per se 2) or the predicates belong in the definitions of the subjects (i.e., per se 1) [more literally, the subjects are belonged-in in their definitions by the predicates] are on account of themselves and by necessity." The passage speaks of "predicates" or "things predicated" (*toîs katēgorouménois*), not "things that belong" (*toîs hupárchousi*). What is *predicated* of triangle is not line but "bounded by lines," and the necessary fact that is relevant to science is that triangles are bounded by lines, not that the term "line" occurs in the definition of triangle (see discussion of per se 1 above).

b. "Without qualification" refers to per se 1: if *A* is predicated in the definition of *B*, then *A* cannot fail to belong to *B*. The word for "without qualification" (*haplôs*) implies a contrast with some kind of qualification. Here the qualification is "the opposites," which stands for the adverbial phrase "with the qualification that the attribute is one of a pair of opposites, one or the other of which must belong to the subject." This is meant to characterize the per se 2 relation.

c. The nature of the opposition excludes intermediates: the absence of one attribute entails the presence of the other, within the relevant class.[23]

d. Among numbers anything that is not odd is even, "insofar as it follows"—insofar as being not odd entails being even.[24]

e. If an attribute must be either affirmed or denied of a subject, and if a pair of attributes (odd and even) belongs per se 2 to a subject (number), such that for any given number, odd must either belong or not belong to it and if it does not belong, then even must belong, then it is necessary for the per se 2 attributes to belong, in that one or the other (but not both) must belong. This view is reaffirmed at I.6 74b8–10.

DISCUSSION

Two features of the passage require discussion: the doctrine that per se 2 predicates come in coordinated pairs of opposites, and the link between per se relations and necessity. The former involves several difficulties. Aristotle was struck by the fact that his principal examples of per se 2 relations are mutually exclusive pairs of attributes that between them cover the entire range of their subject. Even and odd have that relation to number, and straight and bent[25] have it to line. But not all the examples work like this. Equilateral/oblong is a distinction of *composite* number,[26] but these terms are treated in 73a40–b2 as per se 2 to *number*. Further, there are cases of groups of three (acute angle, right angle, and obtuse angle) or more (three-sided figure, four-sided figure, etc.) mutually exclusive and exhaustive coordinate per se 2 attributes, and cases where a per se 2 attribute has no coordinate attributes (e.g., "perfect" number) or where coordinate per se 2 attributes do not exhaust the subject (square number, cube number, etc.).

Since nothing in the account of the per se 2 relation requires the doctrine of "opposites" and since the doctrine fails so badly to pick out attributes that satisfy the account, it should be recognized as a hasty generalization and discarded as an unwelcome excrescence on the theory. In what follows, I shall disregard the "opposites" view of the per se 2 relation. The resulting "purified" theory is more philosophically interesting and viable.

The passage argues that if *A* belongs per se 1 or per se 2 to *B*, then *A* belongs to *B* on account of itself and by necessity, because it cannot fail to belong without qualification or "the opposites." The application to per se 1 relations is clear. Bounded by lines belongs necessarily to triangle because it holds of triangle by definition, and relations asserted in definitions are necessary. Further, since definitions are scientific principles, per se 1 relations hold "on account of themselves."[27]

As for per se 2 relations, even though the doctrine of "the opposites" is irrelevant, Aristotle's point is good. Everything that is even is necessarily a number, and similarly for the other examples. And since these are definitional relations, they are scientific principles and so hold "on account of themselves." Per se 2 predications meet the "by necessity" and "on account of themselves" conditions just as well as per se 1 predications do.

Per Se 1 and Per Se 2: Summary and Comparison

If *A* belongs to *B* and "*A*" occurs in the definition of *B*, then *A* belongs[28] per se 1 to *B*. If *A* belongs per se 1 to *B*, then the definition of *B* yields a proposition in which either *A* itself or some expression containing "*A*" is predicated of *B*. If triangle is defined as a figure bounded by three lines, line belongs per se 1 to triangle. "Line" is not predicated of triangle, but "bounded by lines" *is* and belongs necessarily to all triangles, and the proposition "all triangles are bounded by lines" can serve as an immediate premise of demonstrations. Since the only type of definition envisaged in *APo* is definition by genus plus differentia, and since for *A* to be related per se 1 to *B* requires *B* to be the subject of which *A* (or an expression containing "*A*") is predicated, the per se 1 relation is found only between subjects (as opposed to attributes) in a science's subject genus and their differentiae and genera. I will return to this point below.

If *A* belongs to *B* and "*B*" occurs in the definition of *A*, then *A* belongs per se 2 to *B*. If *A* is predicated of *B* and *A* belongs per se 2 to *B*, then *B* is predicated necessarily of *A*: all *A*'s are necessarily *B* (all odd things are numbers). Since *B* is the subject to which *A* belongs, cases requiring elaborate and ingenious paraphrases (like the line and triangle examples of per se 1) may not arise often. Such cases can be dealt with by observing that it will be possible to find some necessary proposition predicating of *A* an expression that includes the term "*B*." It is worth stressing that the directions of necessity are opposite: if *A* belongs per se 1 to *B* and A is predicated of B, then all B's are necessarily A. If A belongs per se 2 to B and A is predicated of B, then all A's are necessarily B.

Aristotle's emphasis on mutually exclusive and exhaustive coordinate per se 2 attributes can be expressed as follows. There are cases where $A_1, A_2, \ldots A_n$ belong per se 2 to B and are mutually exclusive (if x is A_i, x is not A_j for $j \neq i$) and exhaustive (if x is a B, then x is A_i for some i) of B. His ill-fated doctrine of "the opposites" amounts to a fascination with cases where $n = 2$, so that A_1 and A_2 are contradictories.

If A belongs per se 2 to B, can A be simply an attribute of B, or must it be related in some other way? In the initial discussion of the per se 2 relation, I used the barbarism "belongers" to avoid prejudging this question. Determining the relation between A and B is vital for understanding the role of the per se 2 relation in science. Returning to the examples, we might initially think that odd could be a species or differentia of the genus number, or an attribute of the subject number. There are several reasons against seeing it as a species of number—most obviously the clear statement that odd is not a species but a differentia of number (*Top.* IV.2 122b19–20, b23–24). More generally, if A belongs per se 2 to B, A belongs to B, but I know of no passage that says a species belongs to its genus.[29] Moreover, if A is a species of B, "B" occurs (as the genus) in the definition of A, and so A and B are already related per se 1. There seems no point in their also being related per se 2, and if they were it is scarcely credible that Aristotle would not have mentioned it.

It follows from the *Topics* passage cited above that a differentia can be related per se 2 to its genus. But the examples show that attributes other than differentiae can also belong per se 2 to their subjects. On the assumption that the division of a genus into species is unique, if even and odd are differentiae of number, then prime and composite are not, nor are square and oblong. Therefore any attribute whose subject is mentioned in its definition belongs per se 2 to its subject.

Per Se 2 and Definitions of Attributes

This fact makes the per se 2 relation central to demonstrative science instead of a relation Aristotle takes some pains to define and then puts to little use. The occurrence of a subject in the definition of an attribute locates the attribute in the subject genus. (We have already seen that a subject is located by being defined in terms of another subject [its genus] and by being mentioned in the definitions of other subjects [its species].) In definitions of attributes, the subject term corresponds to the genus term in definitions of subjects. Accepting the argument above that prime is an attribute (not a species or differentia) of number, we find that Euclid's definition of prime number as a number that is measured (i.e., divisible) by a unit alone[30] specifies the subject "number" and the primary[31] criterion for determining whether a number is prime. Similarly his definition of parallel—"straight lines which, being in the same plane and being produced indefinitely in both directions, do not meet one another in either direction''[32]—specifies the subject ''straight lines.'' These cases show that the subject mentioned in the definition may apply more widely than the attribute. In fact, the subject mentioned need not even be the most specific subject to which all instances of the attribute belong. Euclid defines even-times-odd as the product of an even number and an odd number.[33] In this case the subject, which appears in the definition of even-times-odd, is ''number,'' so that even-times-odd is related per se 2 to number; but there is a subclass of numbers, namely, even numbers, to which even-times-odd belongs.[34] Thus, if attribute *A* is related per se 2 to subject *B*, *B* need not be the narrowest subject to which *A* belongs.

Must each attribute, then, be defined in terms of a subject? We would expect not. Equilateral can be defined as ''having equal sides,'' without mention of any subject. But in order to ensure that equilateral is an attribute in *geometry* we need to supply or understand a reference to a specifically geometrical subject, such as figure.[35] Some subject can always be supplied in this way—the subject genus term (spatial magnitude for geometry) if no more specific subject will do. It follows that each attribute belongs per se 2 to some subject. The subject may not occur explicitly in the attribute's definition, but if it does not, it can be supplied, and in fact must be supplied in order to locate the attribute in the science's subject genus.

Per Se 2 and Compound Subjects

Geometry proves properties of parallel lines; arithmetic proves properties of prime numbers, even numbers, and proportional numbers. Since number is a subject, prime an attribute, what is the status of prime number?[36] It is evidently a subject of which further attributes can be proved to hold, but by what right can a compound of subject and attribute serve as a subject, and what determines whether a given compound can serve in that way? These questions must be answered for two types of cases—first, those where an attribute is compounded with the subject found in its definition (to which it is related per se 2), and second, those where it is compounded with a different subject.

For the first class of cases the per se 2 relation between the attribute and the subject legitimates the use of the compound as a subject. In effect, the subject term in such a compound is a relatively meaningless tag enabling the attribute to be used as a subject. Since all odd things in arithmetic are numbers, compounding ''number'' with ''prime'' does not provide any more information than is implicit in ''prime'' by itself; attributes proved to belong to prime numbers belong to prime things. On the other hand, since ''prime'' is an attribute and demonstrations prove that attributes belong to subjects, not to other attributes,[37] there are philosophical as well as grammatical objections to treating ''prime'' as a subject. These objections do not apply to ''prime number,'' which specifies a subclass of the subject ''number,'' which is itself a legitimate subject for attributes.[38]

Once again, the per se 2 relation is unexpectedly important in the structure of sciences. Thanks to it, the distinction between subjects and attributes can be maintained, while the number of legitimate subjects for proofs is increased and along with it the number of necessary per se linkages between subjects and attributes that can be proved.

Equilateral triangle is a typical case of the second kind. Equilateral does not belong per se 2 to triangle, since "triangle" does not occur in its definition and because there are equilateral figures that are not triangles. As above, I shall suppose that figure is the subject either expressed or implied in its definition, so that equilateral belongs per se 2 to figure. Now triangle is a subclass of figure. Figure, in turn, is a subclass of magnitude, and it is possible to speak of equilateral magnitudes too. So the kinds of compounds in question conjoin an attribute and a subject that is either included in or includes the subject to which the attribute belongs per se 2.

The two kinds of compounds behave somewhat differently. When the subject is broader than the subject to which the attribute belongs per se 2 (as in equilateral magnitude), things are straightforward. Since all equilateral things are figures, equilateral magnitude has the same extension as equilateral figure, and the same results that hold for equilateral figures also hold for equilateral magnitudes. Since widening the subject term makes no practical difference, we can treat such cases simply by replacing the wider subject with the subject to which the attribute belongs per se 2.

The interesting cases are those like equilateral triangle. The questions here are, when can a given compound serve as a subject of demonstrations, and by what right? Why can equilateral triangle and equilateral quadrilateral serve as subjects, but not equilateral circle or equilateral line? Line cannot because it is not a figure, but all equilateral things are figures. Equilateral circle cannot because there is no such thing. Equilateral triangle and equilateral quadrilateral are figures that exist (more precisely, they can be proved to exist). These case studies suggest the following doctrine. A compound of a subject (*C*) and an attribute (*A*) may legitimately serve as a subject of demonstrations when *C* is a subclass of a subject (*B*) and *A* belongs per se 2 to *B*, and the existence of *C*'s that have attribute *A* has been proved.[39]

Once again credit for legitimating such compound subjects goes to the per se 2 relation. Since it legitimates equilateral figure as a subject and since triangles are figures and equilateral triangles exist, equilateral triangles can be subjects of proofs just as much as equilateral figures can.

Per Se 3

> **I.4 73b5–10** "Also [I call per se] that which is not said of any other subject, as the walking [thing], *being something different*, is walking and the white [thing], but substance and everything that signifies an individual, which are what they are, not being something different. Things that are not [said] of a subject I call per se, but things that are [said] of a subject [I call] accidents."

DISCUSSION

A thing is per se in a third way if and only if "it is not said of any other subject"; otherwise it is an accident. The first example of an accident is "the walking [thing]," an attribute that describes something else, perhaps a person, which is its underlying subject. Properly speaking, the person is walking, and the description of the person as "the walking [thing]" is misleading. (The point is made more neatly in Greek than in English, since the expression "the walking" does not require a noun like "thing" in order to stand as a grammatical subject.) The second example reads simply "the white [thing]" (*tò leukón*). Since it is another example of something that is not per se 3, it is best understood as corresponding to "the walking [thing]" in its occurrence as an improper subject in the first example.[40] Spelled out more fully, this example would be "the white thing, being something else, is white" or "the white thing, being something else, is walking." These accidents are characterized as [being said] of a subject, in fact a subject different from themselves. By contrast, a per se 3 thing is not said of any *other* subject. Thus the first example of an accident stresses that the walking thing is something else (perhaps a person). The characteristic of not being said of another subject, or equivalently of being what they are without being anything different, belongs to "substance and everything that signifies an individual."[41]

The passage calls "substance (*ousía*) and everything that signifies an individual (*hósa tóde ti sēmaínei*)" per se 3, and defines per se 3 items as things that are not said of a subject (*tà mḕ kath' hupokeiménou*, sc. *legómena*; cf. b5–6). It is easy to read this passage as marking off substances from items in other categories, or primary substances from everything else,[42] so that the per se 3 relation is only marginally relevant to the theory of demonstrative science. I prefer an interpretation that makes it more central. I understand substances and individuals as clear cases of per se 3 items, but I see the definition of "per se 3" as extending more widely, to cover subjects in all sciences, even those sciences, like mathematics, that do not study substances. The distinction between per se 3 items and non-per se 3 items is the distinction between the subjects and attributes in each science's subject genus. In the context of a given science, the subjects in its subject genus satisfy the definition of per se 3: they do not belong to another subject, whereas attributes do belong to them. Moreover, a subject has its attributes not qua being something different, which is not the case for attributes. Triangle has 2R *being a triangle*, but equilateral has 2R *not being what it is*, but *being something different*.

Per Se 4

> **I.4 73b10–16** "That which belongs to each thing on account of itself is per se, and that [which belongs] not on account of itself is an accident. For example, if there was lightning while someone was walking, it was an accident, since it was not on account of his walking that there was lightning, but we say this [just] happened. But if

it happens on account of itself we say it is per se; for example, if something died while its throat was being cut, and died by virtue of the cut, [we say] that [it died] on account of its throat's being cut and that it did not [just] happen to die while its throat was being cut."

DISCUSSION

This passage sits awkwardly in its context. Unlike the other kinds of per se relation, this one does not focus on subjects and attributes and their relations. The examples present pairs of *events* (the occurrence of lightning, someone's walking; *x*'s dying, the cutting of *x*'s throat). The point is that in one case an event (the animal's dying) happened on account of or because of another event (the cutting of its throat), in the other case one event did not happen on account of or because of the other. Since there is no suggestion that any of these events happened on account of or because of itself, what is the point of calling any of them "per se"?

"Per se" is contrasted with "accident" (*sumbebēkós*) or "the accidental" (*katà sumbebēkós*), which in Greek are forms of the verb "to happen" (*sumbaínein*). The passage distinguishes between pairs of events where one takes place on account of the other and pairs where one "just happens" to take place while the other is taking place, calling what "just happened" (*sunébē*) an accident (*sumbebēkós*). The treatment of types of per se relations relevant to demonstrative science begins with the two most important ones (cf. 73b16–18), and then identifies two more by finding different ways of using "accident" (b11, b12; cf. b4–5, b9) and "happen" (b13, b15) and opposing a use of "per se" to each of these ways, whether the phrase makes linguistic sense or not.

In any case, since the per se 4 relation has to do with causal sequences of events, not per se predications, it is not central to the theory of science, however, these causal sequences imply a kind of necessity (any animal of such and such a type whose throat is cut in such and such conditions must die), and underlying the necessary causal structure are per se 1 and per se 2 relations, which *are* the proper objects of science.

Universal

I.4 73b26–74a3 "I call universal[a] whatever belongs in every case and per se and as itself. Thus it is obvious that everything universal belongs to things of necessity. But per se and as itself are the same; for example, point and straight belong to line per se, for also [they belong] to line as a line, and 2R belongs to triangle as triangle, for also the triangle has 2R per se. The universal belongs[b] precisely when it is shown in any chance case[c] of what is primary.[d] For example, having 2R is not universal to figure (although it is possible to show of a figure that it has 2R, but not of any chance figure, and the person does not use any chance figure in proving [it]; for square is a figure, but it does not have 2R). Any chance isosceles [triangle][e] has 2R, but [isosceles] is not primary, but triangle is prior. Now that which is shown to be the chance and primary thing having 2R or whatever else, it is to this primary subject that it belongs universally,[f] and demonstration per se is of this universally, and it is of the others in another way and not per se—nor is it of isosceles universally, but to a wider extent."[g]

NOTES ON TRANSLATION

a. *Kathólou dè légō* ("I call universal") can also be interpreted "I say . . . belongs universally" (taking *kathólou* as governed in sense by *hupárchēi*, b27). *Kathólou* is used both as an adjective (b32, b34, a1) and as an adverb (*hupárchei kathólou*, a1). An attribute is universal if and only if it belongs universally: *A* is universal to *B* if and only if *A* belongs universally to *B*.

b. "The universal belongs" means "*A* belongs universally to *B*" or "*A* is universal to *B*."

c. "In any chance case" (literally, "in the chance case") keeps the translation "in the case of" for *epí* (73a28, a33, b16).

d. A more literal translation than "of what is primary" would be "in the chance first [or primary] case," which is too compact in English if not in Greek.

e. In "any chance isosceles [triangle]" (*tò d' isoskelès . . . tò tuchón*), the ability of Greek to omit a subject that is implied in the context is put to good advantage.

f. The clumsy English of "now that . . . belongs universally" results from fidelity to the Greek (*hò toínun tò tuchòn prôton deíknutai dúo orthàs échon è hotioûn állo, toútōi prótōi hupárchei kathólou*). The point is that if *B* is shown to be the chance and primary thing to which *A* belongs, *A* belongs universally to *B*.

g. The lines "Demonstration per se . . . wider extent" contain a number of difficulties. "This" refers back to "this primary subject" in the previous line, and "this," "the others," and "the isosceles" are grammatically parallel, as are "per se" and "not per se" and as are the occurrences of "universally." Since "this primary subject" is the primary subject (triangle), it follows that "this" is the primary subject, "the others" are failed rival candidates for primary subject (isosceles and figure). "The isosceles" confirms this: in the "nor" clause isosceles is an example of "the others." "The demonstration is of this universally" is shorthand for "the demonstration that has this as the subject of its conclusion has as its conclusion a proposition in which the attribute belongs universally."[43] The last part means "the demonstration with "isosceles" as the subject of its conclusion does not have as its conclusion a proposition in which the attribute belongs universally, but it [the attribute] extends more widely." Since the relation isosceles triangle—2R[44] is per se,[45] and since isosceles is a specimen case of "the others," Aristotle is not saying that in a way there is demonstration of non-per se relations like isosceles triangle—2R. "Per se" modifies "demonstration." "Demonstration per se" means "demonstration by definition" or "it is the nature of demonstration." The passage, then, has the following sense: "The demonstration that has this as the subject of its conclusion has as its conclusion a proposition in which the attribute belongs universally. This is demonstration per se. Demonstrations that have the failed rival candidates for primary subject of the attribute as the subjects of their conclusions [sc., where the conclusion is a proposition in which the attribute does *not* belong universally] are not demonstrations per se; for example, the demonstration that has "isosceles triangle" as the subject of its conclusion does not have as its conclusion a proposition in which the attribute [sc., 2R] belongs universally, but the attribute extends more widely." In plainer language this means: "Demonstration per se proves that the attribute holds universally of this, and demonstration that it holds of the others is not demonstration per se. For example, 2R does not hold universally of isosceles triangle, but extends more widely."

DISCUSSION

This passage introduces an unusual sense of "universal" (*kathólou*), which is important in demonstrative sciences. The first sentence appears to be a definition or at least a necessary and sufficient condition: an attribute *A* belongs universally to a subject *B* (*A—B* is a universal connection) if and only if three conditions are satisfied: *A* belongs "in every case" to *B* (as that expression was defined at 73a28–29), *A* belongs to *B* per se, (it is not stated which of the four per se relations is meant), and *A* belongs to *B* "as itself" (*hêi autó*). Since the first two requirements are already well understood (and also, since per se entails "in every case"), commentators frequently assume that the "as itself" relation introduces what is distinctive about the universal (Ross, pp. 61, 522)—wrongly so, because as the third sentence states, "per se and 'as itself' are the same" (b28–29), a stipulation that the following examples confirm. (Point and straight belong respectively per se 1 and per se 2 to line[46] but do not belong universally to line.) The first sentence therefore provides necessary but not yet sufficient conditions for belonging universally. The second sentence states a consequence of the per se requirement. The distinctive feature of the universal relation first appears in the fourth sentence: "when it is shown in any chance case of what is primary [*prôton*]." "In any chance case" means no more than "in every case," but "what is primary" is a new idea. The example of isosceles shows that if *A* belongs universally to *B* then *B* is the "primary" or "first" (most general)[47] subject to which *A* belongs per se. Primacy (and the allied notion of priority) makes a late appearance in the passage but is prominent thereafter (five occurrences in I.4 and nine in I.5) and carries the weight in the exposition of the star example of a universal relation, "triangle—2R."

The key characteristic of universal connections is that the subject and attribute are coextensive. In effect, the "in every case" relation goes both ways. 2R belongs universally to triangle but not to figure (which is too wide) or to isosceles triangle (which is too narrow). The example is elaborated in I.5, which directs us to strip off successive specifications until the attribute no longer holds "in every case."[48] 2R belongs to the bronze isosceles triangle. When bronze and isosceles are stripped off, it still belongs to triangle, but strip off triangle and 2R no longer holds "in every case" of the next most general term. Since triangle is the last in the series of which 2R holds "in every case," 2R belongs universally to triangle.[49] When our present passage says that "it is possible to show of a figure that it has 2R," we are to think of a person's claiming to prove that all figures have 2R, and using a triangle as his diagram. The proof purporting to show that all figures have 2R employs some features the specimen figure has qua triangle, not simply those it has qua figure. The point is that proofs employing *ekthesis*, a specimen case, must make use only of features of the case that constitute it as a specimen of the desired sort.

The other main feature of universal connections is that they are per se. Triangle is explicitly declared to have 2R per se (b31–32). But in which of the types of per se relations do triangle and 2R stand? To begin with, "triangle—2R" is the conclusion of a proof, not an immediate principle; and further, the discussion and examples of universal relations in I.4–5 make it clear that the universal relation is conceived of as holding typically of provable[50] relations.[51] But the per se 1 and per se 2 relations are defined in terms of definition (*tí esti*),[52] which strongly suggests that only immediate principles can be related in either of these ways. It is also clear that the definition of triangle does not mention 2R (so that 2R does not belong per se 2 to triangle) and the definition of 2R does not mention triangle (so that 2R does not belong per se 1 to triangle). There are two possibilities. Either the per se relation between triangle and 2R is one of the four kinds of per se relations discussed in I.4, or it is not.[53] I will argue that it is.

The strategy of the present passage is first to associate "universal" with "per se" and next to make it clear that per se relations extend more widely than definitional relations, so as to prepare the way for identifying primacy as the distinctive feature of universal relations and expounding "primary" through an example of the nonimmediate, nondefinitional relation triangle—2R. The expression "as itself" helps make the point that the per se extends beyond the definitional. In effect it reminds us of the "natural" meaning of "per se," so that we do not make the mistake of construing the accounts of per se 1 and per se 2 as technical definitions. So, point and straight belong per se to line because (*gár*) they belong to it *as* a line—in virtue of the nature of a line.[54] Similarly, 2R belongs to triangle in virtue of the nature of a triangle, and so belongs to it per se. The explicit reference to examples used in the accounts of the per se 1 and per se 2 relations makes it difficult to accept that Aristotle is introducing a different and as yet unheard of kind of per se relation to hold between triangle and 2R.

Are Conclusions of Demonstrations Per Se?

It is most natural to read "what it is" as meaning strictly "definition," for Aristotle speaks not only of belonging in what it is, but also of belonging in the account stating what it is (*en tôi lógōi tôi légonti tí estin enhupárchei*, 73a36–37, and similar expressions at 76a38, b1–2), and this is a *definition* of "definition" (II.10 93b29, 94a11; cf. II.3 91a1). But a good deal of textual evidence favors a broader interpretation of per se 1 and per se 2. To begin with there are the passages that describe conclusions of proofs as per se relations, beginning with 73b31–32 in our passage (among others, also I.6 75a29–30, a34, I.7 75a41, b1, I.9 76a7, a8). Is it likely that immediately after carefully cataloguing different types of per se relations he would use the expression in a way not covered by the catalogue? If it is different, why not list it as a fifth kind of per se relation alongside the four he discusses?

Second, the argument at of I.6 74b5–12 (discussed at the beginning of the present chapter), where the per se is coextensive with the necessary, asserts that the objects of scientific knowledge (the conclusions of demonstrations) are necessary. Thus, they are per se too. But the per se is viewed simply as per se 1 and per se 2 relations (74b7–10), so the passage somehow includes nonimmediate per se connections in the realm of per se 1 and per se 2 relations.

There is also positive evidence that Aristotle believed that his characterizations of per se 1 and per se 2 apply to such relations. One passage in I.22 reminds us of the importance of per se relations—"For demonstration is of all that belongs per se to things."—and specifies that things are per se in two ways: per se 1 and per se 2 (84a11–17). Since demonstration is of nonimmedate connections, it follows that per se 1 and per se 2 must cover more than just definitional connections.

This passage also argues that the consequence of a certain type of demonstration is a per se 2 relation:

> **I.22 84a14–21** ". . . for example, being odd, which belongs to number, [is per se 2] to number, and number itself belongs as an element in the account of it, and again multitude or discreteness belongs as an element in the account of number. Neither of these can be infinite, either as odd [is predicated] of number. (For again there would be something else [that belongs] to odd, to which it [odd] belongs as an element [in the account of the thing] belonging [to it, i.e., odd]. But if this exists, number will belong first as an element [in the account of those things] belonging to it (odd) . . .)"

DISCUSSION

The unusual expression "belongs as an element [in the account of the thing] belonging" (*enhupárchei hupárchonti*) is used of the per se 2 relation. The point is that if x belongs per se 2 to odd, then, since odd belongs per se 2 to number, x belongs per se 2 to number. Number is not mentioned in the definition of x, or x would belong per se 2 to number without further argument. The per se 2 relation is transitive. Since anything odd must be a number and since anything that is x must be odd, anything that is x must be a number—a consequence that follows because of the natures of x and of oddness specified in their definitions.

FINALLY, IN II.8 ARISTOTLE speaks of eclipse and thunder as attributes of the moon and clouds (93a30–b14). The occurrence of those attributes in those subjects needs to be explained, since the relations are not immediate. The hunt for the explanations is a hunt for definitions of the attributes since a lunar eclipse is a screening by the earth (b7) and thunder is quenching of fire in cloud (b8), and these are the accounts (*lógoi*, b6, b12) of the attributes. On the other hand, the nonimmediate explananda (thunder is a certain noise in clouds and an eclipse is a certain loss of light, a22–23)[55] are called "something of the 'what it is' " (*ti toû tí estin*).[56] Nonimmediate demonstrable relations have "something of the 'what it is' " even if they do not include the term or terms actually occurring in the definition. *APo* II.10 even calls this kind of proposition a kind of definition (*horismós*): the "conclusion of the proof of 'what it is' " (94a13–14; cf. a7–8) as opposed to either the explanation ("screening by the earth"), which is the middle term in the proof of the conclusion, or the "rearranged demonstration" type of definition ("eclipse is loss of light from the moon due to screening by the earth," 94a1–7). Thus it appears that Aristotle can call any demonstrable connection "something of the 'what it is' " and a kind of definition, and claim that it is systematically related to indemonstrable definitions (discussed in II.9), which are the actual principles of the science.[57]

All this amounts to good evidence that Aristotle intended the per se 1 and per se 2 relations to cover more than terms occurring in definitions. But even if the evidence is rejected, it is possible to question the credentials of the opposing view, that only definitional relations are per se 1 and per se 2. The accounts of those two relations at 74a34–b5 are standardly read as definitions—"I call per se . . . "[58] but need not be. The translations above are deliberately less strong, to leave open the possibility that the passages are not defining the per se 1 and per se 2 relations but identifying situations in which those relations are found. Somewhat similarly, 73b26–27 appears to define the universal relation, yet we have seen that those lines do not mention the distinctive characteristic of universal relations and at best give an incomplete definition.

If this evidence is accepted, demonstrations prove per se 1 and per se 2 relations. What kinds of demonstrations prove what kinds of per se relations? *APo* I.22 84a14–21 gives some help. Since the per se 2 relation is transitive, a demonstration will have a per se 2 conclusion if all its premises are per se 2. This may not cover a large number of proofs, and obviously does not cover *APo*'s most prominent example of a provable connection, triangle—2R. I must defer discussion of this matter until ascertaining more about the inner workings of demonstrations.[59]

In summary, there is no doubt that provable connections of subject and attribute are per se relations. Aristotle labels them "per se accidents"[60] (I.7 75b1, I.22 83b19–20).[61] Since there is evidence that he believed them to be per se 1 and per se 2 relations and since it is not certain that he intended to define these types of per se relations in terms of definitory relations, it is best to suppose that the conclusions of proofs *do* state per se 1 or per se 2 relations. On the other hand, since the conclusions of proofs *are* per se relations, anyone who rejects the present account should conclude that this type of per se relation should be discussed along with the other four.

In Every Case, Per Se, Universal, and Necessary

I will conclude by examining the relationships among the "in every case," the "per se," the "universal," and "necessary" relations. *A* belongs per se to *B* if and only if *A* belongs necessarily to *B* (see above p. 84); and if *A* belongs universally to *B*, *A* belongs "in every case" (73b26) and per se to *B* (b27), from which it follows that *A* also belongs necessarily to *B* (b27–28).

Aristotle introduces the definitions of "in every case," "per se," and "universal" as a preliminary to identifying the kinds of principles (I.4 73a24–27). The only obvious link between any of the varieties of principles and the relations defined in I.4 is between definitions and per se 1 and per se 2 relations, which are defined in terms of "what it is." I have argued that "what it is" and "what belongs in what it is" need to be taken as referring more widely than just to definitions and the terms found in definitions, but the notion of definition is clearly there. The nondefinitional per se connections of a subject or an attribute are those it has as consequences of its definition and definitions of other terms in the subject genus. Since necessary relations are per se, and per se relations are dependent on definitions, the necessity of scientific facts is definitional necessity: scientific facts are true by definition. We shall return to this in Chapter 8.

The relations between "per se" and "in every case" and between "in every case" and "necessary" are left unexplained. It is safe to say that a necessary relation holds in every case and likewise a per se relation holds in every case, but the more interesting questions whether relations that hold in every case are necessary and per se are not raised explicitly, nor does anything in the text determine a solution.[62] These issues are irrelevant to *APo*. A scientific demonstration does not just show *that* its conclusion is necessary, but must show *why*. Even if necessity is an extensional concept, it is no explanation of why a connection is necessary to show that in fact the attribute belongs to every individual or every species of the subject for which the attribute holds. An example makes it clear:

> **I.5 74a25–32** "Even if someone shows of each [kind of triangle]—equilateral, scalene, and isosceles separately—by one proof or different proofs, that each has 2R, he does not yet know that triangle has 2R (except in the sophistical way), nor does he know it of triangle as a whole, even if there is no other triangle aside from these. For he does not know [that 2R belongs to its subject] qua triangle, nor that every triangle [has 2R], except numerically. But [he does] not [know that] every [triangle has it] by virtue of its kind, even if there is not any [triangle] that he does not know."

DISCUSSION

Knowing something to hold "numerically" (*kat' arithmón*, a31) of every triangle is the same as knowing it "in every case" of triangles, and this kind of knowledge is not scientific knowledge (although it is entailed by scientific knowledge).

CHARACTERISTIC OF UNIVERSAL relations is that the subject and attribute are coextensive, but this does not mean that "universal" is an extensional concept like "in every case." Since universal relations are per se, a merely numerical approach to them is inadequate. They hold on account of the nature, not just the extension, of their subjects and attributes.

The contrast between the universal and the per se suggests that per se relations need not be coextensive, so that isosceles triangle—2R would be per se, although not universal. (Figure—2R is not per se because 2R is not implicit in the definition of figure; this is obvious since there are some figures that do not have 2R.) This suggestion would surprise those commentators who think of the connection "isosceles triangle—2R" as irrelevant to science, but it agrees with what we know about per se relations. Isosceles triangles have 2R necessarily, and therefore per se.[63]

Scientific Essences

CHAPTER VIII

THE DISTINCTION between principles and provables is basic to the theory of demonstrative science. We have seen that both are per se predications,[1] that per se predications are the same as necessary predications, and that the types of per se predications in question, per se 1 and per se 2, are defined in terms of "what it is" (*tí esti*), which is connected with the notions of definition and essence.[2] In this chapter I will look at the distinction between provable and unprovable predications, their interrelations, the notion of essence on which the theory of science is based, and the relation between essence and necessity.

In view of the huge amount of work that has been recently devoted to Aristotle's metaphysics and his conceptions of necessity and other modal terms, I shall say at the beginning what the present chapter does not do. It does not attempt to defend Aristotle's conceptions of essence, real definition, or necessity, and it does not try to compare them with corresponding modern notions. Also, since science for Aristotle treats universals and not individuals (or treats individuals only incidentally), nothing will be said directly about the vexed issue of individual essences, although a good deal of what is said will be relevant to that issue. Finally, since Aristotle himself says little directly on these issues, there will be much less detailed discussion of passages than in previous chapters. My treatment will be based for the most part on the interpretations already reached, especially in Chapter 7, and will bring out implications of Aristotle's text.

The present chapter will explore the metaphysical foundations of the theory of science. Little attempt will be made to fit these metaphysical theses into the views appearing in such works as *Metaphysics*, *Categories*, and *De Anima*. Many of the analytical tools of those works are absent from *APo*, and the thrust of the present treatment cuts across some of Aristotle's other discussions in ways that would take considerable effort and care to tease out. It will be no obstacle to present concerns if some of the metaphysical views of *APo* are different from those of other works. In fact, an account of the metaphysical foundations of the theory of demonstrative science may help with other Aristotelian material, since it can provide a vantage point from which to view the labyrinth of argument in the Aristotelian treatises primarily devoted to metaphysics.

Scientific Facts

A passage in I.6[3] points to two kinds of facts, per se and per accidens for if it is a fact that *B* is *A*, then *A* is predicated of *B*, and (for a science **S**)[4] *A* is predicated either per se or per accidens of *B* but not both. (Put otherwise: it is a per se or per accidens fact that *B* is *A* and that *A* is related (or belongs) per se or per accidens to *B*.)[5] Per se facts are further divided into definitional and nondefinitional ones. A definitional per se fact (*A* belongs to *B*) is expressed in the principles of **S**. I shall call nondefinitional per se facts derivative facts. Thus, there is a three-way division of facts into (1) definitional facts, (2) derivative facts, and (3) other facts.

Both type (1) and type (2) facts are per se; type (3) facts are per accidens. Aristotle sometimes calls both type (2) and type (3) facts accidents, sometimes only type (3) facts. When type (2) facts count as accidents, type (2) facts are distinguished from type (3) facts as "per se accidents."[6]

Since all per se facts are necessary and all necessary facts are per se, type (1) and type (2) facts are necessary and type (3) facts are nonnecessary or contingent. Further, type (1) and type (2) are scientific facts, type (3) facts are not: type (1) facts are basic facts expressed in definitions, which are unprovable scientific principles, and type (2) facts follow from and are based on the principles; type (3) facts are insusceptible of proof and therefore not scientific facts (I.6 75a18–21). Finally, since Aristotle defines "definition" as a statement of essence, we may call definitional facts essential facts. These observations are summarized on the following chart.

essential	{ (1) definitional facts	necessary, per se scientific (covering (1) and (2))
accidental {	(2) derivative facts	
	(3) other facts }	contingent, per accidens, nonscientific

Further, since each fact is a connection between a subject and an attribute, we may talk of essential, definitional, and derivative *attributes* of a subject.

Insofar as scientific knowledge is based on the definitions of the members (subjects and attributes) of its subject genus, it is based on essences. The definitions express essential facts about the things, and the proofs reveal their consequences. Moreover, derivative and definitional facts are related not merely by logical consequence, but equally importantly by *real* consequence. A derivative fact is necessarily true *because* the essential facts necessitate it. A scientific proof of a derivative fact shows *how* it follows from the essential facts by showing how it is *grounded in* and *due to* them.[7]

Individuals, Intelligibility, and Per Se Facts

To get a better grip on the notion of essence in play, consider the relation between scientific knowledge and individuals. For Aristotle scientific knowledge notoriously is knowledge of universals, not individuals (I.31 87b38–39), and the principles of sciences are universal (I.8 75b21–22).[8] The principles and conclusions are true always and everywhere (I.31 87b32), whereas particulars are here and now (b30).[9] There is no unqualified demonstration of individuals (that an individual has some attribute or exists), only per accidens demonstration (I.8 75b24–26).

But it is not to be scorned that there is per accidens scientific knowledge of individuals. Insofar as an individual falls under universals in a science's subject genus, scientific facts about those universals apply necessarily to it. Insofar as a piece of bronze is a triangle, the fact triangle—2R applies necessarily to it. What counts is that it is a *such* (*toioûde*, I.8 75b34, I.31 87b29), not a *this* (*tóde ti*, *tóde*; I.31 87b29, b30, b31). Sciences treat things qua being of a certain sort or kind (*Met.* K.2 1060b21) and so are concerned with individuals only incidentally, only insofar as they happen to be of that sort.

From the point of view of the science,[10] the particular is only partly intelligible. A bronze triangle is comprehensible partly by geometry, partly by natural science. As a triangle, it has all the attributes geometry proves triangles to have, but its being made of bronze, having a certain density, and so on, features intelligible to other sciences, are unintelligible from the point of view of geometry. A given science, then, treats a particular partially, under one aspect, and ignores the manifold other aspects under which it can be considered.

Suppose an individual, x, qualifies as a subject of science **S** by being an F.[11] Some of x's attributes belong to x qua an F and others do not. This is the distinction between per se and per accidens attributes in the context of **S**. For the purposes of **S**, x *is* an F, and the attributes of x that are comprehensible by **S** are those belonging to x qua an F. They belong to x per se, through x's nature *as an* F. The rest are irrelevant.

By contrast, nothing about its own proper subject matter is irrelevant to **S**. Nothing holding of a triangle qua spatial magnitude is irrelevant to geometry. All its spatial attributes are relevant[12] and all are per se attributes of triangles, facts due to the nature of triangles, that is, the *geometrical* nature of triangles.

A science treats an individual qua being of a certain sort, where "being of a certain sort" means belonging to the science's subject genus. Geometry treats the piece of bronze qua spatial magnitude. In effect, it asks "what is it" meaning "what spatial magnitude is it." The answer "a triangle" locates it in the subject genus of geometry and so determines its geometrical attributes.[13] Similarly, natural science asks "what is it" and means "what kind of natural body is it?" The answer "bronze" locates it in the subject genus of natural science, and its location in that subject genus determines its natural attributes. Since one science treats a certain aspect of an individual and others treat others, the "what is it" question is actually a different question for each science. In each case the question asks where the individual is located in the relevant subject genus—what sort of geometrical, natural, or whatever, thing it is. More specifically it asks what sort of geometrical, or whatever, *subject* it is. In the context of geometry, the right sort of answer to the "what is it" question is "triangle" (or "isosceles triangle" or "figure"), not "2R" (or "a geometrical entity with 2R").

Subjects and Attributes

I suggested above[14] that the per se 3 relation makes this distinction: subjects of sciences are per se things, and attributes are accidents. Moreover, I.22 contrasts statements like "the triangle has 2R," in which the grammatical subject is a subject and the grammatical predicate is an attribute (type I statements) with those like "the [thing with][15] 2R is a triangle," in which the grammatical subject is an attribute and the grammatical predicate is a subject (type II statements) (83a1ff.). Type I statements "predicate," but type II statements "either do not predicate in any way or predicate not without qualification, but incidentally" (83a14–17). Only the former kind is found in sciences since "that is how demonstrations proceed" (83a18–21). Scientific propositions (principles and conclusions) are type I statements.

This important distinction between subject and attribute corresponds to the substance/accident distinction familiar from the *Categories* and *Metaphysics*. When no special context is implicit, the "what is it" question is correctly answered by identifying what kind of substance (as opposed to what quality, quantity, etc.) the individual in question is (*Cat.* 5 2b29–36; *Met.* Z.1 1028a13–18, a36–b1). Accidents are predicated of substances but not vice versa (*Cat.* 5 3a1–8), they depend on substances for their being (*Met.* Z.1 1028a29–31), and so on. The passages from I.4 and I.22 that make the distinction between subjects and attributes use examples that happen to be substances and accidents. However, in those passages we hear more of subjects and predicables (*hupokeímenon*, *katēgoroúmenon*) than of substances and accidents (*ousía*, *sumbebēkóta*). To be sure, if triangles are not substances, then in the context of discussion envisioned in *Categories* and *Metaphysics* "it is a triangle" will never be a proper answer to a "what is it" question, even if that question is asked about a triangular object. But in contexts where the metaphysical primacy of substances is not at issue, "what is it" questions can demand answers in nonsubstance categories.[16] In the context of science **S**, to ask what something is is to ask a question expecting a certain kind of answer, identifying the thing as a subject in the subject genus of **S**. Once we ascertain what it *is*, we may go on to ask further questions about it, namely, what its attributes are. Correct answers here will be type I statements: for example, "it (the triangle) has 2R."

Essence and Necessity

Suppose x is an individual and in the context of science **S** x is an F (x is an F and F is a subject in the subject genus of **S**). Further, suppose $g_1, \ldots g_m$ are the essential attributes of F and $f_1, \ldots f_n$ are the derivative attributes. These are the attributes of F relevant to **S**. They belong to all F's qua F and are treated by **S**. All the other attributes of x are irrelevant to it qua F, and so irrelevant to it qua considered by **S**.[17] They hold of x per accidens (from the point of view of **S**). All the g_i and f_i belong to F (and to x qua F) per se and of necessity. The g_i are the essence of F, and the f_i follow from and depend on the g_i (and the other principles of **S**.) We have seen why all the g_i and f_i belong to F per se and how the distinction between principles and conclusions requires the distinction between g_i and f_i. What remains to be shown is why the g_i and f_i belong to F of necessity.

The short answer is that the f_i belong per se to F, and per se facts are necessary, a consequence of I.6 74b5–12 (discussed in Chapter 7 above). But that passage does not *justify* the claim that per se facts are necessary, nor does I.4 in its account of per se relations. In what follows I will attempt to account for the necessity found in science.

One approach to this problem is to say that the nature of scientific proof requires them to be necessary. Since what is known by scientific knowledge is necessary (I.2 71b12), it follows that principles as well as conclusions (essential as well as derivative facts) are necessary (I.6). If there is scientific knowledge, its principles and conclusions are necessary.[18]

But this argument is a bluff. It will not convince anyone who doubts that *there is* anything that qualifies as scientific knowledge in Aristotle's sense. Why should we suppose that there are necessary truths about things? This is a difficult problem on which the most thorough treatment to date[19] can only make a variety of suggestions, none of them entirely convincing.

A different approach is more closely linked to the *APo* theory of science and to the relation between scientific knowledge and individuals sketched above. As before, let **S** be a science, F a subject in the subject genus of **S**, $g_1, \ldots g_m$ the essential attributes and $f_1, \ldots f_n$ the derivative attributes of F. Each f_i is a consequence of one or more of the g_i (together with the other principles of **S**) and the set of f_i is the maximal set of consequences of the g_i (for subject F). It follows that all g_i and f_i are tightly related, in that if the set of all the g_i or the set of all the f_i were any different, the subject would be different from F. This is trivially true for the g_i,[20] but equally true for the f_i, since any change in the f_i will entail some change among the g_i from which the f_i are proved.

Thus, if in the context of science **S**, x is a particular F, then (1) x has all the g_i and (2) x has all the f_i. (3) If x has all the g_i, then x has all the f_i and conversely (4) if x has all the f_i, then x has all the g_i. Moreover, (5) if a particular x has all the g_i, then x is an F, since having all the g_i is what it is to be an F. It follows from this that (6) if x has all the f_i, x is an F. The following diagram represents these facts.

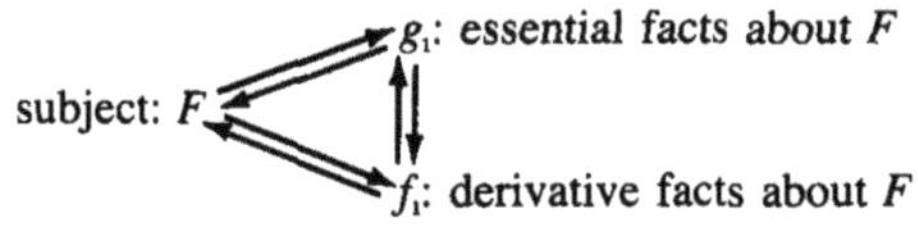

It follows from this that (7) if x is not an F, x does not have all the g_i, (8) if x does not have all the g_i, x is not an F, (9) if x does not have all the f_i, x is not an F, and (10) if x does not have all the f_i, x does not have all the g_i. Conversely, (11) if x is not an F, x does not have all the f_i and (12) if x does not have all the g_i, x does not have all the f_i. And from this it is easy to see that a necessary and sufficient condition for x to be an F is for x to have all the g_i and f_i, all the per se predications of F. If it does not, then *ipso facto* it is not an F. (This does not show that the individual, x, necessarily has the g_i and f_i, but only that *if* x is an F, x must have the g_i and f_i, that qua an F, x must have the g_i and f_i.)

Further, the distinction between per se facts and per accidens facts coincides with the distinction between what is relevant to a given science and what is not. From the given science's point of view, the per accidens facts could change without there being any change in the properties it treats. (The triangle could become a different color or be heated or moved from one place to another, without there being any change in its geometrical properties.) But a difference in any of its per se properties—in any of the g_i or f_i—would mean it was not the same subject of **S**. The per accidens facts can be described in Aristotelian terms as capable of being otherwise, the per se facts as incapable of being otherwise. It follows that per se facts are necessary because Aristotle's basic notion of the necessary is "that which cannot be otherwise" (*Met.* Δ.5 1015a34–36) and this is the kind of necessity found in science. Thus: "Demonstration is of what is necessary, for if it is proved without qualification, it cannot be otherwise. Its cause is the principles, since the things on which the deduction depends cannot be otherwise" (*Met.* Δ.5 1015b7–9).[21] At the level of universals the nature of the necessity is clear. A science treats things insofar as they fall under the universals in its subject genus. If x is an F and F is a subject in the subject genus of **S**, then **S** treats x qua F. And x is an F if x satisfies the definition of F, that is, if it has all the g_i. Each g_i is necessary to F. At the level of universals, the necessity is absolute and unconditional. Triangles must have 2R. It is also based on the nature of the universals in question—the nature specified in the definition.

At the level of particulars the story is quite different. Insofar as a particular falls under a universal treated by a science it must have the necessary attributes of that universal. Insofar as x is an F, x must have all the g_i and f_i. Also, x will have other attributes that are not treated by **S**. (A particular triangle will be made of bronze.) These are per accidens from the standpoint of **S**, and could be otherwise without affecting the fact that qua F, x must have all the g_i and f_i. (The triangle could just as well be made of iron.) All this holds insofar as x is an F, but nothing about **S** forces x to be an F. From the standpoint of the particular, scientific necessity is conditional, not absolute. The nature of science in general does not require that x be an F or that x remain an F once it is an F, nor does it prohibit what is not an F from becoming an F. It does not even guarantee that F's always exist[22]—only that whenever x is F, x has all the per se attributes of F's; if x lacks any of them then x is not an F.

Suppose we have a triangular piece of bronze. For geometry it is a triangle; for natural science it is bronze. Qua bronze it is fusible and ductile, natural attributes irrelevant to geometry. Qua triangle it has 2R, a geometrical attribute irrelevant to natural science. Natural science does not require it to have 2R and geometry does not require it to be fusible or ductile. But geometry does require that qua [bronze] triangle it have 2R, and natural science requires that qua [triangular] piece of bronze it be fusible and ductile. Now suppose that its natural attributes are put to work: it is melted and recast into another shape, say a square. We now have a bronze square, a square piece of bronze with the natural properties of the bronze triangle, but different geometrical properties. The necessity imposed by geometry requires that a bronze triangle have the essential and derivative per se attributes of triangles, but not that the bronze triangle exist. Nor does geometric necessity prevent it from undergoing natural changes, since such changes happen to it qua bronze and not qua triangle. Thus geometry has nothing to do with the conditions under which the bronze triangle continues to exist or when it came into existence in the first place.

This interpretation allows there to be necessary facts dependent on essences, yet for the world to be populated with perishable individuals that have attributes only contingently. It distinguishes two levels: the universal and the particular, such that necessity operates on the universal, not the particular level, whereas contingency operates on the level of particulars. The two levels *are* linked. Universals do not exist independently of individuals, and individuals are not intelligible except *as* falling under universals. Necessity has both an absolute and a conditional character. At the universal level, necessity is absolute: a universal must have its essential and derivative per se attributes. Contingency does not enter into this level except by the back door. Science **S** does not recognize any attributes of its subjects but their necessary per se ones. A triangle may be blue, but its color is irrelevant and indifferent—we might say invisible—to geometry. From this point of view, geometrical triangles have no colors. From another viewpoint they do, since the facts of geometry apply to particular triangles that have physical existence, and hence have physical attributes. But even if particular triangles have nongeometrical attributes, they do not have them qua triangles. Likewise if a particular physical object has geometrical attributes, it does not have them qua physical object. Any changes it undergoes qua physical object (such as being melted and recast into a different shape) may bring with them changes in its geometrical attributes too. Thus scientific necessity, which applies absolutely to universals, applies only contingently to individuals—only insofar as an individual falls under a universal in the science's subject genus do the per se facts about that universal apply to it.[23]

The present chapter has set out the basis of the distinctions between per se and per accidens facts and between essential and derivative per se facts. It has also suggested a way to understand how essential facts (and consequently derivative facts) are necessary. The next two chapters will connect the notion of scientific essence with that of definition, and so establish the place of scientific essences among the principles of sciences.

Definitions, II:
Form and Content

CHAPTER IX

THE DISTINCTION between essential and derivative per se facts is basic to Aristotle's theory. Essential facts, being immediate and unprovable, must be scientific principles. Since the only kinds of principles are definitions, axioms, and existence claims, the definition of a thing must contain all the essential facts about it,[1] from which (together with the other principles of the science) the derivative facts about it are proved.[2] The definition of a subject or attribute, x, will in general be a compound statement of the form "x is A_1 & x is A_2 & . . . & x is A_n,"[3] where each of the "x is A_i" statements represents an immediate per se relation. I shall call this a "fat definition." However, it is not clear how well this notion of definition agrees with Aristotle's notion (occurring in *APo* and elsewhere) of definition by genus and differentiae ("G-D definitions"). But unless G-D definitions contain all the essential facts, it will be impossible to prove all the derivative per se facts, so that demonstrative sciences will fail to perform their function of explicating all the per se relations of their subject genera.

In this chapter I will examine how *APo* treats G-D definitions and show that it conceives of differentiae in two different ways, neither of which is suitable for fat definitions, thus that *APo* contains a fundamental flaw. I will then take up the history of G-D definitions in other works of Aristotle's and sketch a line of development ending with an approach to G-D definitions in *Parts of Animals* that satisfies the demands of demonstrative science.

To begin with, *APo* assumes that definitions will be in the form of genus and differentia:

> **I.22 83a39–b1** "It will predicated as essence, that is, as genus or differentia."

Two passages in *APo* II.13 envisage a procedure for discovering definitions by dividing a genus by successive differentiae. The more important of the two (97a23–34) begins as follows:[4]

> **II.13 97a23–26** "For constructing a definition through divisions, it is necessary to aim at three things: (1) grasping the things predicated in the essence, (2) arranging these in proper order, and (3) [being sure] that these are all."

The following lines expound the three steps. In (1) we find what predicates are essential, in (2) we arrange these in order, and in (3) we check to see if we have done (1) and (2) correctly. If so, we have used up the predicates identified in (1) and need no more: we have got the genus and all the differentiae.

Two aspects of this procedure interest us: first, the proper order of differentiae presupposed by step (2), and second, the method by which essential predicates are determined. Step (2) is expounded as follows:

> **II.13 97a28–34** "The proper arrangement will occur if he takes the 'first,' and this will be the case if what is taken follows[a] all,[b] but all do not follow it[c]—for there must be such a one. When this has been taken, the same method [is used] forthwith for the terms below. For that which is 'first' of the others will be second and what is ['first'] of the subsequent ones is third. For when the thing above is taken away, the next one will be 'first' of the others, and similarly for the rest."

NOTES ON TRANSLATION

a. "Follows" means "is predicated of all." "Footed" is predicated of "two-footed" but not vice versa.

b. "All" means "all the others."

c. "All do not follow it" may mean (in Greek and English) either "none follows it" or "not all follow it." I take it that Aristotle intends the former.

DISCUSSION

This passage assumes without proof that essential predicates of x can be put into an ordered sequence $\{A_1, \ldots A_n\}$, such that for all i,j, if $i < j$ then A_i is predicated of all A_j and A_j is not predicated of all A_i. Each successive differentia of x proves to be a division of its immediate predecessor. A_1 is the genus of x, and the subsequences $\{A_1,A_2\}$, $\{A_1,A_2,A_3\}$, . . . specify increasingly narrower subclasses of A_1 until finally $\{A_1, \ldots A_n\}$ exactly picks out the things that are x's. I shall call this kind of G-D definition a type A definition.

WHY SHOULD WE believe that essential predicates can always be put into a type A definition? It will depend on how the set of essential predicates is obtained—step (1).

II.13 97a26–28 "Step (1) occurs through the possibility of deducing that [a predicate] belongs (as with an accident) and of constructing [the definition] through the genus."

DISCUSSION

This last phrase is a reference to *Topics* IV, which treats methods for determining the genus of a given thing. Ross (p. 660) refers also to *Topics* I.4 101b18–19, which says "the differentiae are to be established by the same *topoi* as the genus." What we look for in *Topics* IV is directions for obtaining genus and differentiae such that the genus and differentiae we obtain (a) are all the essential predicates and (b) can be arranged into a type A definition. But we look in vain.

What we find instead is an array of techniques for refuting or defending assertions that one thing is another's genus, none of which is at all helpful in deciding whether a proposed genus belongs in the essence of the definiendum. Most of the methods are concerned with variations on the theme that a genus (and *its* genus, etc.) is predicated, and predicated univocally of all individuals of the species being defined, of other species as well, and of their subspecies—extensional features that are insufficient to distinguish essential from derivative per se predications. Some methods are useful for distinguishing between genera and other predicables, but even these are not helpful in marking off differentiae from derivative per se attributes—for example, the method of seeing whether the species admits the account of the genus but not vice versa (IV.1 121a13–14). As if to mark the difficulty with the very point we are discussing, the final discussion of *Topics* IV admits that it is hard to distinguish "what always attends something and is not convertible with it" from that thing's genus. It suggests that the person trying to establish a definition should try to *get away with* treating such a predicate as the genus, and then offers a counterexample useful for a person trying to demolish such a claim (IV.5 128a38–b9). Aristotle is virtually admitting that the methods of the *Topics* cannot solve our problem.[5]

THE BEGINNING OF *APo* II.13 also tells "how to look for the things predicated in the essence" (96a22). The method is stated and defended at some length:

II.13 96a24–b14 "Some of the things that always belong to each thing extend more widely [than it], but not outside the genus. I say that those things belong more widely which belong to each thing universally, but also to something else. For example, there is something that belongs to every three but also to [what is] not a three, as being belongs to three and also to [what is] not a number, but also odd belongs to every three and belongs more widely (since it also belongs to five), but [does] not [belong] outside the genus, since five is a number and nothing aside from number is odd. Such things should be taken up to the first point where just so many are taken that each of them belongs more widely, but all of them together [do] not [belong] more widely. For it is necessary that this be the essence of the thing. For example, to every three belong number, odd, and prime in both ways—both not being measured by a number and not being composed of numbers. Now three is precisely this: number, odd, prime, and prime in this way. For each of these—some belong to all odds and the final one also to two, but all [together belong] to nothing [else].

But since we have shown previously that the things predicated in the essence are universal (and universals are necessary), and (1) since the things taken belong to three in its essence, and in whatever other case they are taken in this way, thus (2) three will be these things of necessity. That (3) [these taken together are its] essence is clear from the following. It is necessary, if this is not the essence of three, for it to be a kind of genus, whether or not it has a name. Now this will belong more widely than [belonging just] to three, for let it be supposed that a genus is such a thing as to belong potentially more widely. Now if it belongs to nothing else than the individual threes, this will be the essence of three, for let this be supposed also, that the essence of each thing is the final such predicate that holds of the individuals. And so, there will be a similar relation for anything else shown in this manner and its essence."

DISCUSSION

To begin with, the sample definition of three as "number that is odd, prime, and prime*"[6] is not a type A definition, since the three differentiae do not identify successively narrower sets of things, and since none of them is coextensive with the definiendum. I shall call this a type B definition.

The passage tells us to define x by finding predicates $A_1, A_2, \ldots A_n$ such that each A_i belongs more widely than to x, and x is the widest subject to which all the A_i belong. The second paragraph asserts that (1) each A_i is an element in the essence of x, (2) each A_i found in this way belongs necessarily to x, and (3) the essence of x is the combination of the A_i. It argues for (3) by pointing out that x is the widest subject to which all the A_i's belong, and by postulating that the essence of x is "the final such predicate" (the combination of all the A_i's as a single predicate)[7] that holds of all individual x's. This "final predicate" is coextensive with x, but coextension is not enough, and Aristotle knows it.[8]

The most serious flaw in the passage is that (1) is plainly false. An obvious counterexample is the property 2R, which belongs more widely than to isosceles triangle, without being part of its essence. Even if we grant that the A_i belong necessarily to x,[9] there is not the least guarantee that the A_i will be essential rather than derivative per se attributes, but such a guarantee is needed to solve the problem at hand.

Further, the method described here is different from that for reaching type A definitions, where we take successively narrower differentiae until we reach one that exactly covers the desired class; here we find a set of predicates, each of which is wider than the desired class. (Hence the two approaches are actually incompatible.) There is no indication of any order to follow in selecting them, and no guarantee that the set we reach is unique, only the requirement that the widest class of subjects to which they all apply is the desired class. Therefore, type B and type A definitions are different from each another. Both of them are coextensive with the definiendum, but they approach the definiendum in different and incompatible ways. The passage just discussed has received a good deal of attention,[10] but it is an aberration from Aristotle's usual view that G-D definitions are of type A. I know of no other place either in *APo* or elsewhere where type B definitions appear, and in what follows I shall focus on type A definitions.

WE HAVE OBSERVED the following four facts. (1) The theory of demonstrative science requires a "fat" definition, which includes all the essential predications of the definiendum. (2) *APo* knows only G-D definitions. (3) *APo* does not provide a method of defining that is sufficiently sensitive to the distinction between essential and derivative per se facts. (4) *APo* gives no reason to suppose that a G-D definition is a fat definition.

Indeed there is every reason to suppose that a G-D definition is *not* a fat definition. The purpose of a G-D definition is to distinguish the definiendum from everything else (*Top.* VI.3 140a27–29). Even though the defining terms must be naturally prior to the definiendum and more intelligible than it is (*Top.* VI.4), so that not every description that identifies the definiendum qualifies as a definition,[11] still these requirements do not guarantee that the definition will contain all the essential predicates. In fact, in a type A definition only the final differentia is coextensive with the definiendum; the others are common to it and all the coordinate species of the appropriate genus. It follows that the immediate connection between a definiendum and its final differentia must be the basis for proving all the "universal" connections of the definiendum—a claim badly in need of defending and one that a frequently occurring example of a definition seems to defeat. If man is defined as two-footed terrestrial animal, with animal as genus and terrestrial and two-footed as successive differentiae, and if rational is a universal per se property of man, there should be some way to infer rational from two-footed, or anyway from two-footed terrestrial animal, but this seems most improbable.

The chapters of the *Metaphysics* devoted to essence and definition (especially Z.4–6, 10–12, H.3,6) recognize only type A definitions,[12] and approach the present issue in basically the same way as *Topics* and *APo*.[13]

> ***Met.* Z.12 1037b29–1038a4** "[In such definitions] there is nothing else in the definition except the first expressed genus and the differentiae. The other genera are the first genus and the differentiae taken together with it; for example, animal is the first [genus], the next [genus] is two-footed animal, and again wingless two-footed animal,[14] and similarly if it is expressed through more terms. But in general it makes no difference if it is expressed through many or few, so that neither does it make any difference if it is expressed

through few terms or two. Of the two, the one is differentia and the other is genus; for example, of two-footed animal, animal is genus and the other is differentia."

DISCUSSION

In general, if x is defined as $G, D_1, D_2, \ldots D_n$, where G is the first genus and the D_i are successive differentiae, then G can be regarded as the genus for D_1, G, D_1 can be regarded as the genus for D_2, and so on. At each level (except the first) we have a genus and an immediate differentia of that genus, which combine to form a species of the genus, and this species can serve as a genus for further differentiation into its own species. In a sense, each differentia implies its immediately preceding genus and hence all the previous genera (cf. 1038a29–32).

ARISTOTLE TELLS HOW to take successive differentiae:

***Met.* Z.12 1038a9–18** "It is necessary to divide with respect to the differentia of the differentia. For example, footed is a differentia of animal; next it is necessary to know the differentia of footed animal qua footed, and so winged and lacking wings must not be stated [as differentiae] of footed, if a person speaks well, (but he will do this through inability), but perhaps cloven-footed and uncloven. For these are differentiae of foot, since cloven-footedness is a kind of footedness. And [a person speaking well] always wants to proceed in this way until he reaches things with no differences. At that point there will be as many species of foot as there are differentiae, and [the species of] footed animals will be equal [in number] to the differentiae."

DISCUSSION

This procedure results naturally from the notion of successive differentiation, but it makes the problem we are considering even more conspicuous. To be sure, the process described will end up at differentiae that distinguish each species from everything else, but how can successive differentiation yield all the essential attributes?

ARISTOTLE NEXT ACCEPTS an important consequence of the foregoing, one that was not recognized in *Topics* or *APo*:

***Met.* Z.12 1038a18–30** "If these things are so, it is clear that the final differentia will be the essence and definition of the thing, since it is necessary not to say the same thing many times (i.e., more than once) in definitions, for it is superfluous. But this happens, for when someone says two-footed footed animal he has said nothing but animal having feet having two feet. And if he divides this by its proper

> division, he will say it more times, and equally to [the number of] the differentiae. If, then, it is differentiae that are differentiated [literally, if it is of a differentia that a differentia comes to be], a single one, the final one, will be the species and the essence. If [the division] is in accordance with an accident, for example, if a person takes white and black as a division of what is footed, there will be as many [differentiae] as there are cuts. And so it is clear that the definition is a formula composed of the differentiae, and [is] the final of these, at least if [the division is carried out] according to the correct [method]."

DISCUSSION

A proper definition reduces the definiendum to a single differentia: "the final differentia will be the essence and definition of the thing" in that it entails the prior differentiae and the genus. This is an especially sharp way of inviting our problem, for the science of biology requires more essential facts about humans than information about their feet.

ALTHOUGH ARISTOTLE'S THEORETICAL works do nothing to solve the problem, there are hints of a solution in the first book of *PA*. In the rest of this chapter I will argue that the discussion in *PA* I is applicable to the theory of demonstrative science and show how it points to a satisfactory solution to the problem of reconciling the form and content of definitions with the role they play in demonstrative science.

PA I has undeniable echoes of *APo.*[15] Proof (*apódeixis*) is found in biology. It reveals the necessary properties of its subjects and explains "why" they are necessary in the appropriate sense of "why" (here, the "final cause"). Especially relevant are the statements that correct procedure in biological proofs is to start with the essence of the subject and show that the essence requires the subject to have certain parts, since without them the subject could not be. "Since this is the essence of man, this is why he has these things; for without these parts it is not possible for him to be" (*PA* I.1 640a34–35). It is important to know what (*tí, tí esti*) the subject being studied is (641a16, a23) as the basis for understanding why it has the parts it does. The material nature is less important than the formal nature (640b28–29, 642a17), but earlier thinkers failed to realize this and gave the wrong sort of explanations, "since they lacked the notion of essence and of defining the being [of things]" (*tò tí ên eînai kaì tò horísasthai tèn ousían*, *PA* I.1 642a25–26). The view that the essence of a subject is the basis for demonstrations that prove that the subject necessarily has certain parts, and that explain why it has them, fits well with *APo*,[16] though again, it leaves open the question how to guarantee that definitions will contain all the information a science needs. *PA* I.2 and 3 contain materials for an answer.

Those chapters attack the method of defining species through division by dichotomy, in which the genus is divided into two parts, each of them is subdivided into two, and the process continues until the end products are no longer divisible. At this point each species must be found in exactly one place in the division, and each place must be occupied by exactly one species (I.3 643a7–16). Balme has observed that Aristotle's arguments[17] hold equally against division into more than two differentiae.[18]

In the sequence of differentiae this method yields, each is a differentiation of the previous one. Winged is not divided into tame and wild or into pale and dark, which are accidental to the line of division including winged (I.3 643b19–23). Divisions carried out in this incorrect way fail to achieve the unity that results when a division is done correctly (I.3 643b17–19).

The distance between *PA* and *Metaphysics* appears in their views on the kind of unity a division by successive differentiation produces. *Metaphysics* Z.12 aims to understand the unity of definitions, and concludes that proper definitions are produced by successive differentiation according to a single line of division, so that the final differentia implies the others and makes them redundant (or at least makes it redundant to state them). The unity of the definition consists in the fact that it is reducible to a single term, the final differentia. It is striking that the first of many criticisms of division by dichotomy enounced in *PA* is directed against precisely this feature (*PA* I.2 642b7–9). The criticism is elaborated in the following chapter:

> ***PA* I.3 644a2–8** "Proceeding successively it reaches the final differentia. . . . This is cloven-footed only, or the entire combination, if man is the subject of division; for example, it should form the conjunction footed, two-footed, cloven-footed. If man were *merely* cloven-footed, in that case this would prove to be his single differentia. But as it is, since he is not, there must be many differentiae not under a single line of division."

DISCUSSION

The hypothetical definition of man can be given as "footed, two-footed, cloven-footed," or merely as "cloven-footed," which implies footed and two-footed.[19] Either way, Aristotle objects that man is not *merely* cloven-footed (or indeed footed, two-footed, and cloven-footed). There is more to humans than the nature of their feet!

The right way to proceed is not to take successive differentiations of the same line of division or to begin with one line of division and at some point take a different tack, such as dividing terrestrial animals into footed and footless, but to divide the original genus by many differentiae right from the start (I.3 643b23–24). By making use of many differentiae simultaneously, we can group animals according to their natural kinds, for instance, birds or fish, whereas some dichotomies split such groups—for example, the division of all animals into terrestrial and aquatic groups water birds with fish, an absurd result (I.3 642b10–13; cf. b33–34). In fact, the method proposed is nothing unusual, but what "the many" do in using a number of criteria in distinguishing birds as one class of animal and fish as another (I.3 643b10–13).[20]

The new method of division begins with a genus, such as animal, and finds a plurality of initial differentiae, $A_1, A_2, \ldots A_n$. Each of these differentiae is the beginning of a line of division. For example, A_i is divided into differentiae $A_{1,1}, A_{1,2}, \ldots A_{1,p}$, and each of these is further divided $A_{1,j,1}, A_{i,j,2}, \ldots A_{1,j,q}$. Suppose that the division is fully carried out, and call the set consisting of all the differentiae (including A_1, $A_{1,1}$, and all further subdivisions) the differentia set; and call the set of all the final differentiae reached in each line of division the final set. One such division might appear as follows.

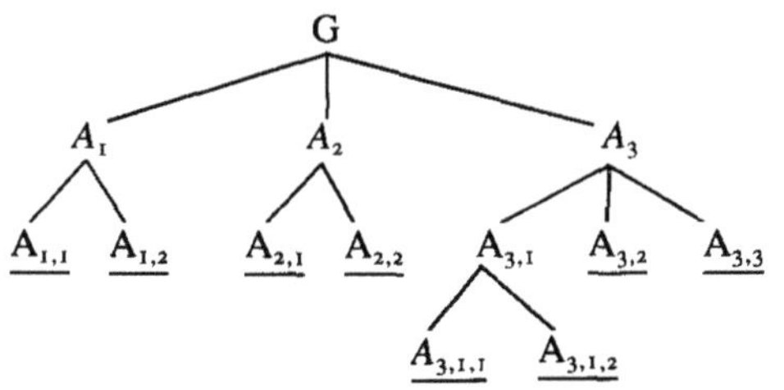

The differentia set will consist of twelve members, all the items below G (the genus) on the diagram. The final set will consist of eight members (those items that are underlined). When the divisions are fully carried out, it will be possible to characterize each species of the original genus, right down to the ultimate species, by a unique combination of the $A_{1,j}, \ldots$, and each member of the final set will occur in one or more of the characterizations. Further, each species will be characterized by n differentiae, one each from the lines of division that start from the initial differentiae $A_1, A_2, \ldots A_n$, though as in the case of dichotomous division, there will be redundant descriptions that mention not only differentiae in the final set but higher differentiae as well. I shall call this a type C definition.

ARISTOTLE MIGHT HAVE changed his mind for practical reasons or theoretical ones. He may have found it difficult or impossible in practice to put his notion of a single line of division to use. There may not be enough kinds of feet to go around, one for each different kind of animal. This kind of complaint may be behind some of his criticisms.[21] But the reasons given for preferring type C definitions are theoretical. "Footed, two-footed, cloven-footed" fails as a definition for man not because it fails as a criterion for identifying man, but because man is not *merely* cloven-footed. If the only purpose of division is to distinguish the definiendum from everything else,[22] there is no force to this objection. But *PA* I maintains that a proper division does more than merely provide criteria to identify the definiendum. Even though 2R identifies triangles among plane rectilinear figures, it is not a differentia of triangle (*PA* I.3 643a28–31). We must divide by essential features, not accidental ones, not even per se accidents (a27–28).[23]

A correct definition, then, will state essential features of the species, both its genus and the essential characteristics that enable us not only to distinguish it from everything else but more. In what follows I will argue that a type C definition contains all and only the essential predications of the definiendum, so as to constitute its fat definition.

That it contains only essential predications (as opposed to both derivative per se and accidental predications) emerges from the passages summarized just above. The elements of the definition must be essential, not per se accidents, as 2R is a per se accident of triangles. Why is 2R not essential to triangles? Because (as we know from *APo*) it is derivative: it follows from the unprovable principles of geometry, including the definition of triangle.

Further, a type C definition contains all the definiendum's essential predications. When Aristotle criticizes the definition of man as "cloven-footed" or as "footed, two-footed, and cloven-footed," saying that this definition would be adequate if man were *merely* cloven-footed, he does not mean that "cloven-footed" is the only thing true to say of man or even that it is the only per se attribute of man. If man were merely cloven-footed, man would still be two-footed and footed.

Since a triangle is *merely* a three-sided rectilinear plane figure, it is rightly defined in this way because all its per se attributes can be shown to follow from this information about triangles plus the other principles of geometry. Man is not *merely* a cloven-footed thing because there are per se attributes of man that do not follow. Since this is so, there must be other differentiae, which come under other lines of division. What constitutes an adequate definition? When we have a complex of differentiae $A_1, \ldots A_n$ coming from different lines of division such that man *is* merely $A_1, \ldots A_n$, we have reached an acceptable definition. And from what we have seen, it seems that we can say that man *is* merely $A_1, \ldots A_n$ when each of man's per se attributes is either included among or entailed by $A_1, \ldots A_n$.

The type C definitions produced by the new method of division therefore state the scientific essence of the definiendum. The definition includes essential predications that are grounds for all derivative per se predications of the definiendum. Since the definitory predicates are essential, moreover, they do not entail any accidental predications. This is precisely the role the theory of demonstrative science requires for definitions. Hence type C definitions are fat definitions.

It is interesting that Aristotle states this theory of division and definition not in *APo* but in one of his scientific works, and equally interesting that much of the theoretical discussion in which it is embedded has to do with topics relevant to the theory of demonstrative science. Reflection on the methods and goals of biology may have led him to spot the shortcomings of the *APo* theory of type A definitions. But even if *PA* I was not written with *APo* in mind, two results stand: the reflections on the nature of definition in *PA* I address a problem that *APo* invites, and the view sketched in *PA* I.3 can be easily extended to solve that problem.

Of course, *PA* is not primarily a theoretical work, and *PA* I contrasts sharply with the remainder. Moreover, the new views on division and definition are not set out very clearly or in an orderly way, but seem more like a series of occasionally inchoate notes than a finished work on biological method. Perhaps Aristotle did not polish these notes because his attention at the time was directed very much toward the particular science of biology. At any rate, he did not integrate their views on division and definition into *APo*, where they would fit very nicely. Nevertheless, in subsequent chapters I shall incorporate these views into the theory of demonstrative science, which will remain an Aristotelian theory, but in this respect will not be identical with the theory of *APo*.

Existence Claims

CHAPTER X

THE FINAL KIND of scientific principle is existence claims. In the present chapter I will discuss what kinds of existence claims are needed, suggest a role for them in the structure of sciences, and consider how well they meet the requirements Aristotle places on scientific principles. Finally I will see whether scientific existence claims are necessary truths.

Along with definitions, existence claims are "proper" principles, specific to a given science. Whereas the definitions of all the subjects and attributes in the science's subject genus are needed as principles, a smaller number of existence claims will suffice. There is no need for principles that posit the existence of attributes, whose existence follows from that of the subjects to which they belong. For an attribute to exist is for there to exist a subject that has that attribute. In the context of demonstrative science, we are concerned only with per se attributes whose existence is necessitated by that of their subjects. The theory requires principles that posit the existence only of subjects,[1] and not even for all subjects.

In Chapter 3 above we saw that *APo* I dithers on the distinction between primitive and derivative subjects.[2] *APo* II.1–2, though, is clear that some subjects are proved, not assumed, to exist, so that a science's principles need to include existence claims for only some subjects. There we find a fourfold division of "things that are investigated." For our purposes, the most important distinction is that between determining whether something exists (*ei ésti*) and whether a subject has an attribute (*hóti*).

> **II.1 89b25–33** "When we investigate . . . whether the sun is eclipsing or not, we are investigating the fact [*hóti*]. Evidence for this is that when we have found that it is eclipsing, our investigation is at an end. . . . Some things we investigate in a different way; for

example, if a centaur or god is or is not. I mean 'if it is or is not' [*ei éstin è mè*] without qualification [*haplôs*], not if it is white or not."

DISCUSSION

This passage distinguishes statements that something "is without qualification" from statements that something (some subject) has an attribute, and does so by a syntactic criterion. The criterion is whether "is" in "if *x* is or is not" is without qualification or with qualification,[3] which corresponds to our distinction between complete and incomplete uses of the verb "to be." This distinction marks off normal subject-attribute propositions from a different kind of assertion, which is best understood as an existence claim.

APo II.2 EXAMINES THE RELATIONS among the kinds of investigations identified in II.1. It gives further examples of things concerning which we might investigate whether they exist—the moon, earth, sun, or triangle (90a12–13)—and identifies these as subjects (a12). Once we know that a subject exists, we investigate "what it is" (II.1 89b34), which amounts to looking for the "middle" of a proof (the nexus of immediate relations that is founded on scientific principles) that it exists (II.2 89b38–90a1). Details of this difficult chapter will be taken up below in Chapter 15. All we need know for now is that II.1–2 not only distinguishes between subject-predicate propositions, but also clearly recognizes proofs of the existence of some subjects as a category of demonstration distinct from proofs of per se relations between subjects and attributes. We must therefore accept a distinction between primitive and derivative subjects corresponding to that between essential and derivative facts. It follows that not all subjects are on an equal footing; some are superior to others from the point of view of priority and intelligibility.

The Role of Existence Claims

Why are existence claims needed? What is their function in science? Someone might argue that sciences can do as well without them as with. If the main interest of science is to prove per se relations between subjects and attributes, and if these depend on the definitions of the terms involved, existence is irrelevant. The distinction between existence claims and per se facts renders existence claims useless in proofs. If the form of a demonstration is "*A* belongs per se to *B*, *B* belongs per se to *C*, therefore, *A* belongs per se to *C*," there simply is no need and no room for existence claims in demonstrations.

I shall respond to this attack on two fronts. First, since existence proofs have conclusions of the form "*A*'s exist," they need premises of the same form. Thus, existence claims are central to one of the main types of proofs recognized in *APo*.

The second response stems from Aristotle's views on essence, which is the basis for per se predication. To put the matter simply, things that do not exist do not have scientific essences. The key passage goes as follows:

> **II.7 92b4–8** "It is necessary for a person who knows what is [*tí estin*] man or anything else to know also that it exists. (For no one knows what the nonexistent is, only what the formula or name signifies, when I say 'goat-stag.' But it is impossible to know what a goat-stag is.)"

DISCUSSION

The difference between "what *x* signifies" (*tí sēmaínei*) and "what *x* is" (*tí esti*) is that to know the latter we must know that *x* exists. The same expression will indicate "what it signifies" to one person and "what it is" to another if the first does not know that *x* exists and the second does. Also, if something, like a goat-stag, does not exist, no one can know what it is. Hence goat-stags are not studied by any science, since if what *x* is cannot be known, neither can any derivative attributes of *x*. Therefore, things that do not exist are irrelevant to science; there is no science of the nonexistent.

THUS THE DISTINCTION between "what it is" and "what it signifies," together with the definition of per se in terms of "what it is," shows the significance of the basic existence claims of a science. A science is the study of things that exist. A subject of a science is a member of its subject genus, and a subject genus is a genus of existing things.[4]

Existence claims prove to be the existential underpinning that gives definitions a grip on reality. Even if they did not appear in any proofs as premises, they would be presupposed in proofs. The links are straightforward: science deals in the per se, the per se depends on "what it is," and the "what it is" presupposes existence. Therefore, some existence claims are needed.

Briefly stated, the function of the existence claims is to identify existing things as subjects for per se predications. Existence claims and definitions cooperate in the task of introducing subjects into a science. The existence claim simply states that the subject exists; the definition informs us what kind of subject it is. Together they legitimize the subject as a member of the subject genus and locate it in the genus. The assertions that lines exist or tigers exist do not tell us what sciences study those subjects, to what subject genera they belong. The definitions do this much and more. Being told only that lines exist, we do not yet know what discipline we are confronting, what context of discourse we are in. But when we are told further that a line is "magnitude continuous in one direction"[5] we know both the subject genus (spatial magnitude) and the specific nature of lines.[6]

We may think of a demonstrative science as telling a kind of evolutionary story, beginning with the simplest, primitive facts and showing how more complex facts emerge from them. "In the beginning" only the indemonstrable principles are available for use in proofs: the axioms, the existence claims for primitive subjects, and their definitions, which state "what it is." The derivative subjects cannot be used, since they are not yet known to exist. Their definitions at this stage have only the status of saying "what it signifies"; only after a derivative subject's existence is proved is its "what it signifies" promoted to "what it is." Likewise for the attributes. Given that a subject exists, the science has to prove that it has certain per se attributes, and thus that these attributes exist. Only at that point do the definitions of the attributes qualify as indicating "what it is." The work of the science consists in proving the existence of derivative subjects and the derivative attributes of both primitive and derivative subjects.

Aristotle's choice of words shows sensitivity to this issue:

> **I.10 76b3–7** "Proper things are (a) those things that are assumed to exist. . . . They assume that these exist and that they are *this*. (b) The per se attributes of these they assume what each signifies."

DISCUSSION

"That they are this" is best taken as an emphatic way of saying that this is their definition ("what it is").[7]

Existence Claims and the Requirements for Principles

The next topic is whether primitive existence claims meet the requirements of scientific principles. *APo* I.2 71b20–22 specifies that principles are true, primary, and immediate, and also more intelligible than, prior to, and grounds for the conclusions.[8] Primitive existence claims obviously satisfy the first three conditions. If there is a demonstrative science and it has existence claims that are principles, they must be true. Being primary and being immediate mean being unprovable, and by specifying that only *primitive* existence claims are principles, we rule out the possibility of proving them on the basis of other *more* primitive existence claims. We have also seen that they cannot be proved from the definitions. As indemonstrable principles, they are more intelligible than, prior to, and grounds for conclusions based on them. More specifically, primitive subjects can reasonably be described as prior to, more intelligible than, and grounds for the derivative subjects whose existence is shown to follow from theirs. There therefore appears to be no difficulty in saying that basic existence claims qualify as principles on Aristotle's criteria.

Existence and Necessity

Finally, are existence claims necessary truths? Does the theory of science require subjects to exist necessarily? Nowhere does *APo* make this demand in so many words. Being necessary is *not* one of the requirements of principles listed in I.2. On the other hand, it is easy to construct arguments to this effect. First, since conclusions of demonstrations are necessary (I.2 71b9–12), principles are necessary (I.4 73a21–24, I.6 74b13ff.). And since primitive existence claims are principles, they are necessary. A second argument for the same conclusion begins with proofs of the existence of derivative subjects. If the existence of a derivative subject is a conclusion of a proof, and if conclusions of proofs are necessary, then the derivative subject exists necessarily. But then reasoning similar to that of I.6 74b26–32[9] will show that the primitive subjects exist necessarily too.

To leave the matter at this stage is to endanger Aristotle's whole enterprise. If the objects of science exist of necessity, then precious few disciplines can qualify as sciences. Even for Aristotle, astronomy and theology will be the only secure sciences, since the stars and god exist eternally and therefore of necessity.[10] The changing sublunary world can be squeezed in only by special pleading. For example, it is the species of animals that exist eternally and of necessity, not the individuals. Geometry does not fare any better, since the existence of spatial magnitudes depends on the existence of things with certain shapes, and most things can perish. The problems this interpretation raises for us today are even worse than the ones it raises for Aristotle, since we no longer believe in the eternity of species. Even those who dispute the evolution of species must admit that species may become extinct. And the view that stars and planets exist eternally is no longer tenable. On this view, there seem to be no scientific disciplines left as serious contenders for being Aristotelian demonstrative sciences.

Before abandoning the theory, it will be worthwhile to look again at the arguments that lead to this uncomfortable result. In the first argument the first premise bears investigation: conclusions of demonstrations are necessary. Since necessary propositions are per se propositions that depend on "what it is," scientific necessity depends on essence. Given that triangles exist and what triangles are, they necessarily have 2R, and this is precisely where *APo* demands necessity. It frequently says that subject *A* must have attribute *B*, never that subject *A* must exist. Further, since necessity is connected specifically with the per se and existence claims are not per se, there is special reason *not* to think that they are necessary. At least the arguments having to do with the necessity appropriate to definitions do not apply to them.

I suggested above that in a proof that an attribute belongs to a subject per se, the subject's existence will not be asserted. It is presupposed in the demonstration, but the premises as well as the conclusion are per se. If the conclusion is necessary, the premises are necessary. But that says nothing about the assertion that *A*'s exist, which serves as an ontological prop for the argument and no more. Since existence claims *introduce* subjects for per se predications and do not assert anything to be true of them, their role in science is very different from the function of definitions. Since this is the only kind of proof clearly envisaged in *APo* I, it is likely that assertions that premises and conclusions of proofs are necessary point specifically to the necessity found in and dependent on definitions.[11] And so, a solution to the problem for this kind of proof will consist in showing that subjects whose existence is not necessary can have necessary attributes.

There remains the second argument, which addresses existence proofs. If the conclusions of all demonstrations are necessary, and some conclusions prove that subjects exist, we have to do with things that exist of necessity. The most straightforward way to deal with this argument is to deny the first premise. *Only some* demonstrations have necessary conclusions, namely those already considered. Existence proofs do not have necessary conclusions. This procrustean move is justified since a proposition is necessary if and only if it is per se,[12] and since the radical distinction between existence claims and definitions implies that existence does not belong to anything per se.

This interpretation has some hurdles to jump. In particular, in a number of places *APo* declares that properly speaking, there is no demonstration or scientific knowledge of perishables (I.8 75b24–25), that the conclusions of scientific demonstrations are eternal (I.8 75b22–23), and that the objects of science are universals, which are "always and everywhere" (I.31 87b32–33; cf. b38–39). This is good presumptive evidence that subjects exist of necessity. For only things that exist necessarily exist always,[13] and unless there is always a subject for the attribute to belong to, the per se predications that the science proves cannot be eternal. In what follows I oppose this argument and offer an interpretation of the eternal truths of science independent of the eternal existence of subjects.

In the first place, these passages do not state that the subjects of sciences are eternal. Since I.8 is interested in the universal nature of scientific propositions, not subjects (75b21–22, b27–28), the claim that demonstration and knowledge are not "of perishables" (things that are not eternal) is meant to hold for conclusions of proofs, not subjects. Similarly, even though much of I.31 applies to universal terms as well as propositions, the examples show that again the universals intended are propositions that can be conclusions of proofs (87b35–36, b39–88a2, 88a14–16).

In the second place, the statement that universals are "always and everywhere" is found in an argument that universals cannot be perceived:

> **I.31 87b28–33** "Even if perception is of what is of a certain kind and not of a particular, it is necessary to perceive a particular at a certain place and time.[a] But it is impossible to perceive that which is universal and applies to all.[b] For it is not a particular or at a certain time.[c] For [if it were] it would not be universal. For we say that what is always and everywhere is universal."

NOTES ON TRANSLATION

a. The Greek translated as "A particular at a certain place and time" means literally "this and somewhere and now."

b. "Applies to all" refers to all the particulars that fall under the universal.

c. The Greek translated as "a particular or at a certain time" means literally "this and now."

DISCUSSION

Does the last sentence of this passage imply that the subjects of science are eternal? The argument does not require it to. First, if "always" means that the subject exists eternally, then "everywhere" should mean that the subject exists in all places—an absurdity. Second, this sentence justifies the claim that the universal,[14] which "applies to all," "is not a particular at a certain time" (or place). And the justification does not require the strong claim that the universal occurs at all times and places, only the weaker claim that it occurs wherever and whenever the relevant particulars do. Third, it is legitimate to understand "always and everywhere" in this way, since the word "always" (*aeí*) has a distributive meaning, "on each occasion," as well as a global one, "always," so the sentence can mean just that the universal occurs "on each occasion" that an appropriate particular instance of it occurs. Likewise "everywhere" (*pantachoû*) can be taken distributively: the universal is found *everywhere that there is a particular*. Aristotle is simply contrasting the particular, which is uniquely locatable in space and time, with the universal, which is not uniquely locatable, since it applies to all the particulars on each occasion (*aeí*) and in each location (*pantachoû*) of their occurrence.

THE SAME INTERPRETATION works for one of the two problematic passages in I.8:

> **I.8 75b33–35** "Demonstrations and scientific knowledge of things that occur repeatedly, such as lunar eclipse, are always, insofar as they [demonstrations and scientific knowledge] are of something of a certain kind, but are 'partial' insofar as they [the things that occur repeatedly] are not always."

DISCUSSION

This passage distinguishes between a way in which demonstrations and scientific knowledge of things that occur repeatedly hold "always" and one in which they do not. They hold "always" in that the particulars to which they apply fall under a single universal (are "of a certain kind") but not so in that the particulars are not "always." That is to say, demonstration and scientific knowledge are "always" in the same way that "universals" are always, and strictly speaking they apply to universals. They apply to particulars too, which are not universal and therefore not "always" (the example given is of especially ephemeral particulars, but the point holds for all), but in that they are conceived as applying to a particular, they are only "partial," that is, taken only with reference to a part of the range of the universal.

THE OTHER PASSAGE may appear more difficult to gloss in the suggested way:

> **I.8 75b21–23** "It is also clear that if the premises on which the deduction depends are universal, also the conclusion of such a demonstration, that is, unqualified demonstration, must be eternal [*aídion*]."

DISCUSSION

The word for "eternal," *aídion*, is an adjective formed from the word for "always," *aeí*, and it may admit the same distributive use. In fact certain parallels between I.8 and I.31 encourage taking it in this way. First, the present passage occurs in an argument that universal premises yield eternal conclusions that apply to particulars only at a particular time and in a way (75b21–26)—the same point about particulars and universals found in I.31.[15] Second, both chapters speak of particulars as belonging to kinds or sorts and contrast the nature of the particulars with that of the sorts (I.8 75b34–35 and I.31 87b28–30, both translated above). Both passages say that particulars can be considered as particulars or as things of a certain kind, as falling under a certain universal. Third, the statement at b22 that the conclusion of unqualified demonstration is *aídion* amounts to the same as the statement at 75b34 that demonstrations and scientific knowledge are *aeí*. Both passages contrast proper scientific knowledge of universals with qualified scientific knowledge of perishable particulars on the grounds that when applied to particulars scientific demonstrations and their conclusions are not taken in their full generality, but with the qualifications needed to adapt them to the limited circumstances of the particulars in question. Unqualified demonstrations, conclusions of demonstrations, and scientific knowledge are described as being *aídion* or *aeí* in order to contrast them with their limited applicability to a given perishable particular. Accordingly, *aídion* in I.8 is used in the same way as *aeí*, and also in the same way as *aeí* in I.31.

So FAR FROM requiring the subjects of sciences to be eternal, I.8 and I.31 are quite clear that the particulars to which scientific propositions apply are not eternal. The universal premises and conclusions of proofs are "always and everywhere," but this means only that they apply to all their individual cases, not that there are individual cases of them in all places and at all times. This interpretation accords with ontological dependency of universals on particulars since it reconciles the "eternity" of universal scientific premises, conclusions, and knowledge with the temporary occurrence of particular instances.

Still, science treats particulars not qua particulars, but qua falling under a universal, and even though the universal is ontologically dependent on perishable particulars, scientific knowledge is primarily of universals and only derivatively of particulars. From the epistemological point of view, particulars are dependent on universals. Regardless of the ontological backing of existence claims in the existence of particulars, then, and regardless of the limited applicability of universal scientific propositions to existing particular cases, we still need to understand just how subjects are introduced into the discourse of a science. We already know that the existence claims and definitions cooperate to introduce a subject *as* a subject of a given science. This is done by asserting that it exists and that it has certain essential attributes (which are the basis for proving that it has derivative per se attributes). We also know that different sciences treat different features as relevant and overlook different sets of features as irrelevant.[16] Since all sciences treat universals, they will all overlook features relevant to particulars *as* particulars. Prominent among these are considerations of time and place.[17] In overlooking these features, they treat their subjects as timeless and placeless. So, in disregarding these temporal features, sciences deal with timeless universals, where the suffix "-less" is a true privative: science deals with its objects irrespective of time.

Now this view of the universal truths considered by sciences fits well with the discussion of what it is to be "in time" in *Physics* IV.12. I quote the relevant sections:

> ***Phys.* IV.12 221b3–5, b23–25, 222a3–6** "Things which are always [*aeí*], insofar as they are always, are not in time, for they are not contained by time, nor is their existence measured by time. . . . It is also clear that not everything that is not will not be in time, for example, all those that cannot be otherwise [sc., than nonbeing], as the diagonal [of a square] is commensurable with the side. . . . Things that are not include all those whose opposites are always; for example, that the diagonal is incommensurable with the side is always, and this will not be in time."

DISCUSSION

These passages say that the fact that the diagonal of a square is incommensurable with its side is "always" and so is not "in time."[18] Also its denial is not "in time." Since this fact is a theorem of geometry, we may understand the claim as holding generally for per se relations.

SCIENTIFIC FACTS ARE always and are not in time, not because they extend beyond any measure of time, so as to be eternal, but because they are timeless. Similarly we may suppose that they are everywhere not because their particulars are found in every possible location, but because universals are not in space. These are consequences of the process of abstraction which yields universals that are legitimate subjects of sciences.

Since scientific proofs are at the level of universals, they prove attributes to belong timelessly in that they belong to universals as such. Moreover, scientific existence claims are timeless as well. Thus there is no need to require infinitely long lasting particulars as objects of scientific knowledge, or even infinitely long lasting universals. What makes a subject suitable for science is the way it is introduced into the discourse of science. Existence claims introduce subjects without reference to time, and science treats them timelessly. Thus nothing in the nature of demonstrative science requires that subjects exist of necessity, only that given their existence (considered timelessly) the attributes proved to belong to them belong per se and of necessity.

But what is involved in considering temporal particulars "timelessly"? What happens when there are no particulars of the appropriate kind? Do the timeless universals wink out of existence, and do propositions about them cease to be true? The question here is not about centaurs and goat-stags, but about eclipses when no eclipse is taking place—a case considered in a passage already discussed (*APo* I.8 75b33–35). There are demonstrations and scientific knowledge of lunar eclipses because they can be considered as a "such," things of a certain kind. It is timelessly true that the earth's screening the sun's light produces a lunar eclipse, and this timeless truth is what applies to and explains each lunar eclipse when it occurs.[19] If eclipses of the moon never occurred, there would be nothing to be explained or applied to, and the proposition would be science fiction, not a scientific fact. It would not be a per se relation and could not be proved.[20]

There may be a residual doubt based on the extinction of species: how is the status of the scientific facts about a species affected when the species becomes extinct? Does the universal wink out of existence when the last member of the species does, and do the necessary, per se truths about that species become false or less scientific? Is there no scientific knowledge of dodo birds and dinosaurs?

The key to the solution is in the existential commitment of timeless universal facts. Must one or more individuals always be existing in time and space for the corresponding universal fact to enjoy its timeless existence? The answer is surely no. The evidence, again, is the assertion that even though there is not always a lunar eclipse taking place, still there is demonstration and scientific knowledge of eclipses. Eclipses qualify as objects of scientific knowledge because they are "of a sort," and *in that way* they can be considered timelessly, in abstraction from their actual particular instantiations. Thus we can have scientific demonstrations and knowledge of lunar eclipses at times when there are no particular lunar eclipses in existence. These demonstrations and this knowledge apply to those particulars whenever and wherever they occur—in the past, the present, or the future—though their pastness, presentness, or futurity is not part of the scientific demonstration or knowledge.

The case of dodos and dinosaurs is relevantly the same. Dodos formed a kind with an essence, and truths about that kind are timelessly true. From this point of view, the fact that there are no dodos now is the same as the fact that no lunar eclipse is now taking place. The sad fact that there will be no dodos again does not affect the issue. The same holds true for dinosaurs.

However, for thoroughly (as opposed to currently) nonexisting things, it turns out differently. If there were never any existing particulars to ground the existence of a universal (recall the goat-stag) then there would be no essence, no per se relations, and no scientific knowledge or demonstrations. But if there ever are such particulars (whether or not there always are), the universal is grounded in them and so is subject to appropriate scientific treatment, and this involves abstraction of incidental attributes, notably temporal and spatial ones.

The preceding discussion suggested two complementary interpretations of the claim that the universal objects of science are "always and everywhere." On the first, it means that scientific facts apply to all the particulars that fall under them. On the second, it means that they are timelessly true. Aristotle would agree with both: the universal facts science deals with are expressed without reference to time and also apply in all relevant cases. The existence claims for subjects of a science satisfy both these demands. They are grounded in the relevant individual cases, and they introduce the subject into scientific discourse as a timeless universal, which is the only kind of subject appropriate for necessary, per se predications.

Therefore, the statements that scientific facts are always and eternal do not entail that science requires eternally existing particulars. Since necessary and eternal scientific facts can apply to non-eternal particulars, there is no need to suppose scientific existence claims to be necessary, and so we may hold that when Aristotle claims that premises and conclusions of demonstrations express necessary truths, he intends them to apply only to per se predications, not to existence claims.

Aristotle's Principles and Greek Mathematics

CHAPTER XI

THE PREVIOUS CHAPTERS have shown that *APo* explicitly recognizes definitions, primitive existence claims, and axioms as scientific principles; they have also explored the nature of each kind and have suggested a role for each in scientific demonstrations. The following chapters will examine different types of proofs and how they depend on the principles. The present chapter will compare Aristotle's principles with the principles of Euclid's *Elements*. The purpose is to consider whether Aristotle's and Euclid's types of principles are, if not identical, similar enough to justify the use of Euclid's proofs as models for Aristotelian demonstrations. This move is crucial to the interpretation of Aristotelian demonstrations in the following chapters, since with the *Elements* in hand it is possible to fill in gaps in Aristotle's exposition and come to an understanding of a number of difficult passages in *APo*. However, since there are differences between Aristotle and Euclid, it is necessary to proceed cautiously. For if the differences are important enough, the strategy of using Euclid to illuminate Aristotle is illegitimate. I proceed by considering Aristotle's and Euclid's relations not only to each other but also to pre-Euclidean geometry in an attempt to show, within the limits of our extremely limited evidence, that Aristotle's classification of principles is based on the mathematics he knew and that Euclid's principles in *Elements* I are organized with Aristotle's classification in mind. The differences are mainly imposed by the fact that geometry uses techniques (constructions) in its proofs that are not found in all sciences, so that Aristotle's general theory of science requires a more generally applicable concept (existence). When appropriate allowances are made for these differences, Euclid's proofs can be used to understand the nature of Aristotle's demonstrations.

Also it has been claimed that "if the only primary premises of a science *apud* Aristotle" are the ones so far identified, "there just are not enough premises to get any interesting science going, nor indeed anything remotely like an interesting science."[1] The present chapter will give grounds for thinking that *Aristotle* believed (even if wrongly) that an interesting science could be based on such principles; summarily, the grounds are that his principles correspond well with those of the geometry he knew, and it cannot be doubted that in the fourth century geometry was a subject of very great interest.

Since the similarities between Aristotle's principles and Euclid's have been discussed a number of times,[2] I take up the issue reluctantly, and only for two reasons. First, previous treatments have tended to focus on terminology rather than the function of the different principles and have not given enough prominence to certain anomalies in Euclid's lists of principles. Second, the interpretation developed in the preceding several chapters of the nature and roles of the different kinds of Aristotelian principles brings a different perspective to the comparison with Euclid. The picture that emerges is complex. The similarities between Aristotle and Euclid are too marked to be the result of sheer accident, yet the differences preclude the suggestion that Euclid simply followed Aristotle in every detail or that Aristotle simply took on board the principles used in the geometry he knew. The evidence suggests influence in both directions, and that Aristotle and Euclid both made adaptations appropriate to their subject matters.

The geometry of the *Elements* is based on three types of principles: definitions (*hóroi*),[3] postulates (*aitē̆mata*), and common notions (*koinaì énnoiai*).[4] The postulates and common notions are as follows:[5]

Postulates. Let the following be postulated:
P1. To draw a straight line from any point to any point.
P2. To produce a finite straight line continuously in a straight line.
P3. To describe a circle with any center and distance.
P4. That all right angles are equal to one another.
P5. That, if a straight line falling on two straight lines make the interior angles on the same side less than two right angles, the two straight lines, if produced indefinitely, meet on that side on which are the angles less than two right angles.

Common notions.
CN1. Things which are equal to the same thing are also equal to one another.
CN2. If equals be added to equals, the wholes are equal.
CN3. If equals be subtracted from equals, the remainders are equal.

H.D.P. Lee's influential interpretation[6] is an attractive approximation: Aristotle's definitions correspond to Euclid's definitions, Aristotle's axioms to Euclid's common notions, and Aristotle's primitive existence claims (hypotheses) to Euclid's P1–P3. The first of these correspondences is so obvious to Lee that he does not even discuss it. The second rests heavily on Euclid's CN3, which is identical to one of Aristotle's standard examples of axioms. Euclid's common notions also satisfy Aristotle's requirement that axioms be common to more than one science. The third relies on the view that in geometry, "the assumption of the possibility of construction is in effect an assumption of existence."[7] P1–P3 are in effect assumptions of the existence of straight lines and circles, the basic subjects of geometry from which Euclid constructs the rest.[8] P4 and P5 are different in form and function from P1–P3, and Lee follows Heath[9] in assigning them to Euclid himself. Thus the correspondence between Aristotle's principles and Euclid's principles is not perfect, but if we allow that P4 and P5 were not principles of geometry in Aristotle's time and admit that constructibility is the same as existence, the fit between Aristotle and the geometry known to him is arguably exact.

According to Lee,[10] "Aristotle's account of the first principles of science in *APo* is an account of the first principles of geometry," and his choice of principles is explained by fourth-century geometry, which recognized only the three "Aristotelian" types of principles. The great success and rapid advances geometry was enjoying in Aristotle's time gave him no reason to look for further sorts of scientific principles and good reason to think that other sciences would be similarly successful if based on the kinds of principles used in geometry.

I call this account an approximation because it moves too quickly over some difficult matters and assumes too easily that the geometry known to Aristotle had principles more or less identical with those of *Elements* I (excluding P4 and P5). While acknowledging that certainty on these matters cannot be obtained, I believe that more can be said and that the picture of Aristotle copying geometry wholesale must be discarded.

There are three main lines of interpretation regarding the relation between Aristotle and geometry. (1) Before Aristotle, the principles of geometry were essentially the same as those of Euclid (except for P4 and P5). Aristotle turned to the *Elements* of Euclid's predecessor Theudius, followed geometrical practice closely, and asserted that other disciplines could also be based on the kinds of principles found in the geometry he knew, although a certain amount of adaptation was needed to generalize the theory adequately—hence primitive existence claims instead of construction postulates. Aristotle's contribution was philosophical discussion and justification of demonstrative sciences based on the kinds of principles found in geometry. Euclid was not influenced by Aristotle's theorizing, but followed established mathematical tradition in his types of principles (again, with the exception of P4 and P5).[11] (2) Before Aristotle, geometers made no attempt to identify and classify unprovable principles. Aristotle's inspiration came not from Theudius, but from philosophical sources—Plato and the Academy. Hence the stress on definitions and the prominence given to the law of the excluded middle and the law of noncontradiction. Hence also the importance of existence claims, a survival of Plato's requirement that sciences study Forms, the only fully existing things. On this interpretation philosophy influenced mathematics, not vice versa. Euclid followed Aristotle's directions in the principles of the *Elements*, diverging from them only where compelled.[12] (3) Euclid's predecessors had gone some way toward systematizing geometry and identifying its principles. Aristotle drew on their work as a source of ideas. He did not blindly follow geometrical practice, though, but reflection on the nature of proofs and principles led him to develop a theory applicable both to mathematics and to other subjects, which set them all on a firmer footing. Euclid's improvements on his predecessors are due to his following Aristotle's advice as far as he could. This is an intermediate view, and is the view I support.

Pre-Euclidean Principles

To begin with, we must presume that some work had been done toward identifying principles of geometry before Aristotle. Although the loss of all geometrical texts from that period makes it impossible to prove the point, Eudemus's statement that Hippocrates of Chios, Leon, and Theudius of Magnesia composed books of elements earlier than Euclid[13] indicates that such work had begun. At the very least, these "elements" would have presented selections from the geometrical knowledge available at the time of writing. In general, it is probable that they were arranged in order, beginning with more elementary and proceeding to more difficult material, although there is no telling how thoroughly this kind of systematization was achieved.[14]

Nothing yet shows that early writers of elements recognized the need for unprovable principles, much less that they classified them in the ways advocated by Aristotle and Euclid. But something may have been achieved in this field too. Proclus praises Euclid for including a wider range of material in his *Elements* than his predecessors had included in theirs, and for improving some proofs,[15] but not for originating the conception of principles or for being first to identify or classify them. If Proclus had any reason to think that Euclid had done so, he would surely have said something, since he is eager to applaud Euclid,[16] and is interested in Euclid's arrangement of first principles.[17]

There is also direct evidence of interest in the foundations of geometry among Euclid's predecessors. Aristotle and other authors give information about definitions of mathematical terms used before Euclid.[18] Further, Euclid defines a number of terms but does not use them (see below), a sign that he took these definitions over from his predecessors and did not successfully integrate them into his text.[19]

Second, the axioms. Aristotle refers to "the common [principles] that are called axioms" (I.10 76b14) and "the things called axioms in mathematics" (*Met.* Γ.3 1005a20) and one of his principal examples of axioms is Euclid's CN3. Also similar principles are stated by Plato at *Theaetetus* 155a.

Third, the construction postulates (granting that P4 and P5 are Euclid's innovations). On this topic there are widely diverging views. Heath,[20] noting that Aristotle gives no instances of geometrical postulates and makes no "allusion recalling the particular postulates found in Euclid," thinks that Euclid formulated all the postulates for the first time. Von Fritz believes that Speusippus knew P1–P3,[21] but his view has been rebutted.[22] Mueller discusses the evidence and concludes, "none of this information seems to me to relate in any specific way to the content of fourth century mathematics.''[23] Szabó[24] gives credit for P1–P3 to Oenopides, who lived in the later fifth century.[25] But his case is weak.[26]

Another piece of evidence is Plato's account of the principles of geometry and arithmetic in *Republic*:

> ***Republic* VI 510c3–d3** ''Hypothesizing the odd and the even and the figures and three kinds of angles and other things related to these in each field, supposing that they know these they make them hypotheses and no longer think that they must give themselves or others any account of these, on the grounds that they are clear to all. They begin from these and, systematically proceeding through the rest, consistently finish up at the thing they set out to consider.''

But the passage leaves it unclear whether the mathematicians' hypotheses are the objects of geometry,[27] or definitions, or existence claims,[28] or whether they both assert existence and define, and no further light is given by the context. The passage would support Lee's view of Aristotle's relation to Greek mathematics if Plato's ''hypotheses'' are existence claims and if the passage accurately reflects the mathematical procedure of his day. But uncertainty on both counts is too great for the passage to be clear evidence that pre-Euclidean mathematics recognized the need for postulates of construction or existence.

THE PRECEDING DISCUSSION points to the following conclusions about pre-Euclidean mathematics. That there was interest in identifying basic principles of geometry is proved by plentiful evidence of pre-Euclidean definitions of mathematical terms and by solid information that mathematicians recognized a kind of principle called axioms, which included Euclid's CN3. I take it for granted that these definitions and axioms were used the same way that Euclid uses them. There is no clear evidence that mathematicians before Euclid saw the need for construction postulates or existence claims, or that they identified a separate class of such principles, but equally there is nothing to rule out this possibility. Even so, there are sufficient grounds to reject interpretation (2) (see above, p. 135.) since it says that pre-Euclidian geometry was unprincipled. Also, since at least two of Aristotle's three kinds of principles bear close relations to pre-Euclidean principles, it seems very likely that he was influenced by the mathematics he knew.

Differences between Aristotle and Euclid

Reexamining the correspondences between Aristotle's and Euclid's principles throws further light on the originality of Aristotle's classification of principles. This time I will look for differences instead of similarities.

Axioms and Common Notions. Aristotle's principle EQUALS is identical with Euclid's CN3, but his other examples, LEM and LNC, are not found in Euclid.

Commentators have ignored the latter examples in their attempts to show that Aristotle's axioms are the same as Euclid's common notions. However, I have argued[29] that Aristotle grouped EQUALS with the logical laws as axioms because the important thing for the purposes of *APo* is not their differences but their similarities: they are common to more than one science and they have a special role in scientific proofs.[30] EQUALS is used in both the arithmetical and the geometrical books of the *Elements*, and licenses the inference of one fact from another in the way suggested for Aristotelian axioms.

What about the absence of the logical laws from Euclid's list of common notions? If Euclid was influenced by *APo*'s discussion of scientific principles, he did not follow Aristotle blindly. The logical laws were probably not explicitly stated by Euclid's predecessors. If he followed earlier lists of definitions even when they do not tally with his own technical vocabulary,[31] it is unlikely that he would have omitted LEM and LNC had they occurred in earlier lists of principles. It follows that Aristotle's including them as axioms is his own contribution. He saw that the roles they play in demonstrative sciences are relevantly the same as those played by mathematical axioms such as EQUALS, but his source in this case was not earlier mathematics but philosophical discussion on the principles of thought.[32] His reflections on the special character and function of mathematical axioms led him to recognize that other principles of this sort apply in all sciences, even nonmathematical ones. In this matter Aristotle did not blindly copy contemporary mathematics, but generalized in a way that went beyond his source.

Hypotheses and Postulates. I follow the interpretation that P4 and P5 are Euclid's own invention in response to a fault in the theory of parallels known to Aristotle.[33] On this interpretation, Aristotle is capable of reflecting on and criticizing geometrical practice, and Euclid's improvement may be due to Aristotle's influence.

As for P1–P3, there are reasons not to identify them too closely with Aristotle's hypotheses, despite claims to the contrary. First, Aristotle says that geometry must assume the existence of points and lines (*APo* I.10 76b5), but P1–P3 do not do so. At most they posit the existence of two *kinds* of lines, not lines in general. Second, the existence of points is *presupposed*, not posited, in P1 and P3[34] (in P3, the given center is a point; cf. defs. 15, 16). P2 presupposes the existence of straight lines. Third and most important, constructibility is not the same as geometrical existence.[35] Although constructibility guarantees the existence of what is constructed,[36] existence does not entail constructibility, let alone constructibility on the basis of P1–P3. For example, it is most implausible that controversies over squaring the circle, duplicating the cube, and trisecting the angle were focused on the question whether there exist squares, cubes, and angles of the required sizes. Assumptions of continuity, such as were frequently presupposed in ancient geometry,[37] will guarantee that they exist. The problem was, rather, whether and how they could be constructed.

The function of Euclid's postulates is not to assert existence, but to introduce geometrical entities in a way suitable for geometry, to allow them to be treated geometrically. Simple assertions of the existence of the basic subjects of geometry may be important in the foundations of geometry, but are of limited help in the business of geometry, in stating and proving theorems and making constructions. If P1–P3 were replaced by assertions that points, straight lines, and circles exist, geometry could not get off the ground. What geometry requires is the license to construct (or a guarantee of the existence of) *certain* circles and straight lines, not lines and circles in general. A geometrical proof might require a line between two given points or a circle with a given point as center and a given distance as radius, because this is how it shows the resulting figures to have the attributes under consideration.[38] Indeed, the fertility of Euclidean geometry—its profusion of theorems based on so small a number of principles—is largely due to its use of constructions.[39]

If P1–P3 are more remote from Aristotle's basic existence claims than they are frequently said to be, does it follow that Aristotle's existence claims have no basis in geometrical practice? I think not. Their function is to introduce things into the realm of discourse of a science, as subjects for per se predications, in a way suitable for the methods and purposes of the science. And *mutatis mutandis* this is the job of geometrical constructions, since the techniques of geometrical argument make constructions, not bare existence claims, the appropriate way to introduce subjects. P1–P3 introduce geometrical entities of determinate sorts, entities that satisfy certain definitions. They are used to construct other geometrical entities of determinate sorts, and anything constructed in ways permitted by P1–P3 becomes a legitimate subject for proofs. Also, the different function in geometry between constructions that yield certain lines and figures, and theorems that prove results about those lines and figures, is highlighted by differences in the way the results to be established are expressed[40] and in certain terminological distinctions[41] (although it would be dangerous to say that these differences are not dependent on Aristotle).

Whether or not construction postulates were formulated before Aristotle, he may have recognized the distinctive function of constructions in geometry, seen the advantages corresponding procedures would bring to other sciences, and generalized accordingly, allowing for existence proofs alongside proofs of per se attributes, and connecting these existence proofs with definitions.[42] Aristotle's primitive existence claims have the same correspondence to P1–P3.[43]

This account leaves open the question whether the need to posit basic entities was first seen by Aristotle or by pre-Aristotelian mathematicians, and consequently does not pronounce definitely on either Aristotle's or Euclid's originality. Either way, Aristotle's theory may have its origin in mathematical procedures, though it does not follow mathematics uncritically. If we believe with Heath[44] that construction postulates were not formulated before Euclid, Aristotle's demand for explicitly stated existence claims for the basic subjects of demonstrative sciences is all the more impressive, and in stating P1–P3, Euclid may have been following Aristotle's advice. But even if pre-Aristotelian mathematicians knew construction postulates, Aristotle still deserves credit for recognizing the significance of such principles and generalizing accordingly. Whatever the state of postulates and constructions before Aristotle, Euclid's treatment of them corresponds quite closely, *mutatis mutandis*, to what Aristotle's theory demands.

Definitions. Here the issue is doubly complicated since *APo* is insufficiently clear whether definitions consist of genus and differentia or contain all immediate per se relations, and it is difficult to see how Euclid conceived of the proper form, content, and role of definitions in geometry. Several features of Euclid's definitions need to be borne in mind. (1) Some of the defined terms are not used in demonstrations, for example, "line," "plane," "surface," "oblong," and "rhomboid" (*Elements* I, defs. 2, 7, 22). (2) Some terms used in the *Elements* are not defined, for example, "parallelogram," "intersection," and "pentagon."[45] (3) Although some of Euclid's definitions are in genus-plus-differentia form, others are not, for example, *Elements* I, definitions 1, 3, 5, 11–14, and 16. (4) Some of Euclid's definitions are overdetermined, for example, *Elements* I, definition 17 ("A diameter of the circle is any straight line drawn through the center and terminated in both directions by the circumference of the circle, and such a straight line also bisects the circle"), and book VII, definition 7 ("An odd number is that which is not divisible into two equal parts, or that which differs by a unit from an even number"). (5) Some of Euclid's definitions do not appear to be definitions at all. *Elements* I, definition 3 ("The extremities of a line are points") is "an explanation that, if a line *has* extremities, those extremities are points" and serves "to connect a point with a line."[46] (Points and lines have just been defined in defs. 1 and 2.) A similar relation holds for definitions 5 and 6[47] and book XI, definitions 1 and 2. (6) Some of Euclid's definitions differ from previous definitions in philosophically significant ways. Thus, the definitions of point and line (*Elements* I, defs. 1–2) as "that which has no part" and "breadthless length" are different from the earlier definitions of a point as the extremity of a line and a line as the extremity of a surface, definitions that Aristotle finds objectionable for explaining the prior by the posterior (*Top*. VI.4 141b19–22).

Do these features of Euclid's definitions tell us anything about Euclid's relation to Aristotle? (1) and (2) violate Aristotle's requirement that all (and only) subjects and attributes of the science's subject genus be defined. Even though *APo* requires definitions of basic subjects (point, line) regardless of whether those definitions are used in proofs, there is no philosophical justification for including definitions of terms that appear nowhere else in the *Elements* or for failing to offer definitions of terms that do. The former discrepancy is most plausibly explained as due to carelessness (see above, p. 136). I suppose the latter is due to carelessness as well. For there is no mathematical justification for these features of Euclid's definitions, and no reason not to think that Euclid intended to define all and only the terms of his science, even if he failed in some cases to do so. I therefore do not regard (1) and (2) as seriously conflicting with Aristotle.

The remaining concerns have to do with the form and content of definitions. (3) Many of Euclid's definitions are in genus-plus-differentia form, but several are not. But since *APo* is unclear on the form definitions must have, Euclid's definitions that are not through genus and differentia do not violate any clear prescriptions. Nor can they be attributed to *APo*'s influence.

(4) More interesting are overdetermined definitions. It is possible to think of these as Aristotelian "fat definitions,"[48] which specify immediate per se relations in addition to those needed to identify the definiendum. Aristotle would doubtless disapprove[49] of redundant definitions like the definition of "odd," which offers alternate specifications either of which implies the other, but definition 17 is not redundant in this way. The diameter of a circle can be identified *either* as a straight line drawn through the center of the circle and terminated in both directions by the circumference *or* as a straight line terminated in both directions by the circumference of a circle and bisecting the circle; there is no need to specify both. Further, the fact that the diameter bisects the circle was treated by earlier geometers as a proposition requiring proof. Thales is reported to have proved it,[50] and Proclus offers an argument for it that turns on the first part of Euclid's definition[51] and is probably the proof Thales gave.[52] The proof involves "superposition"—imagining the circle split on the diameter and one of the two parts fitted on the other. This method of proof may have been common in early Greek geometry, but it has been argued that Euclid tended to avoid it and used it only in those few cases (bk. I, props. 4, 8; bk. III, prop. 24) where "he had not been able to see his way to a satisfactory substitute."[53] In the present case, the substitution consists in appending the proposition about the diameter to the definition, i.e., treating it as an unprovable principle.[54] Book I, definition 17 can therefore be regarded as a fat definition.

(5) The relation between definitions 2 and 3 is the same as that between the two parts of definition 17: definition 2 defines line and definition 3 states an unprovable proposition about lines that is frequently used in proofs. The same holds for definitions 5 and 6 and book XI, definitions 1 and 2. Although these propositions are stated separately and not simply appended to the definition, as occurs in definition 17, not much weight should be placed on this difference, since printed texts may not perfectly reflect Euclid's grouping of definitions.[55] In any case, the fact that he places them among the definitions is evidence that he conceived of definitions as doing more than simply identifying terms—a conception that accords with Aristotle's fat definitions.[56]

(6) The fact that Euclid defines line and surface in ways not open to Aristotle's objections to earlier definitions suggests sensitivity to philosophical issues, and to Aristotle's criticisms in particular.

A FINAL PIECE of evidence for the relation between Euclid and Aristotle is the uniqueness of the threefold classification of principles in *Elements* I. There is no clear evidence that mathematicians before Euclid employed a tripartite division of principles. In fact the only pre-Euclidean mathematical text that survives (Autolycus of Pitane, *On the Moving Sphere*, dated to the late fourth century) does not classify its principles at all,[57] although it states several propositions at the beginning and uses them as unproved principles in subsequent proofs. Nor is Euclid's precise division found in post-Euclidean mathematical works,[58] or even in other works of Euclid himself. Of all the works of Greek geometry, Euclid's *Elements* (in particular book I and the books that depend directly on book I)[59] pays the closest attention to foundational issues. The most obvious explanation is that those were live issues in Euclid's time and he thought it important to take account of them in his basic textbook in order to be up to date and provide his treatment with a solid basis. His more advanced treatises resemble the works of Archimedes in not attempting to state all the principles on which the proofs depend. On the other hand, we have seen that Euclid's principles do not readily fit into his own tripartite classification. Some definitions require special pleading, and P4–P5 are different in kind from P1–P3. Proclus attempts at length to rationalize Euclid's classification,[60] but does not succeed.[61]

We are left with the impression that Euclid set out with the conviction that the principles of geometry fall into three kinds and that he squeezed them until they fitted. If he did not begin with such a belief, it is hard to understand why he did not group his principles in some other way.[62] Since there is no good evidence that he obtained his threefold classification from earlier geometers, and since his classification resembles Aristotle's as closely as it does, it seems inescapable that Euclid was influenced by the doctrine of principles in *APo*.

It is now time to pull together the results of the discussion. The evidence does not allow for certainty, but strongly suggests the following historical sequence. Pre-Aristotelian geometry certainly knew definitions and at least some of Euclid's common notions, and possibly knew construction postulates. Aristotle's views on axioms and existence claims very likely have their roots in geometrical practice (even if geometry had not yet explicitly formulated construction postulates), but were also influenced by both philosophical and methodological considerations, the former having to do with Aristotle's views on essences and the basic laws of reasoning, the latter with the need to produce a theory applicable to all branches of theoretical knowledge, not just mathematical subjects. As for the importance of definitions, Aristotle must have been influenced heavily by philosophical concerns that go back at least to Socrates and continued to be a lively topic in contemporary discussions. But his belief that all the facts that the science proves to hold of its subjects and attributes are due to their definitions may stem from his study of the methods of geometrical proof.

Euclid in turn was strongly influenced by Aristotle's classification of principles, at least when it came to setting out the foundations of geometry at the beginning of his introductory book, the *Elements*. He took over Aristotle's tripartite division, but did not follow Aristotle blindly any more than Aristotle uncritically followed Euclid's predecessors. As Aristotle generalized, so Euclid specified. His choice of wording in certain definitions shows an awareness of philosophical discussion on the appropriate forms of definitions. What I take to be his intention to provide definitions of all the specifically geometrical terms he uses could easily be due to familiarity with *APo*. His list of common notions may omit LEM and LNC for the reason given by Aristotle himself (I.10 76b16–21)—they are too obvious to need stating. Besides, Euclid as a practicing mathematician, even if a mathematician interested in providing sound foundations for geometry, may have thought that the logical laws were insufficiently mathematical in character to be appropriately included among the principles. The first three postulates are related to but importantly different from Aristotle's existence claims, but as was noted above, they presuppose the existence of one of the types of the basic geometrical entities (points) and they, but not existence claims, introduce entities in ways useful in geometrical proofs.

Why then are P4 and P5 classed as postulates while other principles that are not definitions but do not play the role of axioms or basic constructions are placed among the definitions? One answer is that P5 is a postulate because it amounts to an existence claim (that there exists a point of intersection of nonparallel straight lines)[63] and P4 is needed to guarantee the uniqueness of the point of intersection. P4 is adjacent to P5 for the same reason that definition 3 is next to definition 2 and definition 6 to definition 5. P5 is a postulate because it posits existence, even though it is not a *construction* postulate. This is a further reason to think of P1–P3 as related to existence claims and so as a kind of adaptation of Aristotle's hypotheses to the needs of geometry.

The fit between Euclid's and Aristotle's principles is close if not exact—so close that Euclid's axiomatization of geometry suffers. Were it not for Aristotle's strictures he might have seen the advantage of admitting a fourth class of principles to include his definitions 3 and 6, and the second half of definition 17. In Aristotle's terminology, these relations would be immediate and nondefinitional and would include P4 and perhaps P5 as well. Euclid's formalization is inadequate, but far more successful than anything else in Greek mathematics. Possibly both its strengths and weaknesses are due to Aristotle's theory. It would be surprising indeed if an ancient mathematical work satisfied the rigorous standards of formalism that appeared for the first time in the nineteenth century. It would be even more surprising if Aristotle's three kinds of principles proved to be all the kinds of principles required, since the task he set himself was not simply to state rules for formalizing *one* science, but to identify types of principles that are sufficient for *any* science.

These resemblances permit a qualified affirmative answer to our original question, whether Euclid's principles and Aristotle's are sufficiently alike to ground an attempt to use Euclid's proofs to understand Aristotle's theory. In Chapter 12 I shall examine two Euclidean proofs. When allowance is made for their use of construction postulates instead of existence claims and their nonsyllogistic form, they prove to be helpful and instructive examples of several of the features of Aristotelian demonstrations and in particular of the use of the different kinds of principles.

Aristotle's Demonstrations and Euclid's *Elements*

CHAPTER XII

NEAR THE END of *APo* Aristotle says he has explained what demonstration and demonstrative science are and how they come about (II.19 99b15–16). It is disappointing that he never sets out any rigorous demonstrations and analyzes them in detail. His examples are schematic, ad hoc, and frequently sloppy. Nevertheless, an examination of arguments Aristotle would count as demonstrations will provide a clearer idea of what some of his assertions mean, point out shortcomings of the theory, and indicate the changes needed to overcome them.

Chapter 11 ended with the hope that proofs from Euclid's *Elements* can serve as models of Aristotelian demonstrations. I begin with a simple example.

Elements I, Proposition 1

D1. On a given finite straight line to construct an equilateral triangle.

D2. Let *AB* be the given finite straight line.

D3. Thus it is required to construct an equilateral triangle on the straight line *AB*.

D4.1. With center *A* and distance *AB* let the circle *BCD* be described [P3];

D4.2. again, with center *B* and distance *BA* let the circle *ACE* be described [P3];

D4.3. and from the point *C*, in which the circles cut one another, to the points *A*,*B* let the straight lines *CA*,*CB* be joined [P1].

D5.1. Now, since the point *A* is the center of the circle *CDB*,*AC* is equal to *AB* [Def. 15].

D5.2. Again, since the point *B* is the center of the circle *CAE*,*BC* is equal to *BA* [Def. 15].

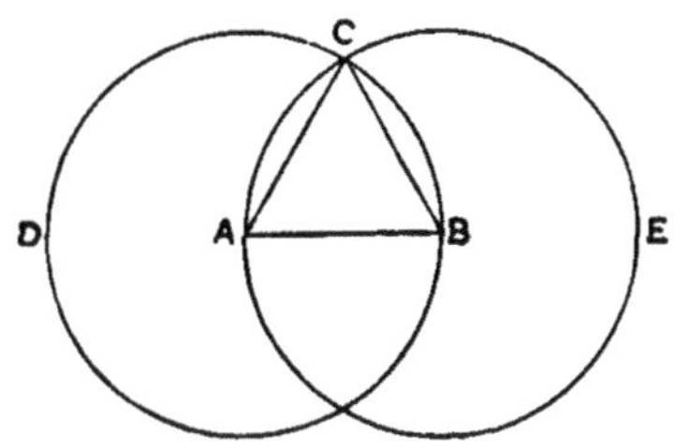

D5.3. But *CA* was also proved equal to *AB*; therefore each of the straight lines *CA*,*CB* is equal to *AB*.

D5.4. And things which are equal to the same thing are also equal to one another [CN1]; therefore *CA* is also equal to *CB*.

D5.5. Therefore the three straight lines *CA*,*AB*,*BC* are equal to one another.

D6. Therefore the triangle *ABC* is equilateral; and it has been constructed on the given finite straight line *AB*. (Being) what it was required to do.[1]

I have indicated the traditional dissection of a Euclidean proof:[2] D1 is a general statement of the proposition to be proved, D2 a restatement of the "given" lines and figures in terms of a specimen case, D3 a specification of what is to be proved in terms of the specimen case, D4 supplementary constructions needed to prove the desired result, D5 the proof proper, and D6 the conclusion, which is "what it was required to do."[3]

Each step of D5 is justified by reference to a definition or a common notion. (In subsequent propositions, previously proved theorems can be used too.) The construction postulates are used in D4, not D5. (In subsequent propositions, previously justified constructions can be used too.) Let us consider the use of each of the three kinds of principles.

Common Notions. D5.1 and D5.2 having shown that $AC = AB$ and $BC = AB$, in D5.3 CN1 licenses the inference to $AC = BC$. Even though it does not contain any information specifically about geometrical entities, it *applies to* the objects of geometry. It states a fact about equal things, and geometrical lines are one type of things that can be equal.

Definitions. D5.1 uses the definitions of circle and of the center of a circle to infer $AC = AB$ from the fact that *C* and *B* are points on a circle with center *A*. Not only do the definitions identify the subjects and attributes of a science, but also the information they contain justifies inferences in proofs.

Postulates. The strategy of the proof requires referring to a point (*C*) not given in D2, but having certain relations to *A* and *B*, the endpoints of the given line, which *are* given in D2. Also, the proof requires three lines, whereas only one is given. The function of the postulates is to license the construction of the additional geometrical entities required by the proof.

Definitions and construction postulates have other uses as well. P1 underlies the whole theorem. Without a guarantee (based ultimately on the construction postulates) that the "given" figures are legitimate, a geometrical proposition does not get off the ground. No proof can begin, "Let *ABCD* be a square circle." The given line *AB* is a legitimate geometrical object since, given points *A* and *B*, P1 entitles us to construct straight line *AB*. Definitions underlie proofs in a different way. For example, in *Elements* III, the definition of circle is frequently used as the basis for proving further properties of circles. In the present case, the strategy in D5 is to show that the figure constructed in D4 fulfills the definition of equilateral triangle. All of these uses to which Euclid puts his principles match the functions of Aristotle's principles as discussed in previous chapters.[4]

THERE ARE THREE big differences between Euclid's practice and Aristotle's prescriptions. (1) Euclid's proof depends heavily on a diagram; Aristotle does not require diagrams. (2) Euclid uses constructions; Aristotle talks of existence. (3) An Aristotelian demonstration is a series of syllogisms; Euclid's proof is not syllogistic. I shall discuss these matters in turn.

Geometrical Diagrams

Geometry's extensive use of diagrams is due to its special nature. Disciplines not concerned with spatial aspects of things may well not need diagrams, and a general theory like Aristotle's need not make a special place for such devices. Aristotle's only reference to geometrical diagrams, a parenthetical comment in his discussion of types of "hypotheses," accords with this view:

> **I.10 76b39–77a2** "Nor does the geometer hypothesize things that are false, as some declared, saying that one should not make use of what is false, but the geometer is saying what is false when he says that the line he has drawn is a foot long when it is not a foot long or straight when it is not straight. But the geometer draws no conclusion from the fact that the line he mentioned is this one, but [from the fact that] the things indicated by these [diagrams are the things he mentioned]."[5]

Aristotle gives a reasonable interpretation of the fact that no harm is done if diagrams are not exact because the proof does not depend on any properties of the actual diagram, but only uses it to indicate relevant features. Proofs depend on the given information, not the diagram, although the diagram may illustrate or suggest certain features appropriate for treatment.

The following considerations support Aristotle's view of the use of diagrams in geometry. First, diagrams have heuristic value in discovering geometrical properties and ways to prove them,[6] but that fact is irrelevant in presenting the results reached by their aid, and it is the organized presentation of such material that Aristotle treats and Euclid displays.

Second, demonstrations use diagrams to assist in visualizing how proofs work. Especially handy for this purpose is the practice of labeling the parts of the diagram on which the proof turns. But neither the labels nor the diagrams are necessary. Instead of referring to the given finite straight line as "*AB*," its endpoints as "*A*" and "*B*," and the point of intersection of the two circles as "*C*," we can (laboriously) give a description each time. For example, D4.3 becomes "Let the straight lines from each of the endpoints of the given line to the point of intersection of the two circles be joined." Labeling apart, diagrams help us see how proofs work by aiding our spatial intuition. When we draw circles with centers *A* and *B* and radius *AB*, our spatial intuition enables us to "see" that lines drawn from *A* and *B* to the point of intersection form the missing lines of the desired triangle, though this fact still needs to be proved. But even though diagrams may *aid* our spatial intuition, they do not replace it, and it rather than the diagrams is fundamental to geometry.

Third, the points, lines, and so on, in diagrams are not intended as particulars, as unique objects, but as specimen objects characterized by the stated relations. In the present case, the conclusion holds not only for the given line *AB*, but for any finite straight line. What interest us are not the particular lines in a diagram but any lines satisfying the description found in D1 and in a correspondingly general statement of D4. Thus, we may reasonably think of a diagram as a representation of a description of geometrical entities and their relations. When Euclid refers to point *C*, he is of course talking about point *C* in the particular diagram, but he is considering point *C* just as "the point of intersection of circles whose centers are the endpoints of a given finite straight line and whose radii are equal to that line." The circles, points, and lines that make up the diagram are considered only as specimen cases of figures that satisfy the descriptions. But it follows from this that the descriptions can be substituted for the diagrams without loss of information. What counts, therefore, is not the diagram itself, but what it represents—it should represent nothing more or less than what is given, and what is said about it must apply equally well to any other particular satisfying the given.[7]

The result of this discussion is that even though Aristotle (quite rightly) does not make the use of diagrams a feature of all demonstrative sciences, nothing in geometry's use of diagrams hinders us from using geometrical proofs as models for Aristotelian scientific demonstrations. Nor is there any reason for him to prohibit their use. Diagrams, if correctly used, are valuable devices, but mainly as an aid to our spatial intuition. They do not make possible any more logical moves in the argument and do not import any more content to the geometrical situation they depict.

Existence vs. Constructibility

Aristotle's primitive existence claims, while not equivalent to Euclid's construction postulates, nevertheless correspond to them in their function of introducing subjects in a way suitable for scientific treatment.[8] The Aristotelian version of the present theorem would be an existence proof showing that the existence of points, straight lines, and circles guaranteed[9] in P1 and P3 entails the existence of an equilateral triangle.

P1 would become: "there exists a straight line from any point to any point"; that is, given two points, there exists a straight line that has those points as its endpoints. P3 would become: "there exists a circle with any center and distance"; that is, given a point and a distance, there exists a circle with the point as center and the distance as radius.[10] D1 becomes: "given a finite straight line, there is an equilateral triangle on that line (an equilateral triangle with the given line as a side)." D2 becomes: "let *AB* be the given finite straight line." D3 becomes: "thus it is required to prove that there exists an equilateral triangle with side *AB*." D4 becomes: "there exists a circle with center *A* and distance *AB* (sc., as radius).[11] There exists a circle with center *B* and distance *AB* (sc., as radius). There exist lines *AC* and *BC*, where *C* is a point in which the circles cut one another."

Euclid does not prove that the circles intersect[12] or that lines *AB*, *AC*, and *BC* form a triangle, but as Euclid constructs triangle *ABC*, so Aristotle can claim that triangle *ABC* exists.

In D4 the primitive existence claims guarantee the existence of the two circles, and therefore of point *C* and lines *AC* and *BC*, and (together with the definition of circle) that certain things are true of these entities. They also guarantee that D2 does not require anything not known to exist.

This second maneuver is therefore straightforward, and it indicates that Euclid's construction postulates play a role in geometry analogous to the role of existence hypotheses in Aristotelian demonstrative science (with due allowance made for the fact that the postulates permit the construction of *certain* lines and circles—nothing Aristotle says about his existence claims points to an analogous use).

The similarities between the roles of Euclid's common notions and Aristotle's axioms and between Euclid's and Aristotle's definitions are even closer. The common notions and axioms are used to license inferences, and Euclid's uses of definitions, both in identifying the entities of geometry and in providing basic information about them for use in proofs, are what Aristotle needs his definitions to do. Also the key to existence proofs is the definition of the thing being proved to exist,[13] just as the key to the constructions and proof of *Elements* I, proposition 1 is the definition of equilateral triangle.

This is an important result. Chapter 11 above showed close resemblances between the *kinds* of principles in Aristotle and Euclid; we now know that there are also close correspondences between the *roles* of different kinds of principles in proofs. The resemblances are therefore not superficial, but reflect a basic similarity in outlook on how sciences and scientific proofs are structured. Euclid's practice may therefore clarify obscure points in Aristotle's theory and even suggest ways to correct and improve the theory as presented in *APo*.

Proofs and Syllogisms

The remainder of the chapter will explore the differences between Euclid's reasoning and Aristotle's syllogistic by seeing what is involved in recasting Euclid's proof into a series of syllogisms. This project addresses two questions. First, can each step of the proof be transformed into a syllogism, and second, can the steps be arranged in the right way?

Does Aristotle believe that proofs are syllogistic? At *APo* I.2 71b18 he says that demonstration is a "scientific syllogism" (*sullogismòs epistēmonikós*), but the word *syllogismós* can be translated simply as "deduction." For in his work on logic, Aristotle initially defines "syllogism" as "discourse or reasoning in which, some things having been posited, something other than the things that have been laid down follows of necessity by the fact that these are the case—that is, follows through these" (*APr* I.1 24b18–21). So, *APo* I.2 71b18 can be taken as saying only that demonstration is a kind of deduction, characterized by the requirements imposed on its premises. On general grounds this is a reasonable interpretation, for as the deductive-nomological view of scientific explanation exemplifies, a theory of scientific reasoning need not specify the logic to be used to draw conclusions.

Still, there is no doubt that the logic of demonstration is the syllogistic developed in *APr*. In the first place the premises and conclusions of demonstrations are in subject-predicate form, the only form appropriate to syllogistic reasoning. Second, the theory of the syllogism is used extensively in setting out details of the theory of demonstrative science. Third, at the beginning of *APr* Aristotle announces that the subject matter of the work is demonstration and demonstrative science (*APr* I.1 24a1–2) and asserts at the beginning of the last chapter of *APo* that he has discussed syllogism and demonstration (II.19 99b15). The works are meant to be taken together, with the theory of syllogism presented as the logic for scientific proof.

I am not concerned with the issue that provoked a debate between Solmsen and Ross and that has been taken up again by Barnes,[14] namely, whether Aristotle originally conceived of the theory of demonstrative science independently of the theory of the syllogism. The subject under discussion is not the genesis of the theory of demonstration but the theory itself as presented in *APo*. And as it appears, it presupposes that the logic by which the demonstrations proceed is the logic of the syllogistic.

Some advantages accrue from this fact. The theory of the syllogism enables Aristotle to state a number of points clearly and argue them convincingly. For example, at I.3 72b37ff. he uses the syllogistic to show how circular arguments proceed.[15] Similarly, the syllogistic is used in I.15 to show that there are immediate (atomic) disconnections (negative predications) as well as connections, and in I.16–17 to analyze errors in arguments, specifically to point out for different forms of the syllogism how a false conclusion follows from one or two false premises.[16] In the analysis of demonstrations with negative conclusions in I.21, Aristotle examines the syllogistic figures in which negative conclusions occur (Celarent, Camestres, Cesare) to show that they cannot introduce an infinite number of negative premises.[17] *APo* I.14 employs results of the theory of syllogism to make a theoretical point of general relevance to the theory of demonstration, that the first figure of the syllogism is the most scientific. In general, the benefits gained from the syllogistic are greater detail and precision in both theoretical points and particular arguments. The syllogistic makes possible a more concrete picture of the way in which demonstrations proceed and how they are best analyzed. The results of *APr* are won by much labor, and Aristotle puts them to work in *APo*.

But there are disadvantages as well. Corresponding to the gains in specificity is a loss in generality. Since the syllogistic is inadequate to represent many kinds of valid deductive arguments, Aristotle is hitching his wagon to a broken-down horse. Before considering the sorts of inadequacies to which the syllogistic is subject, it will be useful to consider how Aristotle should be understood when he talks of proofs proceeding through the syllogism. Some of his statements make the claim that syllogistic proofs are actually used in mathematics. For example:

> **I.14 79a17–22** "Of the figures [of the syllogism], the first is most scientific. For the mathematical disciplines construct their demonstrations through this figure, for example, arithmetic, geometry, and optics, and practically all disciplines that investigate the reason why. For either completely, or for the most part and in most cases, the deduction of the reason why occurs through this figure."

If this were true, *APo* would simply be following the practice of mathematics. But it is false. He cannot be talking about past mathematical practice since obviously no one used the syllogistic before Aristotle invented it. And all our contemporary or nearly contemporary evidence of mathematical proofs indicates that they were not expressed in syllogistic form. Since Aristotle's claim simply cannot be accepted at face value, an obvious way to make sense of it is to take it as meaning that mathematical proofs can be put into syllogistic form, and if they were they would be in the first figure, for the reasons given in I.14.

APr I.23 contains an argument for a thesis entailing the first of these claims: "every demonstration and every deduction takes place through the three figures" (i.e., is syllogistic in form) (40b20–22). The soundness of his arguments for this substantial metalogical result need not be discussed here. What is important for now is that *APr* I.23 argues for this view by claiming not that all deductive reasoning in fact is in syllogisms, but that it can be reduced to syllogisms.[18] It follows that the syllogistic is an adequate logic for the theory of demonstration.

On this view, the program of the *Analytics* is clear. First, develop a logic that is general and powerful enough to formulate all deductions; second, employ it in setting out the theory of demonstration, in order to reap the benefits described above. As Lear points out,[19] the syllogistic can be seen as playing a role analo-

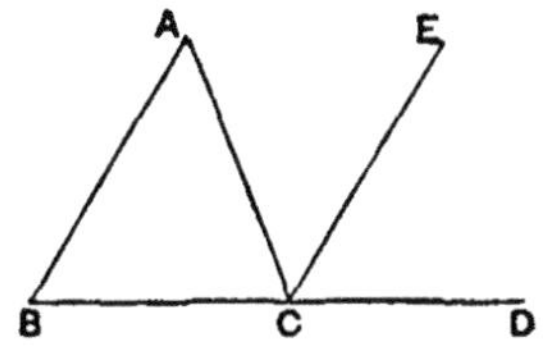

gous to that of the predicate calculus in modern logic. It provides a precise model of reasoning that is supposed to be adequate to represent informal mathematical proofs and that, because of its precision, can be shown to have proof-theoretical results that apply to the mathematical proofs it can be used to formalize, even if no effort is actually made to formalize them. The claim that the predicate calculus is adequate to formalize informal mathematical proofs is substantiated by formalizing some mathematical proofs. But Aristotle takes another tack, the metalogical arguments in *APr* I.23, which provide a theoretical reason to accept the thesis that all deductive arguments and therefore all scientific deductions can be syllogized.

In considering how far it is possible to syllogize a scientific demonstration, I shall discuss Euclid's proof of triangle—2R (*Elements* I, prop. 32), which is either the same as the proof known to Aristotle or quite similar.[20]

Elements I, Proposition 32

D1. In any triangle, if one of the sides be produced, the exterior angle is equal to the two interior and opposite angles, and the three interior angles of the triangle are equal to two right angles.

D2. Let *ABC* be a triangle, and let one side of it *BC* be produced to *D*.

D3. I say that the exterior angle *ACD* is equal to the two interior and opposite angles *CAB*,*ABC*, and the three interior angles of the triangle *ABC*,*BCA*,*CAB* are equal to two right angles.

D4. For let *CE* be drawn through the point *C* parallel to the straight line *AB* [I.31].

D5.1. Then, since *AB* is parallel to *CE*, and *AC* has fallen upon them, the alternate angles *BAC*,*ACE* are equal to one another [I.29].

D5.2. Again, since *AB* is parallel to *CE*, and the straight line *BD* has fallen upon them, the exterior angle *ECD* is equal to the interior and opposite angle *ABC* [I.29].

D5.3. But the angle *ACE* was also proved equal to the angle *BAC*; therefore the whole angle *ACD* is equal to the two interior and opposite angles *BAC*,*ABC*.

D5.4. Let the angle *ACB* be added to each; therefore the angles *ACD*,*ACB* are equal to the three angles *ABC*,*BCA*,*CAB*.

D5.5. But the angles *ACD*,*ACB* are equal to two right angles [I.13]; therefore the angles *ABC*,*BCA*,*CAB* are also equal to two right angles.

D6. Therefore etc. Q.E.D.[21]

The following analysis is based on a curious work of the sixteenth century by Herlinus and Dasypodius entitled *Analyseis Geometricae sex librorum Euclidis*,[22] which puts the first six books of Euclid's *Elements* into syllogisms.[23] In Herlinus's version syllogisms appear only in D5, which proceeds as follows:

The apódeixis. Six syllogisms.

S1A. A straight line falling on parallel lines makes the alternate angles equal to one another.

S1B. The straight line *AC* falls on parallel straight lines *AB*,*CE*.

S1C. Therefore, angle *ACE* is equal to angle *CAB*.

Explanation. The major is proposition 29. The minor is known from the *kataskeuḗ* [part (4)].

S2A. A straight line falling on two parallel straight lines makes the exterior angle equal to the interior and opposite angle on the same side.

S2B. The straight line *BD* falls on parallel straight lines *AB*,*CE*.

S2C. Therefore, angle *ECD* is equal to angle *ABC*.

Explanation as above.

S3A. If equals are added to equals, the things that result are equal.

S3B. Angle *ACE* is equal to angle *CAB*: and angle *ECD* is equal to angle *ABC*.

S3C. Therefore, the whole angle *ACD* is equal to the two angles *CAB*,*ABC*.

Explanation. The major is a common notion. The first part of the minor is the conclusion of the first syllogism. The last part is the conclusion of the second syllogism.

S4A. If equals are added to equals, the things that result are equal.

S4B. Angle *ACD* is equal to the two angles *CAB*,*ABC*. And angle *BCA* is common.

S4C. Therefore, the two angles *ACD*,*ACB* are equal to the three angles *CAB*,*ABC*,*BCA*.

Explanation. The major is a common notion. The first part of the minor is the conclusion of the third syllogism. The last part is known per se.

S5A. A straight line standing on a straight line and making angles, makes those angles equal to two right angles.

S5B. The straight line *AC* standing on the straight line *BD* makes angles *ACD*,*ACB*.
S5C. Therefore, the angles *ACD*,*ACB* are equal to two right angles.
Explanation. The major is proposition 13. The minor is known per se.

S6A. All right angles are equal to one another, and things that are equal to the same thing are equal to one another.
S6B. The three angles *CAB*,*ABC*,*BCA* are equal to the two angles *ACD*,*ACB*. And the two angles *ACD*,*ACB* are equal to two right angles.
S6C. Therefore, the three angles *CAB*,*ABC*,*BCA* are equal to two right angles.
Explanation. Each part of the major is a common notion. [Actually the first part is P4.] The first part of the minor is the conclusion of the fourth syllogism. The last part is the conclusion of the fifth syllogism.

Herlinus then restates relevant parts of syllogisms 4 and 6, and then abbreviates the general statement of the conclusion:

The sumpérasma.
Angle *ACD* is equal to the two angles *CAB*,*ABC* and the three angles *CAB*,*ABC*,*BCA* are equal to two right angles.
Therefore, in any triangle, etc. Q.E.D.

This version of the proof has a number of shortcomings from the point of view of the syllogistic. To begin with, Aristotle states that a premise is an affirmative or negative statement of one thing (*A*) about one thing (*B*).[24] In the case of affirmative statements, *A* belongs to (*hupárchein*) or is predicated of (*katēgoreîsthai*) *B*. Since each syllogistic premise has only two terms, it is not apparent how *any* of the premises in Herlinus's version can be put into correct form. Premise S1A, for example, refers to at least five things in addition to the attribute equal: the straight line, the parallel lines, and the alternate angles. We might attempt to recast the proposition as follows:

S1A′. (Being) equal to one another belongs to alternate angles produced when a straight line falls on parallel lines.

Here the predicate is "(being) equal to one another" and the subject is "alternate angles produced when a straight line falls on parallel lines." Similar machinations will work S2A and S5A. Likewise S3A and S4A might become:

(Being) equal to one another belongs to things that are the results of equals added to equals.

Premise S6A presents special problems, which are perhaps best dealt with by simply omitting its first part and treating its second part the same way as S3A and S4A.

At this point, syllogism 1 becomes:

S1A′. (Being) equal to one another belongs to alternate angles produced when a straight line falls on parallel lines.
S1B′. (Being) alternate angles produced when a straight line falls on parallel lines belongs to angles *ACE* and *CAB*.
S1C′. Therefore, (being) equal to one another belongs to angles *ACE* and *CAB*.

But how do we know that 1B′ is true? Probably the simplest solution is to make it explicit in D4, which could be reworded:

D4′. Let *CE* be drawn through the point *C* parallel to the straight line *AB*. Thus *BAC* and *ACE* are alternate angles, and the interior angle *ABC* is opposite to the exterior angle *ECD*.

Syllogisms 2 and 5 can be treated similarly.

Another problem is raised by 1B′. In first-figure syllogisms, the middle term is subject of the major premise and predicate of the minor premise. But is the middle term a subject or an attribute? The distinction between subject and attribute is basic in demonstrative sciences,[25] yet even the simplest syllogistic arguments appear to violate it. Premise S1B′ preserves grammatical rules by adding the word "being." Can the subject-attribute distinction be preserved as well?

Aristotle does not appear to have reflected on this problem, since he shows no hesitation in speaking of the middle term as being identical in the two premises. A straightforward solution would be for him to adopt a convention for deriving attributes from subjects and vice versa. I suggested that the per se 2 relation legitimates compound subjects like equilateral figure,[26] which are derived from attributes. It is also possible to derive attributes from subjects as "being (or having) alternate angles" is derived from "alternate angles." In general, a term could be marked as an attribute, so that it can legitimately appear as a predicate, by the word "being" or "having," and as a subject by the expression "things that are" or "things that have." (The addition of "being" to straightforward attribute terms such as "equal" also makes for better English, though the issue here is purely one of grammar.) A syllogism in Barbara would appear as "Being *A* belongs to all things that are *B*; being *B* belongs to all things that are *C*; therefore being *A* belongs to all things that are *C*." The dual role of the middle term is justified by the trivial fact that being *B* belongs to all and only things that are *B*.

Premises S3B, S4B, and S6B present a further problem. Syllogism 3 will be rephrased as follows:

S3A′. (Being) equal to one another belongs to things that are the results of equals added to equals.
S3B′. (Being) things that are the results of equals added to equals belongs to angle *ACD* and the angle that is the sum of angles *CAB* and *ABC*.
S3C′. Therefore, (being) equal to one another belongs to angle *ACD* and the angle that is the sum of angles *CAB* and *ABC*.

Premise S3B′. combines the results of syllogisms 1 and 2 with the fact that angle *ACD* equals angles *ACE* plus *ECD* (this could be justified by the principle—which, however, does not occur in Euclid—that the whole is equal to the sum of its parts). But this combination cannot be made within the rules of the syllogistic, since it starts with four terms (the angles *ACE*, *ECD*, *CAB*, and *ABC*) and ends with the single term that forms the subject of S3B'. A similar difficulty arises in syllogisms 1, 2, and 5.

Of these difficulties, all but the last can be met within the strict requirements of the syllogistic, as long as appropriate modifications to D4 are made. But the kind of reasoning involved in going from $M = N$, $P = Q$, and $M + P = R$ to $N + Q = R$, or from $M = N$ and $N = P$ to $M = P$, cannot be represented syllogistically.[27]

A major shortcoming of the syllogistic is its restriction to only one relation between terms, that of "belonging to" or "being predicated of." Since De Morgan, it has been recognized that other relations are susceptible to logical treatment too. In particular, mathematical reasoning uses more relations than the one treated by Aristotle. This is proved true by the examination of "syllogisms" 3, 4, and 6 in Herlinus's version, but is also indicated by the complexity of the gyrations needed to put the mathematically very simple reasoning of the other syllogisms into subject-predicate form.

There remains the question whether Aristotle would have acknowledged the force of this objection. Our question amounts to asking how closely Aristotle felt obliged to stick to the strict demands of the syllogistic. Would he have accepted sloppy syllogisms in proofs? An indication that he would is gleaned from the ways he refers to the conclusion of the proof just studied. He does not say "(being) equal belongs to the interior angles of a triangle and two right angles," or even "(being) equal to two right angles belongs to the interior angles of a triangle," but "two right angles belongs to triangle" (I.4 73b31), "the triangle is equal to two right angles" (b31–32), and "the triangle has two right angles" (b33–40). This lax attitude is also shown in his sample demonstrations. In II.8 he suggests that a demonstration explaining thunder can be constructed from the following terms: cloud (*C*), thunder (*A*), and quenching of fire (*B*) (93b9–10). He says that *B* belongs to *C* and *A* belongs to *B*, so *A* belongs to *C*. Although *C* is connected with *A* via *B*, there are a number of problems in putting it into a syllogism. If we take it as a syllogism in Barbara, the result is as follows:

Thunder belongs to all quenching of fire.
Quenching of fire belongs to all clouds.
Therefore, thunder belongs to all clouds.

Even if we can make sense of the premises and conclusion, all three are false. Suppose we take it as a syllogism in Darii:

Thunder belongs to all quenching of fire.
Quenching of fire belongs to some clouds.
Therefore, thunder belongs to some clouds.

On the appropriate interpretation, the minor premise and conclusion are true, but the major premise is false. Since these are the only likely syllogistic interpretations of the argument, the example is poorly chosen for syllogizing.

Further, the relations between the terms are not appropriate for syllogistic premises. Aristotle justifies the minor premise by saying that fire is extinguished in cloud (b11). But this only identifies the location of the quenching, and provides no better grounds for saying that the relation is one of predication than one would have in saying that man is predicated of the Earth, since that is where men are found. The relation between noise and the quenching of fire is also not one of predication. Noise is *produced by* the quenching of fire, but this relation of cause and effect is not a relation of predication. I suppose we could construct some kind of argument from the premises when they are properly spelled out:

Noise is produced by the quenching of fire.
Quenching of fire occurs in clouds.
Therefore, noise is produced in clouds.

But this is not a syllogism, nor would it be if the premises were quantified. The relations between the terms in the premises are different, and neither of the two relations is the predication relation required for a syllogism.

There are further objections as well, since it is hardly conceivable that either premise is immediate and it is not clear how they would be mediated within the present argument, nor is it clear that either premise is per se. But even though there are reasons why we should not expect to find good examples of scientific demonstrations in *APo*, those reasons do not condone a casual attitude toward syllogisms.[28] And the case discussed is by no means unusual.[29]

An explanation for Aristotle's slackness in these syllogisms can be found in the view he argues for in *APr* I.23, that every demonstration and deduction is syllogistic in form.[30] If he is convinced by metalogical considerations that demonstrations can be syllogized, there is no need to take the trouble to make his examples conform strictly to the requirements of the syllogistic. No matter if the examples are sloppy. It is enough to give the reader a rough idea of how the syllogizing would go. Further, Aristotle need not have thought it a simple task to syllogize a science. The theory of demonstration is programmatic and prescrip-

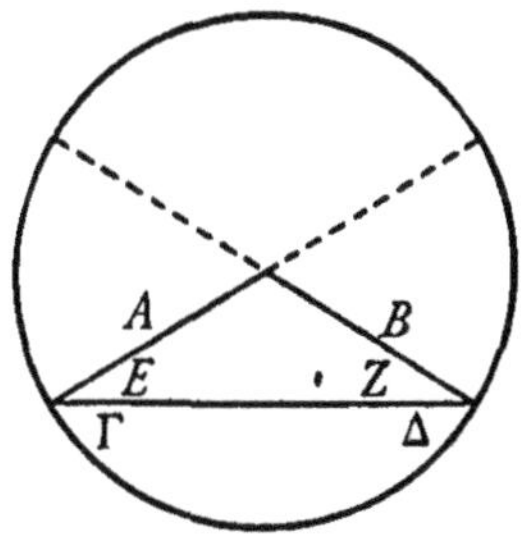

tive, not descriptive. It calls for a radical restructuring of all sciences except the mathematical ones, and even for them, it requires recasting their propositions in a radically different form. In these circumstances, rough "hand-waving" examples are all we can expect.

But one piece of evidence strongly suggests that Aristotle accepted a degree of sloppiness in syllogisms that is incompatible with the strict application of syllogistic theory. *APr* I.24 argues that every syllogism must have (at least) one universal premise and uses a mathematical example:

> ***APr* I.24 41b13–22** "This is more manifest in geometrical propositions, for example, the proposition that the base angles of an isosceles triangle are equal. Suppose the straight lines *A*,*B* have been drawn to the center. Then, if one should assume that the angle *AC* is equal to the angle *BD* without having claimed generally that the angles of semicircles are equal, and again, if one should assume that the angle *C* is equal to the angle *D* without making the additional assumption that every angle of a segment is equal to the other angle of that segment, and if one should then, lastly, assume that, the whole angles being equal and the angles subtracted being equal, the remaining angles *E*,*F* are equal, he will beg the question unless he assumes [generally] that 'if equals be subtracted from equals the remainders are equal.' "[31]

A and *B* are radii of the circle; *AC* and *BD* are the "mixed" angles made by the circumference of the circle with the radii *A* and *B*, respectively; *C* and *D* are the "mixed" angles made by the circumference with the chord of the circle, which is the base of the triangle; and *E* and *F* are the base angles of the triangle.[32] The proof that $E = F$ depends on the premises $AC = BD$ and $C = D$, from which the conclusion follows, since $E = AC - C$ and $F = BD - D$. Aristotle says that the proof requires three syllogisms whose major premises are "The angles of semicircles are equal," "every angle of a segment is equal to the other angle of that segment," and "if equals be subtracted from equals, the remainders are equal." The syllogisms are:

S1A. The angles of semicircles are equal.
S1B. *AC* and *BD* are angles of semicircles.
S1C. Therefore, *AC* is equal to *BD*.

S2A. Every angle of a segment is equal to the other angle of that segment.
S2B. *C* is one angle of a segment; *D* is the other angle of that segment.
S2C. Therefore, *C* is equal to *D*.

S3A. If equals be subtracted from equals, the remainders are equal.
S3B. *E* and *F* are the remainders of equals subtracted from equals.
S3C. Therefore, *E* and *F* are equal.

Since Aristotle is not concerned to state the premises and conclusions in subject-predicate form, it is unclear how he would treat a proposition asserting equality: would S1C be rendered "(Being) equal to one another belongs to *AC* and *BD*," or "(Being) equal to *AC* belongs to *BD*"? There are objections to both choices. In the former case, S1 must be rendered thus:

S1′A. (Being) equal to one another belongs to the angles of semicircles.
S1′B. (Being) the angles of semicircles belongs to *AC* and *BD*.
S1′C. Therefore, (being) equal to one another belongs to *AC* and *BD*.

But S1′B is actually a composite of the two propositions asserting that *AC* is an angle of a semicircle and that *BD* is an angle of a semicircle, and as we saw above, the syllogistic does not, strictly speaking, permit moves from two terms to one. In the latter case, the syllogistic does not allow us to get from the major premise Aristotle explicitly identifies as "the angles of semicircles are equal," whose predicate is "equal," to the conclusion, where the predicate is "equal to *AC*." Similar problems attend S2 and S3, and these are the very kinds of problems identified above in discussing Herlinus's syllogized version of Euclid's proof.

Aristotle did not observe the formal requirements of his logic as strictly as modern standards demand. Even when we would expect him to pay close attention to those concerns, he allows premises that are not in subject-predicate form and can be put into that form only implausibly and with difficulty. He also accepts inferences that are not valid in the syllogistic. Quite likely he simply failed to notice this fact, since they were such simple and manifestly correct moves.

If faced with this discussion, if he were shown that some of his syllogisms are sloppy, would he say that they are not really syllogisms? If so, he would have to admit that the syllogistic cannot represent all valid deductions, and that it is inadequate to represent scientific proofs. Or would he say that they *are* to count as syllogisms and syllogisms can be sloppy? If so, he can still maintain that all valid deductions and therefore all scientific proofs can be represented by syllogistic arguments. One is inclined to prefer the former of these alternatives, although it is not clear to me on what evidence. From the historian's point of view it is important to ask whether the preference does not come from anachronistic modern prejudices about the nature of formal logic. I would prefer to believe that if the objections had been put to him, he would have revised the theory to make it cover a wider field of relations than the subject-predicate relation. But this belief is admittedly due more to historical hindsight than to adequate ancient evidence.

So FAR I have shown that the structure of individual syllogisms makes them inadequate to represent scientific proofs. In what follows I shall consider additional problems, which arise for Aristotle's theory from his view that proofs consist of particular kinds of *series* of syllogisms. He believes that proofs are series of syllogisms in which the conclusion of each syllogism becomes a premise of the next (except for the last syllogism, whose conclusion is the *demonstrandum*). For example, in proving that *E* belongs to all *A* via the middle terms *B*,*C*,*D* we have the following syllogisms:

S1A. *C* belongs to all *B* (immediate premise).
S1B. *B* belongs to all *A* (immediate premise).
S1C. Therefore, *C* belongs to all *A*.

S2A. *D* belongs to all *C* (immediate premise).
S2B. *C* belongs to all *A* (conclusion of syllogism 1).
S2C. Therefore, *D* belongs to all *A*.

S3A. *E* belongs to all *D* (immediate premise).
S3B. *D* belongs to all *A* (conclusion of syllogism 2).
S3C. Therefore, *E* belongs to all *A* (*demonstrandum*).

I shall represent this proof as follows:

A—B—C—D—E,

where *X—Y* represents a connection between *X* and *Y*.

Evidence for this view of the nature of the demonstrations is found in the arguments against the possibility of infinitely long proofs in I.19–23. For example, the beginning of I.19 presents what is evidently intended as a general characterization of demonstrations:

> **I.19 81b15–18** "For it is necessary to assume these and to prove in this way, for example, that *A* belongs to *C* through *B* and in turn that *A* belongs to *B* through another middle and similarly that *B* belongs to *C*."

Likewise, in the account of how to discover proofs, this is the kind of proof that is the goal:

> **I.23 84b31–35** "When it is necessary to prove [that *A* belongs to *B*], we must assume what is predicated first [i.e., immediately][33] of *B*. Let it be *C*. And similarly [let] *D* [be assumed as predicated immediately] of this. And as we continue in this way at each step, no premise is ever taken in the process of proving which is outside of or belongs to *A*, but the middle is continuously being compressed [i.e., the logical gap needing to be filled between *A* and each successive term becomes smaller] until they become indivisible and one."[34]

It should be noted that this is more restrictive than the normal definition of a proof as a finite linear sequence of statements $A_1, A_2, \ldots A_n$, where $A_1, A_2, \ldots A_{n-1}$ are the premises and A_n is the conclusion, and for all i such that $1 \leq i \leq n$, A_i is either primitive or follows from axioms and/or previous statements. Aristotle in effect requires that the propositions be arranged in such a way that for all i such that $3 \leq i \leq n$, A_i is immediate or follows from A_{i-2} and A_{i-1}. In the example just above, the propositions will be arranged as follows:

A1. *B* belongs to all *A*.
A2. *C* belongs to all *B*.
A3. *C* belongs to all *A*.
A4. *D* belongs to all *C*.
A5. *D* belongs to all *A*.
A6. *E* belongs to all *D*.
A7. *E* belongs to all *A*. Q.E.D.

But this requirement is impossibly restrictive for demonstrations, since it implies that no theorem can be used in the proof of subsequent theorems. Consider how *Elements* I, proposition 32, reaches its conclusion. It assumes tacitly that triangles have three angles (a result that can be proved from suitable definitions of triangle and angle) and then argues that in the specimen triangle *ABC*, together with the lines *CD* and *CE* given in parts D2 and D4,

D5.1. Angle *ACE* equals angle *BAC*;
D5.2. Angle *ECD* equals angle *ABC*;
D5.3. Angle *ACD* equals angle *BAC* plus angle *ABC*;
D5.4. Angle *ACD* plus angle *ACB* equals angle *BAC* plus angle *ABC* plus angle *ACB*;
D5.5. Angle *ACD* plus angle *ACB* equals two right angles;
D6. Angle *BAC* plus angle *ABC* plus angle *ACB* equals two right angles. Q.E.D.

D5.1, D5.2, and D6 depend on previously proved theorems and so, in Aristotelian terms, are not immediate. Nor is it simply for the sake of shorthand, convenience, and avoiding repetition[35] that the proofs of those theorems are not repeated here. Their proofs are not *wanted* here. What is wanted and what is used is the theorem proved elsewhere. This is an obvious point, but an important one. For it means that the proof of a proposition may contain nonimmediate links, as long as they are justified by proofs that contain only immediate links, or by proofs with nonimmediate links that are justified in the same way. As an inspection of Euclid's *Elements* or any book of modern mathematics will show, this characterization of proofs does not lead to an infinite regress or circularity, since the initial proof is justified exclusively by the starting points.

Aristotle's simple picture must be altered to cover this situation, where a connection that needs justifying is part of the proof of another connection. If the outline of the proof is

A—*B*—*C*—*D*—*E*,

suppose that *C*—*D* is a previous theorem. According to the simple view of demonstration, the interim conclusion *A*—*D* follows from *C*—*D* and *A*—*C*, which is itself an interim conclusion. But if *C*—*D* is a theorem, it is the conclusion of a proof based on at least two immediate propositions. If the proof of *C*—*D* is inserted in the proof of *A*—*E*, the serial nature of proof cannot be maintained for the proofs of both *C*—*D* and *A*—*E*. In the simplest case the proof of *C*—*D* will have two immediate premises, *C*—*X* and *X*—*D*. We have this:

S1A. *C* belongs to all *B* (immediate premise).
S1B. *B* belongs to all *A* (immediate premise).
S1C. Therefore, *C* belongs to all *A*.

S2A. *D* belongs to all *X* (immediate premise).
S2B. *X* belongs to all *C* (immediate premise).
S2C. Therefore, *D* belongs to all *C*.

S3A. *D* belongs to all *C* (conclusion of syllogism 2).
S3B. *C* belongs to all *A* (conclusion of syllogism 1).
S3C. Therefore, *D* belongs to all *A*.

S4A. *E* belongs to all *D* (immediate premise).
S4B. *D* belongs to all *A* (conclusion of syllogism 3).
S4C. Therefore, *E* belongs to all *A*. Q.E.D.

Syllogism 3 violates the requirement that A_i be immediate or follow from A_{i-2} and A_{i-1}.

Aristotle occasionally recognizes that previously proved propositions have a legitimate use in proofs,[36] but does not seem aware that they conflict with his serial view of proofs. Since it requires only a small modification in his view of the nature of scientific demonstration to accommodate this important feature of proofs, we may regard it as a minor oversight.

If prior results are allowed to justify steps in a proof, the proof of *A—E* as given above may be diagrammed as follows,

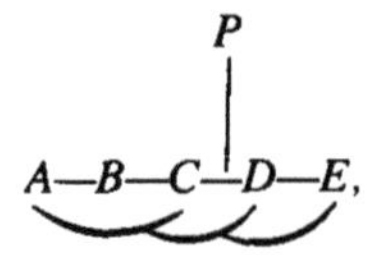

to indicate that link *C—D* depends on previous theorem *P*, and the other links (*A—B*, etc.) do not depend on previous theorems.

More important is the fact that even with this modification, the serial interpretation of demonstrations does not capture the nature of reasoning in the proof of this theorem. In proving that the three angles of a triangle equal two right angles, it shows that they are equal to the two angles made by a straight line set up on a straight line (D5.4), which are equal to two right angles (D5.5, depending on *Elements* I, prop. 13). But in reaching D5.4 (which treats the sum of all the angles taken together), it treats certain of the angles separately and then in combination. D5.1 proves a result about angle *BAC*; D5.2 proves a different result about angle *ABC*; D5.3 combines these results to prove a further result of the sum of angles *BAC* and *ABC*; D5.4, finally, proves a result about the sum of all three angles of the triangle, which follows from the result of D5.3. The line of argument can be represented schematically as in the diagram below.

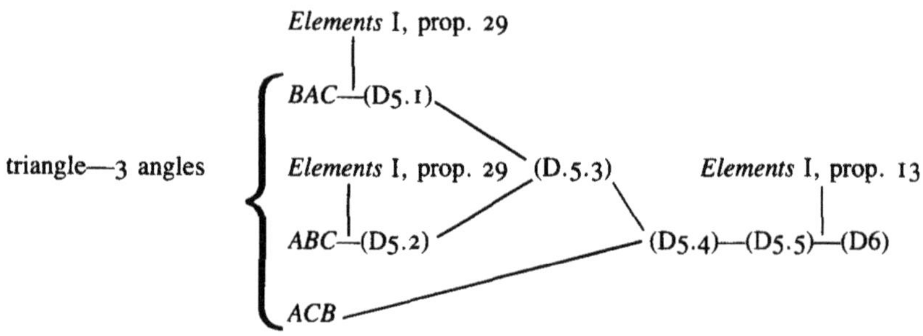

Thus, in the proof of a conclusion it can happen that reference must be made to compounds (such as the sum of three angles) and also to their components. A fact about the compound may depend on one or more facts about its components or vice versa. Indeed, geometrical proofs frequently employ this kind of decomposition and recomposition, and it may occur in other disciplines as well. But when this sort of argument occurs, the conception of proof as a series of syllogisms does not adequately account for the relations between the different steps of the proof. As in the previous case, Aristotle needs to alter his view of the nature of demonstration to allow for the complexity we have noticed in the structure of the Euclidean proof.

Nevertheless, since the needed modifications do not require Aristotle to abandon the substantial result he argues for in I.19–23, that demonstrations must be finite in length, we may consider the simple model of proofs he uses there as no more than a first approximation that contains the basic argument he would use to prove the results for the more complex cases that have been discussed here. There are no theoretical obstacles to prevent him from making the suggested changes.

The present chapter has shown how the proofs of Euclid's *Elements* can illustrate how Aristotelian demonstrations work, what their faults are, and how to overcome them. The most significant change needed, abandoning the syllogistic, does not affect the basic conception of demonstration. (In the light of *APr* I.23, even the insistence on syllogistic reasoning need mean only that proofs must be valid.) Even though Aristotle underestimates the complexity of reasoning involved in mathematical demonstrations,[37] much of what he says can be adapted to take this complexity into account. What is perhaps of greatest interest is that Euclid uses his various kinds of principles in just the ways we would expect Aristotle to apply his (with provisions made for differences between Aristotle's existence claims and Euclid's construction postulates). The next three chapters will draw on these results in considering the three types of demonstrations recognized in *APo*.

The Varieties of Demonstration, I: Universal Subject-Attribute Demonstrations

CHAPTER XIII

DEMONSTRATION IS introduced as one (possibly not the only) way[1] to attain scientific knowledge. A demonstration is a scientific deduction (*sullogismòs epistēmonikós*), a deduction that yields scientific knowledge of its conclusion (*APo* I.2 71b17–18).

After saying this much about demonstration, Aristotle passes on to his account of scientific principles. But what of the conclusions of demonstrations and the way they follow from the principles? In Chapter 12 I took up some aspects of these topics—the roles of the different kinds of principles and the reasoning used in proofs. It remains to consider what kinds of propositions can be proved and in a general way what proofs achieve and how they do it. I will supplement Aristotle's limited remarks on these matters with material from Euclid's *Elements*.

The Nature of Demonstrations

From the opening lines of I.2 we know that the conclusion of a demonstration is necessarily true and that a demonstration is a deduction that proves it is necessarily true and contains the explanation of why it is true. A person who understands the demonstration knows that and why its conclusion is necessarily true, and his understanding the demonstration is the grounds of his conviction that it is so (I.2 72a25–26). Also, when something can be demonstrated, the only way to get unqualified scientific knowledge of it is to have a demonstration of it (I.2 71b28–29).

Since conclusions are necessary, so are the principles.[2] In fact principles and conclusions are per se relations[3] of the two types identified at I.4 73a34–37 and a37–b5,[4] and there is no scientific knowledge of relations that are not per se.[5]

Demonstrations have a necessary middle (I.6 75a13): the middle term must be necessarily related to the subject and the attribute. If

A—B—C[6]

is a demonstration that attribute *C* belongs to subject *A* via middle term *B*, the relations *A—B* and *B—C* must be necessary truths. Also when there is more than one middle term, adjacent terms must be necessarily related.

Since a demonstration explains its conclusion, the premises are the grounds of the conclusion (I.2 71b30–31). In the proof of *A—C* sketched above, *C* is proved to belong to *A* through the middle *B*. *B* is the grounds of *C*'s belonging to *A* (II.2 90a6–7); that is, *C* belongs to *A because C* belongs to *B* and *B* belongs to *A*. Moreover, since the principles of demonstrations are immediate indemonstrable truths, they have no further grounds. They are "uncaused causes" or "ungrounded grounds" (I.9 76a20). *C* belongs to *B*, as does *B* to *A*, "through itself" (I.6 75a35–37). The word *dihóti*, "because" or "why" (sometimes best translated as "the reason why") frequently appears in this connection. "To know the reason why is to know through the grounds" (I.6 75a35); "the most important thing in knowledge is investigating the reason why" (I.14 79a23–24); "the reason why is the middle" (I.33 89a16). Knowledge of the reason why comes from deductions based on immediate premises, since only thus do we grasp the primary grounds (I.13 78a23–26).[7]

Some consequences of these views have been noticed. The terms used in a scientific demonstration must belong to the subject genus of the science;[8] and the conclusion cannot be perishable, that is, a relation that can cease to be true (I.8 75b24–25) and both premises and conclusions must be universal (I.8 75b21).[9]

The picture so far is simple. Principles are immediate, indemonstrable, per se necessary relations. Conclusions are nonimmediate, demonstrable, per se necessary relations that are proved to follow from the principles. Principles are grounds of conclusions, and scientific knowledge of a provable relation comes from knowing the proof whose conclusion is that connection.

Subjects and Attributes

The picture becomes more precise when we graft onto it the distinction between subject and attribute. A scientific proposition asserts that an attribute belongs to a subject.[10] The conclusion of a proof asserts that an attribute (*C*) belongs to a subject (*A*).[11] This means that in the proof

A—B—C,

B and *C* are both attributes of *A*. *B* belongs immediately and *C* nonimmediately to *A*: *A—B* is a definitional or essential fact and *A—C* a derivative fact.[12] *C* belongs to *A* because *C* follows from the essence of *A*, and *B* belongs to *A* just because it is *A*'s nature to have attribute *B*.

This interpretation of scientific proofs poses some problems. Granted that *A*—*B* and *B*—*C* are immediate, and that *A*—*B* is an essential fact about *A*, what is the nature of the relation *B*—*C*? It should be an essential fact about *B*, so that *C* is part of *B*'s definition, and so that part of the definition of one of *A*'s attributes belongs to *A*. This result is subject to two objections. First, it runs afoul of the *Categories* claim that the definition of an attribute is never predicable of a substance (although the name of an attribute is predicable in some places) (*Cat.* 5 2a27–33). Even though the subjects of sciences need not be substances,[13] in some sciences the subjects *are* substances, and in these cases the *Categories* claim should apply. Second, the result is hard to reconcile with the standard example of a demonstrable fact, triangle—2R, since "having angles equal to two right angles" is not part of the definition of any term in the definition of triangle.

I believe these objections can be met. The key to the solution of both is the conception of "fat" definitions.[14] The *Categories* objection proves to be a red herring, since it is based on the concept of "thin" definitions consisting of genus and differentia. Ackrill's comment on the *Categories* passage suggests a solution to the problem: "In saying that usually the name of what is in a subject cannot be predicated of the subject he obviously means more than that, for example, one cannot say 'Callias is generosity.' He means that there is something else which one does say—'Callias is generous'—by way of ascribing generosity to Callias."[15] Likewise, though the thin definition of generosity, "liberality in giving," cannot be predicated of Callias, if Callias is generous, he is liberal in giving. Further, this same move can be applied to the elements in the fat definition of an attribute and to their per se relations as well. If we take it as an immediate per se fact about generosity that it is a mean state between stinginess and extravagance, then Callias is not that mean state, but he is characterized by being in that mean state. Similar points hold for subjects and attributes in scientific propositions. A line is straight, not straightness, and per se relations of straightness can be predicated of the line in appropriate ways. In general, then, there is no objection to inferring that an element of the fat definition of *B* belongs to *A*, when *B* is an element in the fat definition of *A*. A caveat must be issued here, since the various elements of a fat definition may stand in various relations to the definiendum, so that this transitivity need not always hold.[16]

Second, the relation "triangle—2R" can be accounted for in this way if we make certain adjustments.[17] Since the proof[18] turns on properties of the three angles of the given triangle, it is a tacit assumption that triangles have three angles. If the proof is to be stated fully, we begin with the subject of the proof (triangle) and the essence of the subject (closed plane rectilinear figure with three sides). The essence of triangle and the essence of rectilinear angle entail that a triangle has three angles. So far the argument can be represented as follows:

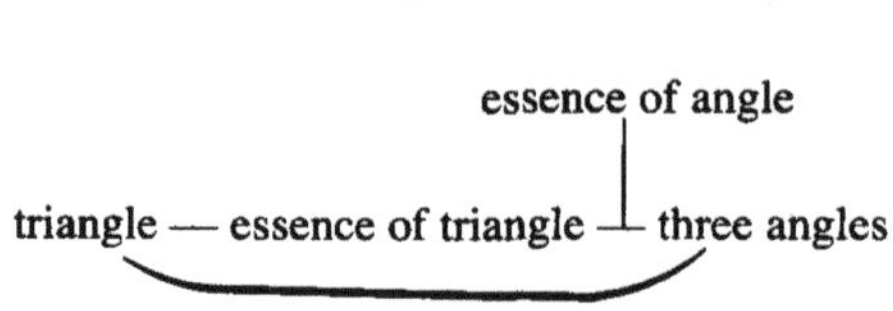

The argument proceeds to consider the angles separately and then in combination. For example, it first shows that angle BAC = angle ACE, a result that follows from *Elements* I, proposition 29, which shows *inter alia* that a straight line falling on parallel lines makes the alternate angles equal to one another. This amounts to showing that the essence[19] of alternate angles entails that they are equal to one another. This result is then applied to the specimen case to give the desired result. Thus, each step in the proof follows from the essence of the subject term of the particular step in question—the triangle or certain specified angles or combinations of angles. And the essence of the triangle underlies every step, since the different angles are defined as angles *of the triangle*, and so the triangle is reasonably seen to be the subject of demonstration throughout.

The situation is more complicated than the simple case where we have

A—*B*—*C*,

with *B* an element in the essence of *A* and *C* an element in the essence of *B*. But the increase in complexity does not produce a difference in kind. Each step in a proof can be described as being either of the form *A*—*B* (*B* is an essential [immediate] attribute of *A*) or of the form

P

A—*B*

(*B* is a derivative attribute of *A*, which is proved to hold by *P*, an immediate relation or a nonimmediate relation that is ultimately based on immediate relations).

An Aristotelian Demonstration

So far I have built up a picture of how an Aristotelian demonstration should work. But how well does this picture fit Aristotle's practice? A look at the most carefully worked out example of a demonstration in *APo* shows that it fits well, but also reveals a further wrinkle in immediate per se relations. *APo* II.17 considers "the grounds" (*tò aítion*), "that of which it is the grounds" (the attribute) (*hoû aítion*), and "that for which it is the grounds" (*hôi aítion*) (99a16ff.). It suggests that the middle[20] is an account (*lógos*)[21] of the first extreme (the attribute) and says that this is the reason why all sciences come to be through definition (a21–23). Then follows the example:

> **II.17 99a23–29** "Being deciduous follows along with the vine and extends more widely, and [it follows along with the] fig tree and extends more widely. But not [more widely] than all, but equally. If, then, you should grasp the first middle, it is the account [*lógos*] of being deciduous. For there will be a first middle in the

other direction, that all are such and such; then [there will be] a middle of this, that the sap congeals or some other such thing. What is being deciduous? For the sap in the connection of the leaf stalk to congeal.''

DISCUSSION

The subjects of the proof are ''the vine,'' ''the fig tree,'' and so on; the attribute is ''being deciduous,'' which extends more widely than each of the subjects, but not more widely than all the deciduous plants taken together. Aristotle assumes that there is a single explanation for the fact that all these plants are deciduous.[22] Slightly perversely, he speaks of ''first middles'' in two directions, by which he means the terms related immediately to the subjects and to the attribute. If we have a demonstration

$$A—B—C—D—E,$$

B is the first middle in the direction of subject A, and D is the first middle in the direction of attribute E. In the present case, we have several subjects $A_1, A_2, \ldots A_n$ that have E. The first step toward discovering a proof of E's belonging to $A_1, A_2, \ldots A_n$ is to determine their common nature, some B such that $A_1, A_2, \ldots A_n$ are all B and such that they are E because they are B. (In practice, it may not at first be apparent which of several common features is the relevant one.) In the present case, the ''such and such'' that all the subjects are is probably broad-leaved.[23] Then we need to find a middle to explain why broad-leaved plants are deciduous—say, that the sap congeals in the connection of the leaf stalk. And, this is the account of deciduousness! The proof is complete. It can be sketched as in the diagram below.

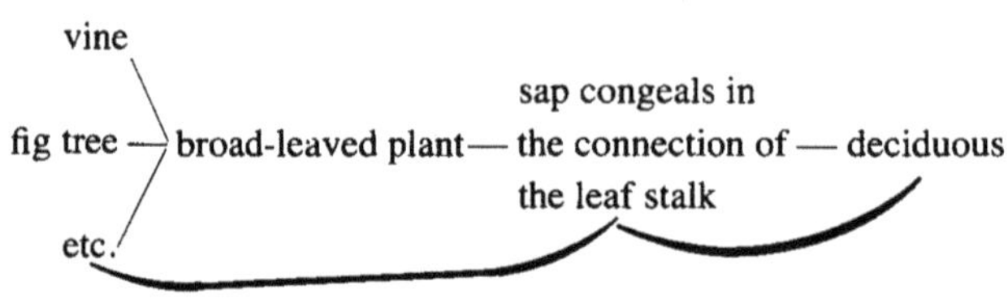

For convenience, assume that the connections between successive terms are immediate. If the vine, the fig tree, and so on, are species of the genus broad-leaved plant, then they are definitionally related on either of Aristotle's views of definitions (fat or thin). If not, being a broad-leaved plant is an element in a fat definition of each. The fact that the sap congeals in the connection of the leaf stalk is neither genus nor differentia of broad-leaved plants, but as an immediate, not further explainable fact about broad-leaved plants, it will be part of the fat definition. Now it happens that the congealing of sap in the connection of the leaf stalk is the grounds[24] of deciduousness, and so, if this relation is immediate, it is part of a fat definition—but definition of what: of deciduousness or of the congealing of sap in the connection of the leaf stalk?

Recall the way we dealt with the problem of the same term's standing as a subject in one premise and an attribute of another (p. 154). In the syllogism

$$A—B—C,$$

the middle term *B* occurs as "being *B*" when it is the predicate of the minor premise, and as "things that are *B*" when it is the subject of the major premise. So here, one of our first (minor) premises will be "being a broad-leaved plant belongs to things that are vines." The second premise will be "being a plant in which the sap congeals in the connection of the leaf stalk belongs to things that are broad-leaved plants." The third will be "being deciduous belongs to things that are plants in which the sap congeals in the connection of the leaf stalk." Aristotle says that the congealing of sap (sc., in the connection of the leaf stalk) is the account (*lógos*) of being deciduous (a25–26).[25] Since it is not the definition (either the genus or differentia) of being deciduous, Aristotle presumably believes it is an immediate attribute that is an element of the fat definition of being deciduous.

On the other hand, is there not precisely equal reason to say that being deciduous is part of the fat definition of plant-in-which-the-sap-congeals-in-the-connection-of-the-leaf-stalk? And similarly at each step of the proof. Being a broad-leaved plant will be part of the fat definition of plant-in-which-the-sap-congeals-in-the-connection-of-the-leaf-stalk, and if broad-leaved plant is the genus of vine, and so on, they will be immediately related to it too.[26] This is to say that immediate relations cut two ways: if *A*—*B* is immediate, then *A* is in the fat definition of *B* and vice versa.

If this move is allowed, Aristotle's example fits my account of demonstration perfectly. Each premise is immediate and due to the essence of both its subject and its predicate. A proof that subject *A* has attribute *D* proceeds by showing that by virtue of *B*, an element in *A*'s essence (fat definition), *A* has attribute *C*, where *C* is an element in the essence (fat definition) of *D*.

Per Se 1 and Per Se 2

The preceding permits us to address an issue left over from earlier discussion.[27] Since conclusions of proofs are necessary, they are per se, but are they per se 1 or per se 2? Aristotle offers little help on this score, simply saying that they are per se, or both per se 1 and per se 2.[28] The matter is not clear-cut since, for example, neither the subject triangle nor the attribute 2R is mentioned in the definition of the other. In what follows I will argue that typical subject-attribute proofs prove that the attribute belongs per se 1 to the subject.[29]

A proof that subject *S* has attribute *A* proceeds by showing that it is a consequence of *S*'s nature, as specified in *S*'s fat definition, that *S* has attributes that satisfy the definition of *A*. Each successive step in the proof has one of the two forms identified at the end of the section "Subjects and Attributes" earlier in this chapter, and for the sake of convenience I will assume that each step is immediate. Thus, the proof will be of the form

$$S\text{—}A_1\text{—}\ . \ . \ .\text{—}A_n\text{—}A,$$

where adjacent terms are so related that each is in the fat definition of the other, and so belongs per se 1 to the other. The per se 2 relation permits the A_i to function as attributes in one premise and subjects in the next. Thus, A_1 belongs per se 1 to S, A_2 belongs per se 1 to A_1, and so on. It is a natural extension to regard the syllogistic consequences (such as S—A_2) of such premises as per se 1, not per se 2.

Further, if the consequences are per se 2 we have the anomalous situation that per se 2 conclusions result both from arguments with two per se 1 premises and from arguments with one per se 1 and one per se 2 premise. For example, in the argument

$$S\text{—}A_1\text{—}A_2\text{—}A_3,$$

the first step,

$$S\text{—}A_1\text{—}A_2,$$

has two per se 1 premises, but the second, whose premises are S—A_2 and A_2—A_3, has one of each type. This anomaly does not arise if the conclusions of arguments all of whose premises are per se 1 are also per se 1.

Also considerations of quantification require the conclusions to be per se 1. If the conclusion of a proof is a universal affirmative relation "A belongs to all S," it must be proved from universal affirmative premises "A belongs to all A_n," . . . "A_1 belongs to all S." But in per se 2 relations the universal affirmative relation is the other way round: S belongs to all A (all odd things are numbers) and not vice versa.

Another general argument for the same point goes as follows. The nature of the per se 1 relation is such that if attribute A belongs per se 1 to subject S, it follows that if S' is an S (where S' is either an individual S or a subcategory of S), A belongs to S'. The nature of the per se 2 relation is such that if A belongs per se 2 to S, it follows that if A belongs to S', then S' is an S. The entailments go in opposite directions. But the entailment found in subject-attribute proofs goes in the direction of per se 1, not per se 2 relations.

Further, Aristotle recognizes cases where an attribute belongs to different subjects for different reasons (II.16 98b25–31, II.17 99a30–b7). In these cases it is especially clear that the entailment relation between subject and attribute must be the one characteristic of per se 1, not per se 2, relations.

Further, even if the per se 2 relation is transitive and some subject-attribute proofs exploit this fact to give per se 2 conclusions,[30] that does not show that all subject-attribute proofs have such conclusions, and in fact is a rather special kind of proof. An example of this sort of argument is this: odd belongs per se 2 to number, number belongs per se 2 to multitude, so odd belongs per se 2 to multitude. This process of moving to wider and wider subjects yields universal affirmative relations characteristic of per se 2, but not per se 1, relations (where *S* belongs to all *A*, but *A* does not belong to all *S*), but it is hard to believe such arguments will have much theoretical interest.

Other considerations point in the same direction. Aristotle's ideal in subject-attribute proofs is for the relation between the subject and attribute to be "commensurately universal," as 2R is to triangle. But Aristotle tends to think of per se 2 attributes as coming in disjunctive pairs: a number is even or odd, a line is straight or bent (I.4 73b18ff.). But then neither odd nor even is commensurately universal to number.

Since every number is even or odd, the compound attribute "even-or-odd" is commensurately universal with number. Even so, Aristotle does not show any interest in proofs of this sort of conclusion.[31] And as to proofs that such and such numbers (e.g., all prime numbers greater than 2) are odd, the arguments above indicate that even though odd belongs per se 2 to number, it will belong per se 1 to the category of numbers in question.

Commensurate Universals

The last issue is the so-called commensurately universal nature of proofs. A frequently recurring theme of *APo* I is the superiority of the proof of triangle—2R to the proof of isosceles triangle—2R.[32] The superiority consists in the fact that triangle is the "primary" subject for which the attribute 2R belongs to any individual case taken at random.[33] The attribute 2R belongs to any isosceles triangle taken at random, but it extends more widely (I.4 74a3), whereas triangle is prior to isosceles triangle (73b39) and in fact is primary (the superlative of "prior"). The method prescribed for locating the primary subject (I.5 74a37–b4) suggests that the primary subject of an attribute is the most general term to which it can be proved to belong.

These passages suggest that the attribute and subject are coextensive, so that the occurrence of either guarantees that of the other. This is so for triangle and 2R and also for rectilinear figure and 4R (having exterior angles whose sum is equal to four right angles).[34] But when Aristotle considers the point more closely, he sees that coextension is not required. He discusses the matter abstractly at II.16 98b25–31 and II.17 99a30–b5, and his example at 99b5–7 makes the situation quite clear. The attribute of being long-lived belongs to various species of quadrupeds, for example, horse and ass,[35] and to various species of birds, for example, parrot and peacock. Those quadrupeds are long-lived because they are bileless, and those birds are long-lived perhaps because they are dry—in any case, for a different reason from the one that accounts for the longevity of the quadrupeds. Now the class of bileless quadrupeds is wider than each of the bileless species of quadrupeds, and the class of dry birds is wider than each of the dry species of birds. And we may assume that there is no common explanation of why both dry birds and bileless quadrupeds (not simply dry things and bileless things)[36] are long-lived. By the criteria of I.4, it follows that dry birds and bileless quadrupeds are *both* primary subjects of the attribute long-lived, since there is no subject to which the attribute belongs that is prior to either of these subjects. In such cases, where there are more grounds than one for the same attribute's occurring,[37] the same thing is predicated of more than one primary subject,[38] with the proviso that the primary subjects are not things of the same kind—either individuals of the same species (in which case the same explanation would necessarily apply to them all) or species of the same genus (where the genus is the primary subject of the attribute). In the example, horse and ass are species of one genus, parrot and peacock are species of another. Thus, an attribute need not be coextensive with its primary subject.[39]

The attribute is universal or belongs universally to its primary subject[40] in the use of that word at I.4 73b26ff.[41] The universal relation is characterized in terms of the conclusions of demonstrations, and demonstration per se proves universal relations. The proof of triangle—2R is a demonstration per se, but proof of isosceles triangle—2R is not (74a1–3). A proof of a universal connection is a proof that the attribute belongs to the subject through (*diá*) or in virtue of (*katá*, *hêi*) or because of (*hóti*, *dihóti*)[42] that subject and not through, in virtue of, or because of anything else.

The following passage tells how to locate the primary subject:

> **I.5 74a32–b4** "So when does he not know universally and when does he know without qualification?[a] It is clear that [he would know universally] if being a triangle and being an equilateral triangle were the same, either for each or for all.[b] But if they are not the same, but different, and it belongs in virtue of triangle, he does not know. Does it belong in virtue of triangle or in virtue of isosceles? And when does it belong in virtue of this primary [subject]? And of what universal is the proof?[c] It is clear that [the proof is universal] when [the attribute] belongs to the primary [subject] when things are subtracted. For example, 2R belongs to the isosceles bronze triangle, but also [it belongs] when the bronze is subtracted and the isosceles. But not when figure or boundary [is subtracted]. But they are not primary.[d] Then what *is* primary?[e] If it is triangle, in virtue of this it also belongs to the others, and the demonstration is of this universal."[f]

NOTES ON TRANSLATION

a. "Knowledge without qualification" implies a correlative, qualified knowledge. We can have qualified knowledge of equilateral triangle—2R, the qualification being that it depends on something further. It is qua triangle, not qua equilateral triangle, that the relation holds. So, here "without qualification" is the same as "universally." The question is, when do we know universally and when do we not?

b. For suggestions as to the meaning of "either for each or for all," see Ross (ad loc.) and Barnes (ad loc.). Mure's interpretation ("for each or for all equilaterals"), although less interesting, may be correct.

c. On this translation the sentence asks, What is the primary universal subject to which the attribute can be proved to belong? It can also be translated "And of what is the proof universally?"—of what subject does the demonstration prove the attribute to belong universally? The two versions amount to the same thing.

d. The genitive case of "primary" is due to the genitives of "figure" and "boundary." The sense is "but neither figure nor boundary is the primary subject of 2R," the first subject in the series: bronze, isosceles, triangle, figure, boundary, such that 2R belongs to it and no longer belongs when it is subtracted.[43]

e. The genitive case of "primary" may be explained as before, but it could also be short for the genitive absolute *tínos prõtou aphairethéntos oukéti hupárxousi dúo orthaí*, "What is the first subject such that when it is subtracted, 2R no longer belongs?" Again, the two questions amount to the same.

f. On this translation, the sentence *toútou kathólou estìn hē apódeixis* states that this is the primary universal to which the attribute can be proved to belong. It can also be translated "the demonstration is universally of this"—the demonstration proves that the attribute belongs universally to this. The two versions come to the same thing. See note c.

DISCUSSION

The passage asks under what conditions a proof of equilateral triangle—2R will give universal knowledge. The answer is that, since triangle—2R is a universal relation, equilateral triangle—2R could be universal only if the essences of equilateral triangle and of triangle were identical.[44] But they are different; in fact equilateral triangles have 2R in virtue of being triangles: "it belongs in virtue of triangle" (a35). This is why the proof does not give universal knowledge of equilateral triangle—2R. Now 2R *does* belong to equilateral triangle, and Aristotle does not deny that this can be known. But it does not belong (and therefore cannot be known to belong) universally. After telling how to locate the primary subject of an attribute, the passage concludes by saying, "if it is triangle, in virtue of this it also belongs to the others, and the demonstration is of this universal" (b2–4). The second part of this sentence answers the question "Of what universal is the proof?" (a37). The first part answers the question "When does it belong in virtue of this primary subject?" (a36). 2R belongs to isosceles triangles and isosceles bronze triangles in virtue of triangle, in virtue of their being triangles. It does not belong to figure or limit in virtue of triangle, since it does not belong to them at all, or belongs only accidentally, so "the others" must refer to

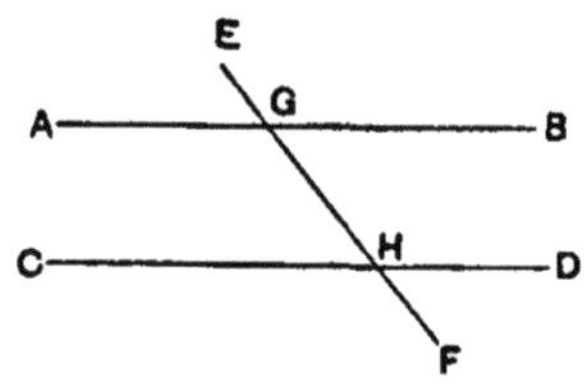

the specifications of the subject that can be subtracted without the attribute's ceasing to belong.

2R BELONGS TO triangle in virtue of itself and to equilateral triangle in virtue of something else. It belongs to equilateral triangle because it belongs to triangle and equilateral triangles are triangles. To argue that it belongs to equilateral triangle directly, and not via triangle, is misleading, as another passage shows:

> **I.5 74a13–16** "If anyone should show that right angles do not intersect,[45] the proof of this would seem to be an account of the fact that it applies in the case of all right angles. But it is not, since this does not occur because they are equal in this particular way, but in that they are equal in any way."

DISCUSSION

Elements I, proposition 28, states: "If a straight line falling on two straight lines make the exterior angle equal to the interior and opposite angle on the same side . . . the straight lines will be parallel to one another." The proof depends on the equality of angle *EGB* to angle *GHD*. A person who proves the theorem for the case where the two angles are both right angles will prove a true conclusion, but the proof will not make use of their being right angles, only their being equal. Aristotle objects to the proof not because its conclusion is false but because it gives the impression that the conclusion holds *because* the angles *EGB* and *GHD* are both right angles, when in fact it holds because the two angles are equal. It *is* proper to ask why perpendiculars to the same line do not intersect, and the proper answer is that they do not intersect because the angles they form with the given line are equal, not because they are right angles.

ELSEWHERE ARISTOTLE STATES that we can know connections like isosceles triangle—2R:[46]

> **I.24 85b37–86a3** "We know best when this no longer belongs because [it is] something else. Now when we know that the exterior [angles of something] are equal to four [right angles] because [it is] isosceles, [the question] still remains, Why [does] the isosceles [have exterior angles equal to four right angles]?—because [it is a] triangle, and this because [it is a] rectilinear figure. But if this [has

> exterior angles equal to 4R] no longer because [it is] something else, then we know best. And then [we know] universally. Therefore universal [demonstration] is superior."

He admits here that "*x* has 4R because *x* is isosceles" is an explanation (although not the best explanation) and that we can know that *x* has 4R for this reason (although our knowledge is not the best that can be had). The point recurs at I.24 85b5–15.

THESE PASSAGES COMPARE two sorts of cases. (1) *x* is a bronze isosceles triangle; *x* has 2R; *P* knows a proof that isosceles triangles have 2R; *P* holds that *x* has 2R because *x* is an isosceles triangle; therefore, *P* knows that *x* has 2R. (2) *x* is a bronze isosceles triangle; *x* has 2R; *Q* knows a proof that triangles have 2R; *Q* holds that *x* has 2R because *x* is a triangle; therefore, *Q* knows that *x* has 2R. Aristotle says that *Q*'s knowledge is better than *P*'s since triangles have 2R in virtue of their own nature as triangles, not because they are anything else, while isosceles triangles have 2R not because they are isosceles triangles but because they are triangles—not in virtue of their own nature, but in virtue of something else's nature:

> **I.24 85a23–24** "We know each thing better when we know it by virtue of itself than [when we know it] by virtue of something else."

The contrast between "by virtue of itself" (*kath' hautó*) and "by virtue of something else" (*kat' állo*) is quite different from the distinction between per se (*kath' hautó*) and accidental (*katà sumbebēkós*), but is a distinction among grounds for knowledge of relations that are per se in the familiar way. The attribute 2R belongs to triangle because of the nature of triangle. It belongs to isosceles triangle not because of the nature of isosceles triangle, but because of the nature of triangle. This does not mean that isosceles triangle—2R is an accidental relation; it is not. It means only that isosceles triangle—2R is proved by virtue of triangle, which is something else than (has a different essence from) isosceles triangle. This point is also made at 85b5–15. Nor does this mean that triangle—2R is self-explanatory. As a nonessential per se relation, it needs explanation, but the explanation depends on the essence of triangle (having three sides and therefore having three angles, etc.), not on the essence of any other subject to which the attribute 2R belongs.

When Aristotle says that demonstration per se is of universal relations (I.4 74a1–2), and that we have scientific knowledge in a nonaccidental (i.e., per se) way when we know that an attribute belongs to a subject by virtue of that by virtue of which it belongs (I.9 76a4–5), he is stressing the importance of proving conclusions in the right way—establishing the relation between an attribute and its primary subject—since this yields the best kind of knowledge. I therefore find no good evidence for the view that there can be no scientific knowledge of relations like isosceles—2R.

I have said nothing so far on the question whether Aristotle thinks all the premises of a proof of a universal relation are convertible, a doctrine that Barnes (p. 247) calls the "most extreme form" of "the doctrine of the Commensurate Universal." The first thing to point out is that Aristotle never asserts this view. Second, his recognition of cases in which an attribute has more than one primary subject and his analysis of the demonstrations found in these cases[47] prove that this view cannot be maintained generally. Third, the analysis of the proof of triangle—2R shows that the premises need not convert. On the other hand, in a *syllogistic* proof whose conclusion is "All *A*'s are *B*," all premises must be universally affirmative (it must be proved by a series of syllogisms in Barbara), and so, if *A* and *B* convert, each of the premises must convert. On a plausible interpretation of the argument that broad-leaved plants are deciduous because the sap congeals in the connection of the leaf stalk[48] the premises are both convertible: in all and only broad-leaved plants does the sap congeal in the connection of the leaf stalk, and all and only the plants whose sap congeals in the connection of the leaf stalk are deciduous.[49]

The Varieties of Demonstration, II: Application Arguments

CHAPTER XIV

THE CHIEF TYPE of demonstration (considered above in Chapter 13) proves that an attribute belongs universally to its primary[1] subject (I shall call this a USA proof). A USA proof establishes, for example, triangle—2R and not isosceles triangle—2R. But what is the status of isosceles triangle—2R?[2] It is either per se or accidental. If it is accidental it does not belong of necessity. But since every isosceles triangle must have 2R, it is not accidental. Therefore, it is per se. But if per se relations are either unprovable principles or conclusions of proofs, and if proofs establish universal relations, then since it is neither a principle nor a universal relation, it is not per se either.

The paradox is only apparent; there is a proof of isosceles triangle—2R, which consists in applying the universal relation triangle—2R to the specific case of isosceles triangle. The argument proceeds thus: isosceles triangle—2R because isosceles triangles are triangles and triangle—2R.

2R — triangle — isosceles triangle

I shall call such proofs application arguments.[3] Aristotle warns against proving isosceles triangle—2R *directly*, as if isosceles triangle had 2R by virtue of its own specific nature. It does not. It has 2R by virtue of having the generic character of triangle. This is the force of the passages discussed in the previous section.

Applications to Species

Application arguments apply universal conclusions of USA proofs (*A* belongs to *S*) to species, subspecies, or other subdivisions of the subjects of the original proofs, or even to individuals belonging to the species, and so on. They can also apply immediate per se relations as in the argument

having three sides — triangle — isosceles triangle.

Application arguments prove per se, non-universal relations, and they use immediate premises. The relation between genus and species is immediate, and a nonimmediate connection between a genus and a subspecies is mediated by intermediate species, so as to form an immediate nexus. For example, in the argument

animal — bird — pigeon,

supposing that animal is the proximate genus of bird and bird of pigeon, then bird mediates the connection between animal and pigeon, but there is no such mediation between bird and animal or between pigeon and bird. Moreover, since bird will be defined as animal plus certain differentiae, and pigeon as bird plus certain differentiae, the immediateness of these connections is revealed by the definitions. Similar arguments establish the same point for other subclasses of a genus than species. As long as they are specifiable as types falling within the genus, the formulas specifying them can be supposed to indicate their immediate connection with the genus or one of its species.

Application arguments therefore have an honest claim to being demonstrations. Their conclusions are per se relations. Their major premises are per se relations that are either immediate or the conclusions of USA proofs. Their minor premises are per se relations that are immediate and definitional. Their middle terms are the grounds for their conclusions: isosceles triangles have 2R because they are triangles and triangles have 2R. The case is similar when the application requires more than one step, as from figure to triangle to isosceles triangle. Once a science is organized with genus/species trees of its subjects, there is no difficulty in constructing such arguments. This is the point of the following passage:

> **II.14 98a1–12** "With a view toward getting the problems,[a] it is necessary to select *both* the anatomies and the divisions[b] and to do so in the following way. (1) Posit the genus common to all (e.g., if the things under investigation are animals), (2) select what kinds of things[c] belong to every animal. (3) When these have been grasped, next select what kinds of things attend[d] the first of the remaining things,[e] for example, if this is a bird, what kinds of things attend every bird, (4) and always in this way [to select the kinds of things that belong] to each subject in order. For it is clear that we will immediately be able to say why the attributes belong to the subjects that fall under the common genus, for example, why they belong to man or horse. Let *A* stand for animal, *B* the things that attend every animal, *C*, *D*, *E* particular [kinds of] animals. It is clear why *B* be-

longs to *D*, for it does so through *A*. And similarly for the others. And the same account always holds in the case of the remaining subjects."[f]

NOTES ON TRANSLATION

a. "Getting the problems" means identifying the subject-attribute relations the science will demonstrate.

b. In "both the anatomies and the divisions," the "divisions" will be a taxonomic arrangement of subjects into genera and species. Aristotle tells us to "select" subjects that form such an arrangement.[4] Ross (ad loc.) gives reasons for treating "anatomies" as equivalent to "divisions"; Barnes (ad loc.) gives reasons against. If they are not equivalent, what information do the "anatomies" provide for the construction of a science? Mure suggests that "anatomies" refers to a process of identifying attributes—an attractive idea, but a usage without parallel and one that Aristotle would need to explain. (Barnes appears to accept something like Mure's suggestion.)

c. "What kinds of things" means "what kinds of attributes."

d. "What kinds of things attend" means "what attributes belong."

e. "The first of the remaining things" refers to the most general of the remaining subjects in the taxonomy, one of the proximate species of "the genus common to all." Aristotle's language at a5–7 should be modified to make explicit that there is more than one species of each genus.

f. "In the case of the remaining subjects" means literally "in the case of those below"—"below" on the taxonomic chart. The same point is made if for *kátō* (in manuscript n) we read *állōn* with the rest of the manuscripts.

DISCUSSION

The first part of this passage (a1–6) prescribes a method for determining relations that a science will subsequently set out to prove: set out a genus/species chart for the subjects in question (the "divisions"), somehow identify all their attributes (the "anatomies"?), and then match each attribute with the most general subject to all of whose members it belongs.[5]

WHAT IS IMPORTANT for now is the second half of the passage (a7–12), which tells how to apply the results obtained by the method of the first half to the species and subspecies of the subjects to which the attributes have been seen to belong. If we know that attribute *A* belongs to subject *S* by the method of a1–6, a7–12 tells us that if S' is a species of *S*, then *A* belongs to S', and the reason why is that *A* belongs to *S*. Thus, if *S*—*A* is the conclusion of a USA proof, S'—*A* is the conclusion of an application argument with middle term *S*. This point is easy to generalize to cover the subspecies and other subclasses of *S*.[6]

The following passage takes up a question left over from II.14:

II.17 99b7–II.18 99b14 "But[a] if they do not at once reach the indivisible[b] and the middle is not only one, but more, also the grounds are more. But which of the middles is the grounds of the particulars[c]—the one that is first in the direction of the universal, or the one [that is first] in the direction of the particular?[d] Now it is clear that it is what is nearest to each thing for which it is the grounds. For this is the grounds of the first thing's belonging under the universal;[e] for example, *C* is the grounds of *B*'s belonging to *D*. Now *C* is the grounds of *A*'s belonging to *D* and *B* [the grounds of *A*'s belonging] to *C*, and [*A*] itself [the grounds of *A*'s belonging] to this [to *B*]."

NOTES ON TRANSLATION

a. The introductory sentence, "But . . . more," belongs to II.17.

b. "The indivisible" means an immediate relation.

c. "The grounds of the particulars" means the grounds for *A*'s belonging to *D*.

d. "The one . . . the particular" (b9–10) means that *B* is first in the direction of *A*, and *C* is first in the direction of *D*. *A* is "the universal," *D* "the particular."

e. In "of the first thing's belonging under the universal," the "first thing" is *D*, "the universal" is *A*.

DISCUSSION

Aristotle considers the relation *A*—*D* where there are two middle terms *B*, *C*, such that *A*—*B*, *B*—*C*, and *C*—*D* are immediate relations. The argument works both where *D* is an individual and where *D* is the narrowest species in the sequence *A*,*B*,*C*,*D*. In a way *B* and *C* are *both* grounds of *A*'s belonging to *D* ("also the grounds are more"), but which is "the" ground? *C* is uncontroversially "the" ground of *B*—*D* and *B* of *A*—*C*. *A* is called "the" ground of *A*—*B*, which as an immediate connection is sometimes called ungrounded and sometimes self-grounded, and holds solely because of the natures of its subject and attribute as stated in their (fat) definitions. As for *A*—*D*, Aristotle identifies *C* as "the" ground. Consider the application argument

4R—rectilinear figure — triangle — isosceles triangle.

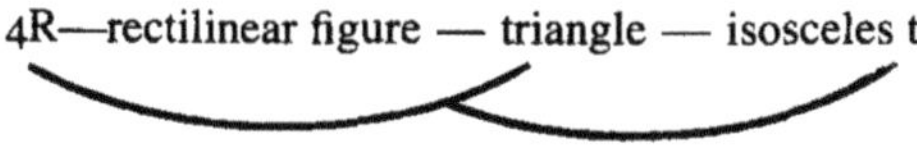

Aristotle says that "the" ground for isosceles triangle—4R is triangle. This may seem an odd choice. We might think it more explanatory to say that isosceles triangles have 4R because they are rectilinear figures. But in Aristotle's view, they have 4R because the proof consists in applying a conclusion of a prior application argument (triangle—4R) to a subspecies of the subject term of the prior proof. This conception is based on the model of the taxonomic chart described in II.14, and requires us to move the minimum possible number of steps up or down.

Whereas USA proofs are of the form

$$\underbrace{S\text{—}A_1\text{—}A_2}$$,

where S is the subject, attribute A_1 belongs per se to S, and attribute A_2 belongs per se to A_1 by being an element in the fat definition of A_1, application arguments are of the form

$$\underbrace{S_2\text{—}S_1\text{—}A}$$,

where attribute A belongs per se to subject S_1, and subject S_2 is related per se to subject S_1.

Applications to Individuals

Application arguments work for individuals too. Universal conclusions of USA proofs can be applied to individual cases to yield scientific knowledge of them. Applications to individuals are importantly different from applications to species. Consider the following two arguments.

A1. All triangles have 2R;
A2. All isosceles triangles are triangles;
A3. Therefore, all isosceles triangles have 2R.

B1. All triangles have 2R;
B2. This is a triangle;
B3. Therefore, this has 2R.

Whereas A1 states a necessary, per se fact about triangles, and A2 and A3 necessary, per se facts about isosceles triangles, neither B2 nor B3 states a necessary, per se fact. There are no necessary per se geometrical facts about this qua an individual. B2 is contingent since any individual triangle can (qua physical) undergo changes, but *if* an individual is a triangle, then qua triangle it must have 2R.[7] The necessary, per se connections of universals carry over to individuals in this derivative way, and so although there is no unqualified scientific knowledge of individuals, there is a *kind* of demonstration and scientific knowledge of individuals nonetheless.

The ambiguous status of this knowledge of individuals is described in the following passage:

> **I.8 75b24–30** "Therefore there is no demonstration or scientific knowledge of perishables without qualification, but as if[a] incidentally, because it does not apply to it universally[b] but at a time and in

a way. Whenever there is [such demonstration and knowledge] one premise must be non-universal and perishable—perishable because the conclusion will hold too when it does, and non-universal because it will apply to one [of the individuals] to which [the major premise applies], but not to another[c]—so it is impossible to deduce [a] universal [conclusion], but [it is possible to deduce] that [the conclusion holds] now."

NOTES ON TRANSLATION

a. *Hoútōs hṓsper* ("as if") is frequently translated "only," for instance by Mure and Barnes, but *hṓsper* means "quasi" (Bonitz 872b48–54). Ross is better (p. 532): "only in a qualified sense."

b. I read *kathólou*, with the manuscripts.

c. I read *tôi . . . tôi*, with Ross.

DISCUSSION

The passage is stated in terms of perishable individuals, but what it says about perishables' not being universal holds for all individuals, not for perishables qua perishable. It admits a derivative, second-rate kind of demonstration and scientific knowledge of individuals, depending on application arguments in which a universal major premise (either a principle or a conclusion of a USA proof) is brought to bear on an individual case through a minor premise whose subject is the invidividual in question. The kind of knowledge such an argument yields is "as if incidental," which is not to say that the fact that an individual triangle has 2R is incidental (if it were, there could be no scientific knowledge of the fact at all), but rather that the necessary scientific fact triangle—2R applies incidentally: not to that *triangle* but to that *individual*. It applies only so far as that individual happens to be a triangle.

ANOTHER FEATURE of application arguments regarding particular subjects is discussed in the following passage:

I.1 71a17–29 "When a person previously knows some things, it is possible [for him] to come to know [something else] at the same time that [he is] getting knowledge of others, that is, all the things that happen to fall under the universal of which[a] he has the knowledge. For he previously knew that every triangle has 2R, but as soon as he recognized[b] that this figure in the semicircle is a triangle, he came to know [that it has 2R]. (For the learning of some things happens in this way and the final subject[c] does not come to be known through the middle[d]—[that is], all the things that happen to be individuals and are not [said] of any subject.)[e] But perhaps it should be said that before he recognizes it [as a triangle] or forms the deduction, in a way he has scientific knowledge,[f] but in another way he does not. For regarding what he did not know without qualification to exist, how did he know without qualification that this has 2R? Clearly, in one way he has scientific knowledge, because he has scientific knowledge universally, but he does not have scientific knowledge without qualification."

NOTES ON TRANSLATION

a. In "of which" I read *hoû* with Ross.

b. I defend "as soon as he recognized" as an interpretation of *háma epagómenos* in McKirahan 1983.

c. In the example, "The final subject" would be the individual triangle.

d. "The middle" refers to the immediate nexus that forms the proof of the universal proposition.

e. "Not [said] of any subject" proves that Aristotle is talking about individuals, not species.

f. Regarding "has scientific knowledge" (*epístasthai*), see note a, on I.2 71b9–12 in Chapter 2 above.

DISCUSSION

This passage treats a partial exception to the rule propounded at I.1 71a1–11, that all intellectual teaching and learning depend on prior knowledge. Here we learn one thing *simultaneously* with learning another when we previously know a third. In the example, we already know *A*: triangle—2R, and then as soon as we recognize that *B*: this figure in the semicircle is a triangle, we know that *C*: this figure in the semicircle has 2R. In such cases we apply *A* (a conclusion holding universally) to one particular falling under it and do not need to prove *C* by the same proof used to prove *A*. Such learning depends on two moves: first (a24), recognizing *B*, and second (a25), forming the deduction

A—*B*—*C*.

Aristotle puzzles about whether a person who knows *A* also knows *C* before he knows that the particular triangle exists, and characteristically concludes that in a way he does and in a way he does not. He does not know *C* without qualification, but since he knows *A*, in some sense he knows *C*. Aristotle says that he knows *C* "universally." Then, when he learns *B* and forms the deduction, he knows *C* without qualification. The passage indicates that there can be unqualified (*haplôs*) knowledge of individuals—a doctrine in apparent conflict with I.8 75b24–25, which only recognizes knowledge of individuals that is "as if incidental"; unqualified scientific knowledge is reserved for universals. I suggest that I.1's unqualified knowledge that this triangle has 2R is the same as I.8's "as if incidental" knowledge that this triangle has 2R—the best kind of scientific knowledge of individuals that can be had. The "universal" knowledge of I.1 is a worse way, an even *more* qualified way of knowing that this triangle has 2R, since we have it without being aware of this triangle.

The function of these arguments is to explain facts about particulars by reference to universal per se connections, which are the proper objects of demonstration. In general, these arguments will be of little interest from the point of view of scientific theory, and little is said about them, but they are the key to applying demonstrative science to the world.

Applications to Natural Phenomena

Application arguments also explain natural phenomena by showing them to be cases of connections proved in full generality. They need not be species. If it is generally true that fire makes a noise when quenched, the phenomenon thunder is explained if we show that it is the noise of fire's being quenched in clouds. Similarly for eclipses: if it is generally true that an illuminated object suffers loss of light when an opaque object is interposed between it and its source of light, and if we then show that an eclipse is just such a loss of light, we then have an explanation of eclipses. (Thunder and eclipses are not species of quenching of fire or of "screening," they are rather kinds of cases.)

This kind of application argument raises interesting issues for Aristotle's theory. The argument regarding thunder starts with what we may take to be the result of a USA proof, "a certain noise is produced when fire is quenched." The other premise must specify a certain range of cases of fire's being quenched. It is hard to see just how the argument would go,[8] but Aristotle clearly intended his theory to cover such cases. The significant feature of the second premise, whether it is "fire is quenched in clouds" or "quenching of fire belongs to quenching of fire in clouds" or "quenching of fire belongs to clouds," is that it does not state a necessary or per se fact about either fire or clouds. It is not part of the nature of fire to be quenched in clouds or of the nature of clouds to have fire quenched in them, nor is all fire quenched there, nor does fire quenching take place in all clouds. Likewise for the first premise. Not all fire is quenched, and so not all fire makes the certain noise. How, then, can such an argument be related to demonstrative science?

I offer the following suggestions for dealing with such cases. The first premise is a per se fact about fire, but one of a type not found in mathematics. As a natural object, fire is by its nature subject to change, and part of natural science will consist in studying what kinds of changes it undergoes and *under what conditions*. If such and such conditions occur, fire will change to earth; if such and such other conditions occur, fire will change to water. This is not to say that all fire becomes earth, much less that it *always* becomes earth or *always* becomes both earth and water. Rather, it is to say that it is a per se fact about fire that *if* such and such conditions occur it becomes earth.[9] This fact can be quantified universally: All fire[10] becomes earth if such and such conditions occur.

More significant for present concerns is the second premise. I suggest that it should be treated like the case of individuals discussed in the previous section. It is a contingent matter whether fire is quenched in clouds, and therefore what happens when fire is quenched in clouds cannot be a necessary, per se fact about fire. Nevertheless, if the first premise is interpreted as stating a necessary, per se fact about fire, then *if* the second holds, the conclusion must hold. *If* fire is quenched in clouds, then the noise must be produced.

This argument therefore takes a necessary, per se fact that holds always, and applies it to a contingently occurring type of case. The conclusion holds whenever and wherever such a case occurs. A distinction useful to our present concerns is introduced at I.8 75b33–35.[11] This passage suggests that thunder can be considered either as a contingent phenomenon that does not occur always but only repeatedly, or as "of a certain kind," as a special case of a universal, necessary, per se connection that does hold "always." According to the former way of thinking, the demonstration of thunder does not hold always, because thunder does not always occur. In that case, we do not have full scientific knowledge of thunder. But thunder is not a merely contingent event either. If it were, it would not be treatable by science in any way. But the point of the passage is that in some way science *can* treat it. The demonstration applies *whenever* thunder occurs; it can do so because thunder is a special case of a universal, necessary, per se fact that holds always. This second way of considering thunder focuses on the relevant universal scientific facts that cover thunder and on the general classifications under which thunder falls. In effect, it considers the first premise in the demonstration, regarding thunder as a specific range of cases of noise produced when fire is quenched. This way of regarding thunder connects it with a universal, necessary, per se connection about fire that holds always. Such knowlege is "partial" in that it is an application of a general truth (the first premise) to a specific range of cases (specified in the second premise).

What scientific knowledge do we have of thunder? There are two possible answers, depending on our focus. In one way, it is the first premise: insofar as thunder is "of a certain kind," noise produced by the quenching of fire, the first premise is the relevant scientific knowledge. But it is possible to know that premise and not even know that thunder exists. Hence the need for the application argument. From this point of view, the relevant knowledge is the conclusion of the argument.

As with application arguments concerning individuals, those concerning types of cases have little theoretical interest for the theory of demonstration. But they do have great practical importance. To judge by Aristotle's examples, many phenomena and facts will be brought within the scope of demonstration and scientific knowledge by being classified as types of cases of universal, necessary, per se truths, so as to be explained by application arguments of the sort just discussed.[12]

Application arguments provide a way to meet the objection that *APo* does not distinguish between knowing why thunder occurs and knowing why thunder is now occurring,[13] the former requiring a general explanation of thunder, the latter an explanation of a particular thunderclap.[14] The preceding discussion indicates that both facts are the subjects of application arguments that have the same major premise but different minor premises. In the general explanation, the minor premise is, perhaps, "fire is quenched in clouds," while in the particular explanation it is, perhaps, "fire is now being quenched in clouds." These different cases fall under the second and third types of application arguments considered, and the passages cited for those types provide the tools to distinguish between the two.

Knowing why thunder occurs can be "always" (qua knowing that noise is produced by the quenching of fire) or "partial" (qua knowing that noise is produced by the quenching of fire in clouds) (cf. I.8 75b33–35); and knowing why thunder is occurring, which depends on a non-universal, perishable minor premise, is not knowledge without qualification, but "as if incidental" knowledge (cf. I.8 75b24–30). It would have been useful for Aristotle to discuss the two cases in relation to each another and not just in relation to full, universal scientific knowledge, and for him to say more about the two types of cases. But it is unfair to charge him with complete failure on this count. Further, however important these two types of cases may be to us, Aristotle's focus in *APo* is understandably elsewhere, and in view of the logically trivial nature of application arguments it is unsurprising that he pays them little attention.

Applications in the Preliminary Stages of Science

Application arguments are also useful in the preliminary work of organizing sciences. Suppose we are following the directions of II.14 98a1–6, translated and discussed earlier in the present chapter. The kind of thought characteristic of application arguments will be useful at the stage of matching each attribute with its most general subject. It is also useful at the stage of determining what subjects are genera and species of what. *APo* I.5 74a33–b4[15] is relevant here. If we notice that a certain attribute (*A*) belongs to some subject (*S*), we may find *A*'s primary subject by abstraction and generalization through successively wider subjects S_1,S_2, and so on. If the abstraction and generalization are done correctly, at each stage (S_i) we will be able to prove that *A* belongs to *S* by an application argument beginning with major premise "*A* belongs to S_i."

Application Arguments and Syllogisms

Although application arguments have little general importance for the theory of demonstrative science, they are specially suitable for syllogistic treatment. In the argument

2R—triangle—isosceles triangle,

though the major premise triangle—2R cannot be proved by syllogistic reasoning,[16] once this result is achieved by whatever means, we can proceed in an application argument along strict syllogistic lines. The case for applying general facts to subclasses that are not species and to individuals is similar. It is presumably this feature that sometimes leads Aristotle to speak as if all demonstrations work like application arguments.

An instructive case is I.22, which contains three arguments that proofs can have only a finite number of steps. Although the difficult and unsatisfactory[17] nature of the arguments makes interpretation uncertain, application arguments are the basis of the chapter. The first and third arguments are based on the principles that definitions by genus and differentia are intelligible, and so contain only finitely many terms. According to the first argument (which extends to 83b31) a demonstration that a subject has an attribute can have only finitely many middle terms because (a) either the attribute is in the definition of the subject, or (b) the attribute is an accident. If so, it is either (1) a per se accident or (2) an accidental accident. Case (2) need not be further considered because there is no scientific knowledge or demonstration of such accidents. If (1), then (i) either it belongs directly to the subject or (ii) it belongs to the subject by virtue of belonging to one of the elements in the subject's definition, in which case there is only a finite number of steps connecting the subject with that element. If this reading of the argument is right, Aristotle overlooks the USA proof that the attribute belongs to the subject or to the element in the subject's definition in virtue of which it belongs to the subject, and concentrates on application arguments, the kind of proofs for which the model of definition by genus and differentia is best suited, and which move through the elements of the subject's definition to the subject.

The third argument (84a7–28) is so firmly based on this view of definition as to maintain that the only provable conclusions are relations between a subject and things appearing in its definition (which here as before includes both its immediate and its higher differentiae and genera) or those of its attributes in whose definition it appears (its per se 2 attributes and their per se 2 attributes, etc.). In both arguments Aristotle's views on the nature of definition allow him to argue concretely for claims that would be powerful results if generally true, but precisely because they are based so narrowly on his view of definition they fail to cover the chief type of demonstration.

The Varieties of Demonstration, III: Demonstrations of Existence

CHAPTER XV

THE TYPES OF demonstration so far considered, the only ones clearly envisaged in *APo* I,[1] prove that attributes belong to subjects. *APo* II opens by identifying four objects of scientific knowledge:

> **II.1 89b23–35** "The objects of investigation are equal in number to the objects of scientific knowledge.[a] We investigate four things: the fact [*tò hóti*], the reason why [*tò dihóti*], if it is [*ei ésti*], and what it is [*tí esti*]. For when we investigate whether it is this or this, putting it into number,[b] for example, whether the sun suffers eclipse or not, we are investigating the fact. . . . When we know the fact, we investigate the reason why; for example, when we know that it suffers eclipse or that the earth moves we investigate the reason why it suffers eclipse or moves. Now these things [we investigate] in this way, but some things we investigate in another way, for example, if [a] centaur or god is or is not. I mean 'if it is or not' without qualification, and not if it is white or not. When we know that it is, we investigate what it is, for example, what is [a] god or what is [a] human?"

NOTES ON TRANSLATION

a. The word I translate as "scientific knowledge" (*epístasthai*) is meant in its strong sense (cf. note a on I.2 71b9–12 in Chapter 2 above). It is significant that this is the only use of the word in this passage (see just below).

b. For suggestions about what the curious phrase "putting it into number" means, see Barnes. Could it mean "quantifying"? Part of this kind of investigation is determining whether the attribute belongs to all, some, or none of the subject. This sort of work will be a part of the procedure described at II.14 98a1–6 (translated and discussed in Chapter 14 above).

DISCUSSION

We may speak of four questions corresponding to the objects of investigation this passage identifies. The first two questions are: "Does attribute *A* belong to subject *S*?" and "Why does *A* belong to *S*?". The second two are: "Does subject *S* exist?" and "What is *S*?" Likewise, we may speak of four kinds of investigations, for example, the investigation that seeks to answer questions of the form "Does *A* belong to *S*?" I shall call these investigations "I1," "I2," "I3," and "I4" respectively.

According to the present passage, I1 and I2 investigate subject-attribute relations: (I1) investigates whether the attribute belongs to the subject, and (I2) investigates why it belongs. Unlike *APo* I, which mainly describes the structure of finished sciences, this passage discusses the research phase of a discipline, when we are determining what the facts are and whether certain facts can be shown to follow from the principles. The situation of the formally presented science is different. There we do not attack the question of fact separately from the question of the reason why. There is no independent confirmation of the fact. The conclusion emerges from the principles, and the proof shows how. The present passage does not say that each of the investigations gives scientific knowledge of its results. A successful I1 investigation leads to discovery that the attribute belongs to the subject. At this point we know that it belongs, but do not have *scientific* knowledge that it belongs. A successful I2 investigation then leads to discovery of why it belongs, how it follows from the principles. At this point we have scientific knowledge of both why it holds—since we have the proof—and that it holds—since we know it as the conclusion of the proof.

The third and fourth questions introduce new material. The examples (god, centaur, human) make it clear that they treat subjects in their own right, not in their relation to attributes. Aristotle attempts to capture this distinction by saying that the third asks whether things are without qualification (*haplôs*), as opposed to the first which asks whether they are something (*ei ésti ti*) (II.2 90a3–5) or whether they are "partially" (*epì mérous*) (89b39, etc.). In addition the third deals with subjects (*hupokeímenon*, 90a12). Further examples are the moon, earth, sun, and triangle (which are all subjects in their sciences, even if not all are substances). The first deals with attributes (*tôn huparchóntōn*, a33). One of the examples is 2R. It is generally agreed that the third question amounts to the question whether there are *x*'s or whether *x*'s exist, and I shall accept this interpretation. The close parallel between the first and second pairs of questions and investigations (their similarity is the theme of II.2) suggests that the following account holds for the second pair. First, we ask (and so, investigate) whether *x* exists, where *x* is a subject in a science's subject genus. When we know that *x* does exist, we ask (and so, investigate) what *x* is. With this knowledge we can construct a demonstration that *x* exists, and this demonstration gives us scientific knowledge of *x*'s existence and of what *x* is.

I believe that Aristotle is committed to this account. But there are several problems of interpretation. P1: how does investigating whether subjects exist fit into the structure of science? P2: how is it possible to investigate whether something exists if we do not even know what it is? How will we know what to look for? How will we recognize it if we find it?[2]

THE SECOND SENTENCE of II.2 raises other problems.

> **II.2 89b37–90a1** "When we investigate the fact or if it is [sc., I1 and I3] we investigate whether there is a middle of it[a] or not; and when we know either the fact or if it is (whether partially or without qualification)[b] and in turn investigate the reason why or what it is [sc., I2 and I4], then we investigate what the middle is."

NOTES ON TRANSLATION

a. "It" refers to the fact that the attribute belongs to the subject or that the subject exists.

b. "Whether partially or without qualification" gives alternate specifications of the question "if it is." APo II.2 90a2–5 makes it clear that the question "if it is" when taken "partially" is the question whether *A* belongs to *S*, and when taken "without qualification" is the question whether *S* exists.

DISCUSSION

This passage brings the following problems. P3: How can investigating I1 be the same as investigating whether there is a middle, since the latter investigation should presuppose that the former, which determines whether the attribute belongs to the subject, is already completed? P4: How can we find that there is a middle without finding what the middle is? P5: How can we look for the middle before knowing what the middle is? P6: Likewise for I3, how can investigating whether something exists be the same as investigating what the middle is? What is the meaning of "middle" in this context? P7: In what sense is the middle the definition or essence ("what it is") of the subject?

The solution to P1 depends on distinguishing between primitive and derivative subjects. Primitive subjects cannot be proved to exist any more than immediate relations can be proved to hold.[3] But if the subject matter of a science is such that some subjects can be proved to exist given the existence of others, this distinction applies. Derivative subjects can be proved to exist, just as nonimmediate connections of subject and attribute can be proved to hold, and in both cases we need to investigate what the proofs are. The context of investigating I3 (like that of investigating I1) is the context of scientific enquiry. The point of I3 is to determine whether something exists qua subject of the given science, and this amounts to determining whether it can be proved to exist.

The key to P3 is understanding when I1 and I2 are appropriate. The "reason why" (90a1) is a scientific explanation via a middle: a demonstration through an appropriate nexus of immediate relations. In investigating the reason why, we are looking for a proof of a fact we already know. The context of I1 is the special context of scientific inquiry. If asking whether attribute *B* belongs to subject *A* is tantamount to asking whether there is a scientific explanation of *A—B*, then I1 does not simply investigate whether *B* belongs to *A*, but whether *B* belongs to *A* *and* whether *A—B* can be proved from the principles of the science in question. In effect, I1 investigates whether the question "Does *B* belong to *A*?" is a scientific question.[4]

P6 yields to this same approach. I3 does not investigate simply whether something exists, but in addition whether its existence can be proved from the principles of the science. And as with I1, for there to be a proof is for there to be a middle.

P4 and P5 present methodological riddles. How can we find that a middle exists without discovering what it is, as I1 and I3 require us to do if I2 and I4 are separate and subsequent investigations? And how can we look for something without knowing what it is we are looking for, as I2 and I4 require us to do? How, that is, can I1 and I2 be kept distinct from each other, and likewise I3 and I4?

The answer I offer is that the confusion arises not in practice but in hindsight. In practice, when we investigate the fact (I1), we try to find out not only whether the relation holds but also whether it is a *likely candidate* for a conclusion of a proof in the given science. If we determined that it *is* a conclusion of a proof, it is hard to see how we could fail to be in a position right then to identify the middle,[5] but if we only determine that it is a likely candidate for a conclusion of a proof, there is room for I2 to take hold. We still need to discover what the middle is, and until we discover the middle and so form the demonstration, it is early to say that we have proof that the relation holds. The case of I3 and I4 is similar. In investigating whether *x* exists, we try to find out not only whether *x* exists but also whether it is likely that *x*'s existence follows from that of the primitive subjects of the science. At that point I4 becomes pressing: what *is* the middle that proves that *x* exists?

The solution to P4 and P5 therefore consists in softening the description of I1 and I3. In pursuing those inquiries, we are not yet determining that there is a middle, but only determining that it is likely that there is one. Then, in investigating I2 and I4 we discover how to prove the desired result, thus determining what the middle is. In retrospect, this natural progression of thought appears different. Even if I1 and I3 identify likely candidates for conclusions of proofs, the only ones of interest are those for which proofs are actually found. Once a proof is found it is natural enough to say that I1 and I3 located not just a likely candidate but a *successful* one. That is, they identified facts that *do* have middles, and then I2 and I4 proceeded to find what the middles are. It is natural to focus on successful cases, and easy for us to see how this perspective leads Aristotle to speak in riddles.

Solving P2 and P7 evidently depends on giving a reasonable sense to the claim that the middle of a demonstration that *x* exists is what *x* is. For if what *x* is is the middle in the proof that *x* exists, P2 amounts to asking how we can investigate I3 without knowing what the middle is, and this just means how can

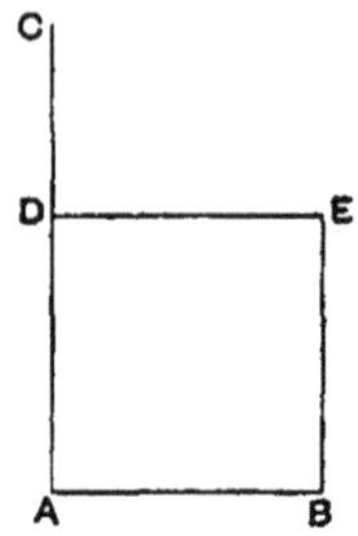

we investigate I3 before pursuing I4, a question already answered (cf. P6, P4, P5).

I WILL APPROACH these questions by examining the structure of a Euclidean construction proof. Recall that constructibility corresponds to existence and that a proof that something can be constructed, which is based on the construction postulates (P1–P3 of *Elements* I) corresponds to a proof that it exists, which is based on the existence of the primitive subjects points, straight lines, and circles.[6] Consider *Elements* I, proposition 46.

D1. On a given line to describe a square.
D2. Let *AB* be the given straight line;
D3. thus it is required to describe a square on the straight line *AB*.
D4. Let *AC* be drawn at right angles to the straight line *AB* from the point *A* on it, [I.11] and let *AD* be made equal to *AB*; through the point *D* let *DE* be drawn parallel to *AB*, and through the point *B* let *BE* be drawn parallel to *AD*. [I.31]
D5.1. Therefore *ADEB* is a parallelogram;
D5.2. therefore *AB* is equal to *DE*, and *AD* to *BE*. [I.34]
D5.3. But *AB* is equal to *AD*; therefore the four straight lines *BA*,*AD*,*DE*,*EB* are equal to one another; [CN 1]
D5.4. therefore the parallelogram *ADEB* is equilateral. I say next that it is also right-angled.
D5.5. For, since the straight line *AD* falls upon the parallels *AB*,*DE*, the angles *BAD*,*ADE* are equal to two right angles. [I.29]
D5.6. But the angle *BAD* is right; therefore the angle *ADE* is also right. [CN 3]
D5.7. And in parallelogrammic areas the opposite sides and angles are equal to one another; [I.34] therefore each of the opposite angles *ABE*,*BED* is also right.
D5.8. Therefore *ADEB* is right-angled. And it was also proved equilateral.
D6. Therefore it is a square; and it is described on the straight line *AB*. Q.E.F.[7]

The proof proceeds in four stages.

S1. It constructs figure *ADEB*. This is justified through the postulates and previously proven construction theorems.
S2. It shows that *ADEB* is a rectangular equilateral quadrilateral (REQ).
S3. It observes that since these are the essential attributes of squares, the figure *ADEB* is a square.
S4. It concludes that the desired square has been constructed.

The proof can be recast into an Aristotelian demonstration[8] that squares of any given dimension exist.

S1. Given that the line *AB* exists.[9] (A) The perpendicular *AC* to *AB* at the point *A* exists (*Elements* I, proposition 11). There exists a segment *AD* of *AC* that is equal to *AB*. Euclid does not justify this step. The point is determined by the intersection of *AC* with the circle of radius *AB* and center *A*, whose existence is guaranteed by P3. The existence of *D* can be proved by the axiom of continuity, which Euclid uses but does not state.[10] (B) The existence of the parallel to *AB* through point *D* is guaranteed by *Elements* I, proposition 31, as is the existence of the parallel to *AD* through point *B*. Euclid does not prove that the lines parallel to *AB* and *AD* intersect, although it follows from P5. Thus, if we may take the previous theorems of *Elements* I as having been proved by Aristotelian demonstrations, it follows that there exists a closed rectilinear figure *ADEB* with the following attributes: *AD* is perpendicular to *AB*, *AD* equals *AB*, *DE* is parallel to *AB*, and *BE* is parallel to *AD*.
S2. It follows from these attributes of *ADEB*, via regular subject-attribute proofs (either USA proofs or application arguments), that *ADEB* is a REQ; that is, being rectangular is a per se attribute of *ADEB*, as are being equilateral and being a quadrilateral.
S3. It follows from the fact that *ADEB* has these attributes that *ADEB* is a square.
S4. Since *ADEB* exists and *ADEB* is a square of the given dimension, it follows that squares of any given dimension exist.

In S1 we ascertain that *ADEB* exists. In fact, the procedure for ascertaining this fact gives us scientific knowledge that *ADEB* exists. (Since *ADEB* proves to be a square, we are in fact ascertaining that a square exists, but at this point it is proleptic to say that we have scientific knowledge that it is a square.) Moreover, the way *ADEB*'s existence is guaranteed (the method of its construction) makes

ADEB a likely candidate for being a square. S2 is the proof that it is a square, the nexus of immediate relations that shows that it is a square. And the attributes the proof shows *ADEB* to have are those that constitute the definition of square. In fact, the claim that the definition of square is the (or anyway, a) middle term in the proof is supported by the following sketch of the demonstration:

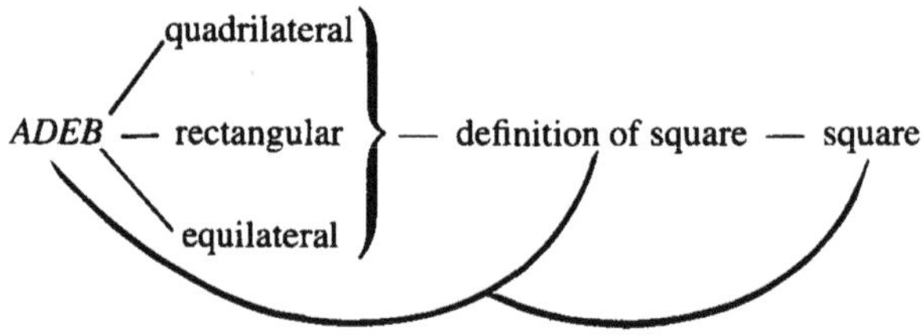

ADEB is a square because the definition of square is predicated of it, and the definition of square is predicated of it because it is an REQ. (There are further middle terms in the proofs of its being rectangular, equilateral, and a quadrilateral.) This is the solution to P7.

The proof does not prove that squares are REQ's. That is known already, an indemonstrable definition, one of the principles of geometry. It would be nonsense to set out to prove this.[11] The object of proof is to show that certain attributes belong to something known to exist. That something turns out to be a square, since the attributes are the essential attributes of squares. But the crucial part of the proof does not show the tautology that the attributes in the definition of square belong to squares, but that the attributes in the definition of square qua attributes (not qua definition) belong to something known to exist (and not known to exist qua a square).

Of course, Aristotle does not say that such proofs prove the definition, only that the definition is the middle in the proof of existence. But then, if we already know the definition, how can we look for it at all? Aristotle *does* say that the definition ("what it is") is the object of inquiry I4. And if we need to know what it is in order to discover the proof that it exists, we are in violation of Aristotle's demand that I3 precedes I4. The solution to these difficulties lies in the distinction between "what it signifies" (*tí sēmaínei*) and "what it is" (*tí esti*) (II.7 92b4–8).[12] The definition of a derivative subject is an indemonstrable principle, to be sure, but it tells "what it signifies" (where "it" is a name or expression), not "what it is" (where "it" is now the actual thing). We begin by knowing the significance of the word "square," but not knowing if any squares exist. When squares are proved to exist, the significance of the word becomes the essence of the thing: a square *is* an REQ.

The proof that *ADEB* is a square can be understood as showing what *ADEB is*, establishing what its essence is: its essence is the essence of square—it *is* a square. Again, this is a proleptic description of the case. In proving that *ADEB* is rectangular, equilateral, and quadrilateral, we are proving per se attributes of it, just as if we were proving that the line *AE* bisects it. Of course the strategy of the proof determines which attributes we will pay attention to, but only in the end, when *ADEB* has been shown to be a square, does it become reasonable to say that we were investigating the "what it is" of *ADEB*. We begin with the definition ("what it signifies," not "what it is") of square, and in constructing a proof that squares exist we investigate the definition ("what it is") of square. But what we ask is whether the existing figure *ADEB* has the attributes contained in the "what it ['square'] signifies," and if it does, then the "what it signifies" proves to be "what it is."[13] This is the solution to P2.

The general form of an existence proof is as follows:

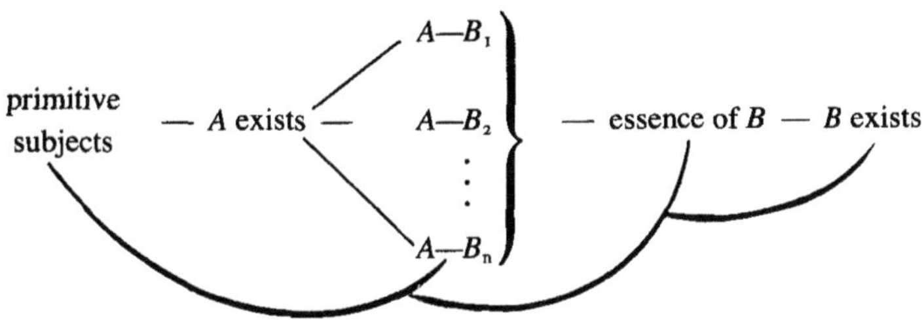

S1. The existence of primitive subjects is shown to entail the existence of derivative subject A.
S2. A is shown to have per se attributes $B_1, B_2, \ldots B_n$.
S3. $B_1, B_2, \ldots B_n$ are the essence of derivative subject B.
S4. It follows that B exists.

A few points remain to be examined. First, how well do existence proofs satisfy the general account of demonstrations? They have as much right as other demonstrations to be called scientific deductions (I.2 71b17–18) and to entitle the person who knows them to a claim to scientific knowledge of their conclusions. The kinds of principles they use will be the same kinds as those used in subject-attribute demonstrations: primitive existence claims (since the purpose of these proofs is to show that the primitive existence claims entail derivative existence claims), definitions (since their strategy is to show that something that exists satisfies the definition of the derivative subject),[14] and common axioms (since the inferences in these proofs may be justified by this kind of principle just as much as those in subject-attribute proofs).[15]

APo II.2 makes it clear that existence proofs give the explanation or grounds of their conclusions (90a6–11, a14–15). The other chief requirement of demonstrations is that their conclusions are necessary, per se propositions. The application of this view to existence proofs can be made, but not wholly straightforwardly, since existence is not a per se attribute, and Aristotle does not require the subjects of sciences to exist of necessity.[16] On the other hand, given that the primitive subjects exist, the proof shows that the derivative subject must exist. Also, the proof proceeds by showing that an existing subject has certain per se attributes. Step S2 is no different from one or more subject-attribute proofs, whose conclusions have just as much a claim to be necessary and per se as do the conclusions of subject-attribute proofs. Further, if it is per se to A that it has attributes $B_1, B_2, \ldots B_n$, and these are the essence of B, it is per se to A that the essence of B belongs to it. Thus, although the conclusion of an existence proof is not a per se and necessary proposition, the penultimate proposition and the propositions derived before that are per se and necessary.

Moreover, there is nothing frail about the conclusions of existence demonstrations as compared with those of subject-attribute proofs. Existence proofs depend on the (nonnecessary) existence of primitive subjects, but subject-attribute proofs also depend on the (nonnecessary) existence of the subject in question. Existence claims are likely to play a more prominent role in existence proofs than in the other kind,[17] but even if they do not appear in the other kind explicitly, they are presupposed. And subject-attribute proofs of per se relations *require* the subject to exist. For per se attributes are either parts of the "what it is" or are implied in the "what it is," and for a thing to have a "what it is" (as opposed to "what it signifies") it must exist.

A further similarity between the two types of proofs comes from the fact that subject-attribute proofs can be read as existence proofs of the attributes, since for an attribute to exist is for it to belong to a subject.[18] Thus Aristotle says that we assume the definitions of attributes and prove that attributes exist (I.10 76b6–10, cf. a32–36). If we look at subject-attribute proofs in this way, the strategy of both types of proofs is seen to be the same. Euclid proves that lines are parallel by proving that if produced indefinitely in both directions they do not intersect (e.g., *Elements* I, proposition 27). Since parallel lines are defined as lines that do not intersect,[19] his strategy is to show that the lines in question satisfy the definition of parallel lines. Thus, just as an existence proof may be sketched as follows:

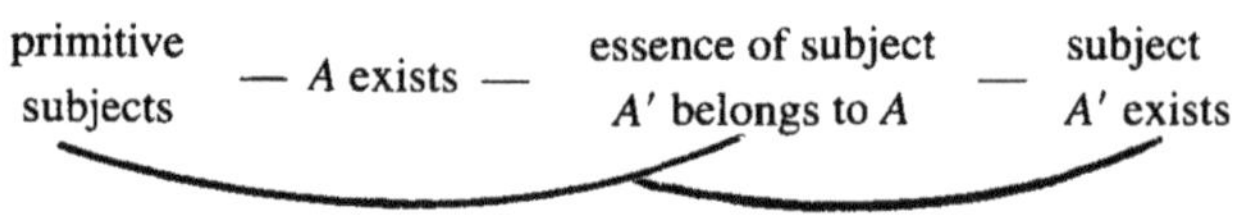

so a subject-attribute proof may be sketched as follows:

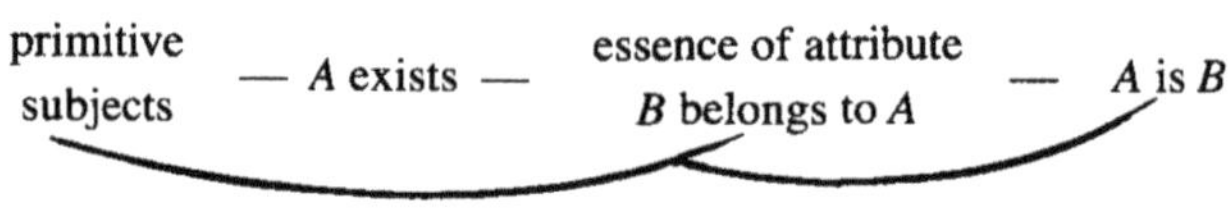

In the same way that the "what it is" of subject A' can be called the middle of the demonstration that A' exists, the "what it is" of attribute B can be called the middle of the demonstration that A is B.[20] This is half the truth. On the other hand, the proofs also depend on the "what it is" of subject A. We have seen how the essence of the subject as well as that of the attribute is used in subject-attribute proofs.[21] What would count by analogy as the essence of A is used in the same way in the proof that A' exists. For the essence of A (qua A) is what is known about it from the way it follows from the primitive subjects. In *Elements* I, proposition 46, what is known about figure *ADEB* is that *AD* is perpendicular and equal to *AB*, *AD* is parallel to *EB*, and *AB* is parallel to *DE*. These facts entail that *ADEB* is an REQ and therefore a square.

The double role of essence in demonstrations is preserved. In subject-attribute demonstrations, the essences of both subject and attribute are crucial to the proof, and both have an equal right to be called middle terms. *A* is *B* not only because *A* is the sort of subject it is and not only because *B* is the sort of attribute it is, but for both reasons. Similarly in existence proofs, A' exists both because A' is the kind of subject it is and because something (*A*) whose existence follows from that of the primitive subjects is the kind of subject *it* is.

In this chapter I have shown a surprisingly close fit between existence proofs and subject-attribute demonstrations. General assertions about the nature and structure of proofs, although made with subject-attribute proofs in mind, apply with only minor alterations to existence proofs, which thereby occupy a rightful place in demonstrative sciences.

Demonstration of Essence: Another Form of Scientific Proof?

CHAPTER XVI

AFTER TREATING subject-attribute proofs, existence proofs, and their relations (II.1–2), Aristotle plunges into a lengthy aporetic discussion mainly on the question whether essence can be demonstrated (II.3–II.8 93a15). This is a surprising question to ask at all, since a cornerstone of the theory of demonstrative science is that essential attributes are indemonstrable, and proofs take as their starting point the essence of their subject (and attribute). *APo* II.3 presents these arguments and others and ends with the wholly negative result we would have anticipated.

So far so good. The initial question is not expected because the answer is so obvious. But at least II.3 gives the obvious answer, and discusses the issue systematically and thoroughly, bringing up a variety of considerations stemming from the theory of demonstration, all of which point to the same negative answer.

We can even see a reason for raising the question. *APo* II.1–2 in effect introduces a kind of demonstration, the existence proof, at best dimly anticipated in *APo* I. Aristotle is seeing how widely the concept of demonstration can be applied. If *APo* I did not (clearly, anyway) see that existence can be proved, maybe it also failed to see that essence can be. The topic is worth discussing.

Even more surprisingly, II.4 and the following chapters resurrect it and pursue it vigorously. *APo* II.8 takes up again the questions what is definition and whether there are demonstration and definition of essence in any way or not at all (93a1–3), and concludes by summarizing the results it has achieved:

> **II.8 93b15–20** "Now it has been said how the essence is grasped and comes to be known, and so there proves to be no deduction or demonstration of the essence but [it comes to be] clear through deduction and demonstration. And so, it is not possible to know the

essence without demonstration in cases where the explanatory grounds is other, nor is there demonstration of it, as we stated also in [our treatment of] the puzzles."

What *has* the chapter achieved? And how well does this description fit? Its principal contribution is a method for investigating the reason why (*dihóti*) an attribute *C* belongs to a subject *A*. Following II.2, it says that this work is a hunt for the middle, which is the grounds of *A*—*C* (93a31–36). We are to find candidates and pick the best one, the one that best explains why *C* belongs to *A* (b3–7). The best explanation will be a term *B* immediately related to *C* (a35–b3). This kind of explanation is best because immediate relations are "uncaused causes" and hold in virtue of themselves. They have no further explanation. On the other hand, if there is a further explanation why *B* belongs to *A*, we must turn to the other candidates and pick the one that best explains *this*, and so on until we have a nexus of immediate relations reaching from *A* to *C* (b12–14).

Two things are clear. First, we are dealing not with finished demonstrative sciences but with techniques for discovering and organizing facts. Second, there is no new type of proof. We are in effect discovering subject-attribute proofs. (An analogous story can be told for existence proofs.) And one thing is not clear: what does this discussion have to do with essences? After all, subjects have essences, and attributes have essences in a secondary kind of way, but here we are dealing with *relations between* subjects and attributes, and nothing has prepared us for the idea that *they* have essences. In fact, II.1–2 stresses to the point of tedium that subjects differ from subject-attribute relations in that subjects have essences and what corresponds to essences in the case of relations is "the reason why."

The two main examples in II.8 (eclipse and thunder) are subject-attribute relations. Both are unusual in that they have names and in that the attributes belong to only one kind of subject: thunder is a noise that occurs only in clouds, eclipse (as Aristotle conceives it here) is a loss of light that happens only to the moon. Since they have names, there seems nothing wrong in asking "what is" thunder or eclipse. There are two kinds of answers: a preliminary answer by which we can identify an occurrence of the relation (thunder is a certain noise in clouds, where we specify what that noise is; thunder is not just any noise in clouds), and a deeper answer that accounts for the relation (thunder is the noise in clouds produced by quenching of fire). The preliminary answer turns out to be simply a statement of the relation that "thunder" names. If "*N*" is the name of the relation *A*—*B* then the preliminary answer to the question "What is *N*?" is simply "the relation *A*—*B*."[1] The additional element in the second answer states the grounds of thunder's occurring and can be represented as the middle in a demonstration that thunder exists (that the noise in question occurs in clouds).

Aristotle says that the first answer contains "something of the thing itself" (93a22) or "something of the essence" (a29). The extra element in the deeper answer is the account of eclipses or thunder (93a30–33, b6–7, b12), and is what they are:

> **II.8 93b6–7** "This [sc., the correctly chosen middle] is the account of the other extreme, *A* [sc., eclipse],[2] in the present examples. For eclipse *is* screening by the earth."

This suggests that "loss of light from the moon" is not part of the essence of eclipse, that is, that a statement of the essence of eclipse does not mention that eclipse is loss of light from the moon. I shall call the preliminary account ("loss of light from the moon") the "superficial essence" and the distinctive part of the second answer ("screening by the earth") the "deep essence." That eclipse occurs (exists) follows from the fact that the moon suffers that kind of loss of light, and we prove that it suffers that loss of light from the fact that screening of the earth occurs. The superficial essence follows from the deep essence, and the existence follows since the superficial essence holds.

We can now assess Aristotle's statement of what II.8 has achieved. The deductions and demonstrations referred to are the arguments that show that (and why) the superficial essence holds, which is equivalent to proving that the thing whose essence we are investigating exists. The deep essence is what 93b15–20 (reasonably) calls "essence."

Aristotle has sketched a method for grasping the deep essence, and the deep essence is not the conclusion of a deduction or proof, but comes to be clear through deduction and demonstration since we discover it by constructing a proof with it as a middle term immediately related to the attribute of the superficial essence. The procedure works only when the middle is "other," different from the attribute of the superficial essence.[3] On this reading, the chapter does what Aristotle says.

There are still some problems. In what way are these essences of subject-attribute relations? For this is what eclipse and thunder are. How is the deep essence an essence? And how is the superficial essence? These are general problems for the theory. There are also specific ones for the way Aristotle presents it. In the same breath he calls eclipse "a certain loss of light" (93a23) (an attribute) and thunder "a certain noise of clouds" (a22–23) (a subject-attribute relation). And just after calling eclipse "screening by the earth" (b7) (the middle), he calls thunder "quenching of fire in cloud" (b8) (a middle belonging to a subject). In the specimen argument for thunder, he calls the major term first "thunder" and then "noise" (b9, b11). Also, in the specimen argument for eclipse it is necessary to read "loss of light" for "eclipse" as the major term in order to make sense of the argument, which suggests that for thunder "noise" should be the major term, but it is further evidence that Aristotle is more careless with his examples than makes for comfortable reading.

Before tackling these questions, it will be useful to discuss II.9 and 10, which give some much-needed perspective. The brief chapter II.9 asserts that there are things that are their own grounds (93b21). *APo* II.8 deals mainly with demonstrable relations, and II.9 asserts that there are also indemonstrable ones.

The essences of these latter things are immediate and are principles (b22). We may wonder[4] how this fact distinguishes the essences of these things, since all essences are immediate. The answer is that *not* all essences, as Aristotle has recently been using the word, are immediate. In II.8 it was necessary to distinguish the deep from the superficial essence, where the latter proved to be a nonimmediate relation. Aristotle's point in II.9 is that there are cases where the distinction does not apply. The initial grasp of the relation cannot be improved. There is no middle to account for it. These relations are immediate, and are the principles on which the deep essences of the others are based. If *A*—*C* is the superficial essence of *N* and we find that *N*'s deep essence is "*A* is *C* because of *B*," we can rearrange this into a syllogism:

A—*B*—*C*.

If we then consider *A*—*B* and *B*—*C* in the same way, either they have a deep essence or they do not. If they do not, they are immediate and function as principles for *N*. If one or both have a deep essence, we go on to treat each of the relations used in accounting for them, and continue in this way until we reach immediate principles. *APo* II.9 assures us we will reach them:

> **II.9 93b23–24** "In these cases (since it is impossible to find further grounds to explain that they exist or what they are) we must posit these facts or make them clear in some other way."

The meaning is partly determined by II.1–2, which tells us to investigate whether subjects exist, and when we know they do, to investigate what they are, which amounts to determining the middle that accounts for their existence. Primitive subjects have no middle, and we must just posit (or make clear in some other way, but not by proof) that they exist. The claim that we do the same for "what they are" is explained by the distinction between two kinds of essences. If our initial conception (corresponding to the superficial essence) cannot be further grounded, it cannot be accounted for via demonstration and must be accepted as immediate. The distinction between superficial and deep essences breaks down, and in a way they can be said to coincide.[5] As in II.8, here the doctrine applies somewhat differently for subjects and for subject-attribute relations. For the latter Aristotle is saying that there are self-explanatory relations. These are immediate and principles. For the former, there are subjects whose existence cannot be proved; we posit it, since we cannot prove that their definition applies to something whose existence is grounded in that of more primitive subjects.[6]

APo II.10, which brings together the results of II.8 and II.9, is Aristotle's final word on these issues. The chapter comprises four parts: first (93b29–37) an account of a kind of definition not previously discussed at length; second (93b38–94a10) a list apparently of four kinds of definitions, including a summary of the results of II.8; third (94a11–13) a list of three kinds of definitions in abbreviated form; and finally (a14–19) a summary of the problems dealt with in II.3–10.

The types of definitions identified in the second section are:

> **II.10 93b38–94a10** "(2a) One definition of definition is the aforementioned (at 93b29–37). (2b) Another definition is the account revealing why it exists. And so the former (2a) signifies but does not show, while the latter (2b) clearly will be a sort of demonstration of what it is, differing in order from the proof. . . . Also (2c) . . . is the conclusion of the demonstration of what it is. (2d) The definition of immediates is an indemonstrable positing of what it is."

The third section states:

> **II.10 94a11–13** "Therefore, (3a) one definition is an indemonstrable statement of what it is, (3b) one is a deduction of what it is, differing in arrangement from a demonstration, and a third (3c) is a conclusion of the demonstration of what it is."

Of these three kinds of definitions, (3a) is identical with (2d), (3b) with (2b), and (3c) with (2c). But what of (2a)? We are told that "it signifies but does not show" (b39–40) and that it has been discussed (b38). It is discussed in the first part of II.10, which calls it "an account of what the name or other namelike expression signifies" (93b30–31). "Eclipse" is a name, and "other namelike expression" covers cases where there is no convenient name for a subject[7] or attribute, or subject-attribute relation, so that the thing being investigated can be described by a noun phrase, for example, "the fact that triangles have 2R" or "the possession of 2R by triangles." The first part of II.10 also says that once we know the definiendum exists we investigate why it exists (93b32), and adds, with a reference to II.8 93a19ff.:

> **II.10 93b32–35** "it is difficult to grasp in this way things we do not know to exist. The reason for the difficulty has been stated previously: we do not even know whether they exist or not except incidentally."

It is clear that (2a) is the description we have on setting out to investigate whether a thing (subject or subject-attribute relation) exists. After learning that it exists, we use (2a) as the superficial essence as we attempt to find a middle (a deep essence) to account for the thing's existence.

Definition (2a) is not the same as definition (3b), the deep essence (though there is no distinction between (2a) and (3b) for immediate relations and primitive subjects). Nor can it be the same as (3a), which has to do with the immediate essences discussed in II.9. But what about (3c)? "Noise in clouds" is a definition of thunder that is a "conclusion of the demonstration of what it is." This definition is contrasted with the "rearranged demonstration" kind of definition, "thunder is noise in clouds produced by the quenching of fire." If "quenching of fire" is the middle in the demonstration of the fact that the noise occurs in clouds, when it is removed, what remains is like the conclusion of the proof in that it contains the extreme terms but not the middle. Definition (3c) proves to be (2a) in a new guise. If (2a) is an account that allows us to learn in a nonincidental way[8] that the thing exists, then in knowing (2a) we know "something of the thing itself." This gives us a basis (the superficial essence) for investigating the thing's explanatory grounds. If we find the grounds, we will be able to construct a (3b) type definition. The difference in content between (2a) and (3b) will be the middle. But since (3c) also differs from (3b) in omitting the middle, (2a) and (3c) are identical. The difference between them lies in the stage of inquiry at which they are found: (2a) before we know that the definiendum exists and (3c) after. Definition (2a) becomes (3c) when the account expressing "what it signifies" is promoted to the status of "what it is" by the discovery that the definiendum exists. Thus, strictly speaking, the basis for investigating whether it exists is (2a), and (3c) is the basis for investigating why it exists (i.e., what it is). Definition (2a) is known before (3b) is known, but is not known *as* the conclusion of the demonstration until after (3b) is known. This leaves us with three forms of definitions: (2a = 2c = 3c) "thunder is a certain noise in clouds," (2b = 3b) "thunder is the noise due to the quenching of fire in clouds," and (2d = 3a) the immediate definitions. The first kind mentions subject and attribute; the second mentions subject, attribute, and middle. No definition mentions only the middle or only the attribute.

This relatively clear picture of the kinds of definitions and their forms enables us now to solve the remaining difficulties in II.8 (see above, p. 200). To begin with, II.10 indicates how to regiment the variations in the definitions of eclipse and of thunder. According to II.10, the superficial essence of eclipse is "a certain loss of light from the moon" (not "a certain loss of light," 93a23), and the superficial essence of thunder is "a certain noise in clouds" (as at 93a22–23). The (3b) type definition of eclipse is "a certain loss of light from the moon due to screening of the earth" (as at II.2 90a15). The subject is "moon," the attribute is "a certain loss of light" (*not* "eclipse," as at 93a30, a37), and the middle is "screening of the earth." The (3b) type definition of thunder is "a certain noise in clouds due to quenching of fire." The subject is "clouds," the attribute is "noise" (as at 93b11; *not* "thunder" as at b9) and the middle is "quenching of fire." There is no proper definition of eclipse that states simply that it is screening of the earth (as at b7) and no proper definition of thunder that states simply that it is quenching of fire in clouds (as at b8).

If we take the examples of II.8 so as to conform to the views on definition given in II.10, they make excellent sense in their context. Substituting "a certain loss of light from the moon" for "a certain loss of light" at 93a23 makes for no difficulties. In fact, the specimen argument (a30ff.) makes better sense if *A* stands for "a certain loss of light" instead of "eclipse." The same is true for the specimen argument regarding thunder. At b9, *A* should stand for "noise."

The only things yet to reconcile with II.10 are the explicit statements that eclipse is screening of the earth and that thunder is quenching of fire in clouds, where the definiendum is identified with the middle term alone. I suggest that Aristotle is speaking a little carelessly, and he means to say that *the accounts of* eclipse and thunder are "screening of the earth" and "quenching of fire in clouds," where "account" means not definition, but grounds. Defense of this interpretation begins in the discussion of the sample argument dealing with eclipse:

> **II.8 93a31–33** "Now to investigate whether it suffers eclipse or not is to investigate *B*—does it exist[a] or not. And this is no different from investigating whether an account of it exists,[b] and if this exists [holds] we say that that [sc., eclipse] exists [occurs] too."

NOTES ON TRANSLATION

a. "Exist": that is, occur. Asking if eclipses occur is tantamount to asking whether *B* occurs (whether the earth screens the sun's light).

b. "Exists": that is, occurs, holds. If asking if an account of eclipse exists is no different from asking if *B* occurs or holds, and if *B* is the account, then "exists" means "occurs" or "holds."

Investigating whether an account of eclipse holds is no different from investigating whether *B* occurs, where *B* is the middle in the proof that eclipse occurs, that is, the grounds of eclipse.[9] Thus "account" too should mean this, rather than "definition."[10] And the next sentence in the text is given a natural interpretation:

> **II.8 93a33–35** "Or, of which member of the pair of contradictory propositions is the account—of having 2R or of not having 2R?"

This means that some accounts explain why an attribute belongs to a subject and others explain why an attribute does *not* belong to a subject.[11]

APo II.8 93b3ff. returns to the topic of how we come to know what it is once we know that it is. If we know that *A* belongs to *C*,

> **II.8 93b4–7** "still, why it belongs is a matter of investigating what *B* is. . . . This is the account of the other extreme—*A* in the present examples."

As before, *B* is the account, the middle or grounds, of *A*. Only at this point do problems arise:

> **II.8 93b7–8** ". . . for eclipse *is* screening of the earth. What is thunder? Quenching of fire in cloud."

We do not want a definition of eclipse here, only an answer to the question why does eclipse occur (b4), which we have been told will be the account of term *A*, the label that Aristotle has given to eclipse. What is expected and wanted is a statement of the account of eclipse, which is *B*, the middle term. And we know from 93a31–33 that the account is "screening of the earth," and this is precisely what eclipse is here said to be! Aristotle has said what he should have said: what the account of eclipse is. It is better to understand him as doing this than as giving a new definition of eclipse that proves to be a definition of a kind otherwise unknown and not found in the thorough summary of II.10. Eclipse is not screening by the earth, but the loss of light from the moon due to screening of the earth. Screening of the earth is the account (not the definition) of eclipse. And the same is true for thunder. The account of thunder is "quenching of fire" (*B*) (b12), and this is what thunder is said to be at b8.[12] But what comes immediately after shows this apparent definition of thunder to be the account:

> **II.8 93b8–9** "Why does it thunder? On account of the fact that fire is quenched in the cloud."

Aristotle's interest here is to identify *why* it thunders, not what thunder is, although the account of thunder becomes part of the (3b) type definition of thunder: noise due to the quenching of fire in clouds.

The examination of 93a31–35 and b3–12 has revealed no good reason to recognize a kind of definition that consists of the account or explanatory grounds of the attribute. *APo* II.8 can be read in a way consistent with and supportive of II.10, and doing so makes not only the examples but also the theory considerably more clear. It amounts to the following.

In investigating the nature of *x*, we begin by knowing what the name "*x*" signifies (a (2a) type of definition). If this contains "something of the thing itself" (93a22), we can determine whether *x*'s exist (occur). In the examples of eclipse and thunder, subject-attribute relations, the initial accounts identify the relations. Once we know that *x* exists, we investigate what *x* is, which is the same as investigating why *x* exists. The reason why (*lógos*) is the middle term of a demonstration of the fact that *x* exists. We proceed by locating candidates (possible explanations of *x*'s existence) and then determining which is the actual explanation. If there is no further explanation for why the candidate we choose holds, our investigation is complete. If there is a further explanation, we must find it in the same way. If the original list of candidates contains all possible explanations, the explanation why the chosen one holds will be one of the remaining candidates. The demonstration can be rearranged so as to form a (3b)-type definition of *x*, and the middle term of the demonstration can be removed from this definition, leaving the (3c)-type definition.

At this point we may answer the questions raised above: in what sense are the superficial essence and the deep essence essences? And in what sense do subject-attribute relations have essences at all? The answer to the first two questions is the same: since a definition is a statement of essence (II.10 93b29) and is of the essence (II.3 90b3–4, b16),[13] and these two "essences" are stated in definitions, it follows that they are essences. The superficial essence is stated in a (3c)-type definition, the deep essence in a (3b)-type definition. In proving that a thing whose essence is in question exists, we proceed by proving that its superficial essence holds. Thus the superficial essence is the conclusion of the demonstration (recast into appropriate form: "eclipse is loss of light from the moon" rather than "the moon suffers loss of light"), and the deep essence is the recast definition (including the explanatory middle or account). This account makes good sense of Aristotle's calling the superficial essence "something of the essence," a variant on "something of the thing itself." It *is* part of the deep essence, requiring the explanatory middle to complete it.

But how convincing are these answers? They work best for demonstrable subject-attribute relations such as the two cases discussed at length, in which the thing whose essence is being investigated is a demonstrable subject-attribute relation with a name.[14] The superficial essence of such a thing is simply a rewording of the assertion that the relation obtains. From "a certain loss of light belongs to the moon" we get "eclipse is a certain loss of light from the moon." To be sure, this is a kind of definition of eclipse, but it cannot do the work required of definitions as principles of sciences, since it is itself the result of a demonstration. The deep essence fares not much better. Of course it is not provable, being a recast form of a proof. But equally it cannot serve as a principle (although the immediate relations it encapsulates will be principles).

The two kinds of essences introduced in II.8 deserve to be called essences only by courtesy, if at all. But sometimes courtesy should be extended. In fact, it turns out that extending the notion of essence in these ways brings some advantages, and is a reasonably natural extension to make.

After II.1 distinguishes two pairs of scientific questions, II.2 begins to assimilate them. After being told (II.1) that questions of the fact and of the reason why belong together and are separate from questions of whether it exists and what it is, we learn (II.2) that the first member of each of these pairs asks if there exists a middle and the second asks what the middle is. Questions of the fact are brought into even closer relation with questions whether it exists when they are both said to ask if it is—only the former ask that question "in part" and the latter ask it "without qualification" (89b39, 90a2–5, a9–13). As early as 90a15 we have the question "what is eclipse" (a "what is it" question) although we have been told to expect "why does eclipse occur" (a question of the reason why). This is allowed, however, because "in all these cases it is obvious that the 'what is it' and why does it exist (occur) are the same" (a14–15). The questions "what is eclipse" and "why does eclipse occur" are placed side by side in order to show that the answers are the same ("loss of light from the moon due to screening of the earth" and "because the light fails since the earth screens"). Though not identical, they contain the same information. And so, even though II.1 says that "what is it" is asked only of subjects, II.2 shows that it can also be asked of subject-attribute relations. We should not be surprised when II.8 also asks both "if it exists" and "what is it" of eclipse and thunder.

And what is wrong with asking these questions? Eclipse and thunder are phenomena studied by sciences. They can be precisely defined and can be explained through the principles of those sciences. The two kinds of definitions given can reasonably be considered definitions of such things. Even if we do not find many such cases in geometry, they may play an important role elsewhere, and the role they play is neatly sketched here. They should be treated not like subjects or attributes, but in the way appropriate to them. The first thing to do is identify the subject and attribute concerned. Next is to explain the relation between the subject and the attribute, which we do by proving the relation by a familiar subject-attribute proof. The new feature, then, is the first step, which need not be trivial.

Finally, does the prescribed manner of treatment apply more widely? An analogous way to treat subjects or attributes would be to start with a preliminary account (superficial essence), use it to discover that the subject or attribute exists, and then go ahead to discover the deep essence, the immediate facts that can explain the preliminary account. For example, we might start out knowing that gold is a soft, malleable, ductile, yellow metal with specific gravity 19.32, melting point 1063°C, and boiling point 2966°C, and after locating the material with these attributes go on to discover that its atomic number is 79 and its atomic weight 197, and that the other attributes follow from these two facts. In that case the outline of a proof would be as shown in the diagram below.

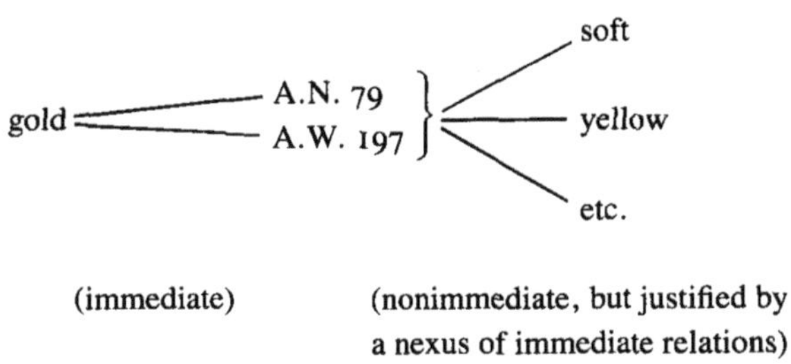

A brief inspection reveals that this case is not very similar. In cases like eclipse the original account (superficial essence) is part of the full essence. But the present account offers no justification for calling the description of gold as a soft yellow metal an essence of any kind. This situation corresponds to cases where we begin by knowing an identifying sign of eclipse (such as the inability to produce a shadow at the time of the full moon when nothing is between us and the moon, II.8 93a37–b3) that does *not* prove to be part of the essence, and it is regrettable that Aristotle says no more about such cases.

However, he suggests a different approach to the essence of subjects: if we grasp that man is a certain animal and soul is a thing that moves itself (93a23–24), we grasp "something of the thing itself," something of its essence (a22, a29). He supposes that these cases are similar to that of grasping that thunder is a certain noise in clouds. In fact, since animal is the genus of man, in grasping that man is a certain kind of animal, we *are* grasping part of its essence. "Animal" occurs in a complete definition of man just as "noise in clouds" occurs in a complete definition of thunder. The cases are different in that "noise" occurs in the discussion not as the genus of thunder but as an attribute of clouds, and also in that the "middle" of the final definition provides the explanatory grounds of the noise in clouds, whereas "rational" or "two-legged" is not the grounds of man's being an animal, but rather the type of animal man is. Even so there is an analogy. For although "rational" does not *explain* animal, it entails it: anything that is rational, or two-legged,[15] is an animal.[16] And if we begin our investigation knowing that a human is a certain sort of animal and so can recognize humans and know that they exist, we can go on to investigate what a human is, which amounts to ascertaining what kind of animal a human is. This is the same as discovering the immediate, unprovable facts about humans that constitute the scientific essence. The parallel with the cases of eclipse and thunder is not exact (which may be the reason why the examples of human and soul are dropped), but at least the method II.8 prescribes for locating essences can in some way apply to the things that *APo* primarily conceives of as having essences.

Scientific Explanation

CHAPTER XVII

TWO REQUIREMENTS of having scientific knowledge of something are that (1) we know its *aitía* and (2) we know that its *aitía is* its *aitía* (I.2 71b9–12).[1] *aitía* is rendered "cause" by Mure, Ross, Tricot, Tredennick, and Heath, and "explanation" by Barnes, but neither of these words captures the sense of *aitía*[2] required by the theory of demonstration. I prefer to translate it as "grounds" or "explanatory grounds," which come closer to the mark.[3]

The importance of knowing the grounds is often emphasized, as in the following passages, which all contain the key word *dihóti*, "the reason why": "the most important thing in knowledge is considering the reason why" (I.14 79a23–24); "to have scientific knowledge of the reason why is to have scientific knowledge through the grounds" (I.6 75a35); "scientific knowledge of the reason why is in virtue of the primary grounds" (I.13 78a25–26).

Middle Terms and Immediate Premises

Since demonstration is the vehicle for attaining scientific knowledge of demonstrable facts (I.31 88a9–11) and since demonstrations are deductions (I.2 71b17–18), Aristotle says that a scientific demonstration is a deduction that displays the explanatory grounds or the "why" (I.24 85b23–24). If *A—B—C* is a proof of *A—C* via middle term *B*, we can regard it as explaining why *C* belongs to *A*: because *B* belongs to *A* and *C* belongs to *B*. What are the explanatory grounds of *A—C*? The unprovable facts *A—B* and *B—C*. If we know this proof of *A—C* (*as* a proof of *A—C*), then (1) we know *A—B* and *B—C* and (2) we know that *A—B* and *B—C* are indemonstrable principles and premises of the proof of *A—C*.

Thus, in knowing the demonstration of *A—C*, (1) we know the *aitía* of *A—C* and (2) we know that the *aitía is* the *aitía*.

But granted that *A—B* and *B—C* are principles of the science and that they entail *A—C*, why should we think that they are the cause, explanation, or grounds of *A—C*? Aristotle does not face this question directly, so far as I know, but repeatedly insists that it is so. In I.2, he states, "We say that we know through demonstration" (71b17); this unproved assertion launches the discussion of the nature of scientific principles, one of the requirements of which is that they are grounds of the conclusion. In I.6, he insists that "a person who lacks an account of why, when there is demonstration, does not have scientific knowledge" (74b27–28): if you do not have an "account of why" (*lógos toû dià tí*), you do not have the demonstration that provides scientific knowledge. And in I.24, "a scientific demonstration is a deduction that displays the explanatory grounds of the 'why' " (85b23–24).

The present chapter will investigate the way unprovable principles of a science are the cause or grounds of demonstrable conclusions and in what sense a demonstration is an explanation. There are good reasons for thinking of principles and proofs in these ways, although the notions of cause or grounds and explanation turn out to be somewhat different from more familiar notions of these concepts. They need to be tailored to fit the needs of demonstration, but they can be made to fit very well.

The picture so far is simple: scientific knowledge of a fact is knowledge by a proof that shows how that fact follows from the principles, which are the grounds of the fact. But other passages express this point differently. The same chapter that says that knowing the reason why is knowing through the grounds also says that proof must be through a necessary middle, or else it does not yield scientific knowledge either of the reason why or of the necessity of the conclusion (I.6 75a13–14). Where we expect to find "grounds" we find "middle." This identification of the middle of a proof and the grounds of the conclusion is repeated elsewhere (II.2 90a6–7: "for the grounds is the middle"; also a9–11, II.8 93a7–8, b4–5, I.13 78a28–30). And I.6 makes it clear that the middle is the middle term in a demonstration.[4]

Now it is a different way of looking at a demonstration

A—B—C

to identify the grounds of *A—C* as *B* rather than as *A—B* and *B—C*. This view brings us closer to a familiar use of the word "cause," and perhaps farther from what we would ordinarily call an explanation. But it is also clear that the two ways may amount to the same thing. If to say that *B* is the grounds of *A*'s belonging to *C* is no different from saying that *A* belongs to *C* because of *B*, and if by this we mean that *A* belongs to *C* because *A* belongs (immediately) to *B* and *B* (immediately) to *C*, then calling *B* the grounds amounts to calling *A—B* and *B—C* the grounds. Indeed, to call *B* the necessary middle is to imply that it is necessarily related to other terms, and to conceive of *B* as a middle implies the other terms for which it is a middle.

In fact this is all Aristotle means. He considers three cases:

> **I.6 75a1–11** "(a) Now when the conclusion is of necessity nothing prevents the middle through which it was shown from not being necessary (for it is possible to deduce what is necessary from nonnecessary things, just as it is also [possible to deduce] what is true from not true things);
> (b) but when the middle is of necessity, the conclusion is of necessity too, just as from true things [the conclusion is] always true (for let *A* belong to *B* of necessity and this to *C* [sc., of necessity]: now it is necessary also for *A* to belong to *C* [sc., of necessity]);
> (c) and when the conclusion is not necessary, neither can the middle be necessary (for let *A* belong to *C* not of necessity, but [let *A* belong] to *B* and this to *C* of necessity; therefore also *A* will belong to *C* of necessity. But it was laid down that [*A* did] not [belong to *C* of necessity])."

In case (a) he begins by speaking of the middle as being nonnecessary, and then in the parentheses shows that he is thinking of premises (the "nonnecessary things," like the "non-true things" can only be premises). In (b) and (c), for the middle to be of necessity is for it to belong of necessity to the minor term in the syllogism and for the major term to belong of necessity to it. I conclude that when Aristotle says that the middle is the grounds, he is emphasizing the importance of the middle term in the premises of the demonstration that are the grounds, and this is the familiar view that the grounds are the principles of the proof. The middle term in a demonstration can always be cashed into immediate premises.

The analysis of sharpness (*anchínoia*) in I.34 confirms this interpretation: "Sharpness is a kind of good guess at the middle in an imperceptible time" (I.34 89b10–11), and "on seeing the extremes, he recognized all the grounds, which are the middles" (b14–15). In the examples given, the sharp person spots the middle term, but the justification for the middle term's being the grounds is its relations to the extremes.

> **I.34 89b11–20** "A person on seeing that the moon always has its bright side facing the sun quickly conceived why this is so—because it is illuminated by the sun. . . . Let *A* stand for the bright side facing the sun, *B* for being illuminated by the sun, *C* for the moon. Now *B* (being illuminated by the sun) belongs to *C* (the moon); and *A* (the bright side facing the source of illumination) [belongs] to *B*; and so *A* belongs to *C* through *B*."

B is the grounds of *A*—*C* if and only if *B* is the middle term in the demonstration of *A*—*C*, and this can happen if and only if *A*—*B* and *B*—*C* are immediate principles. Moreover, if we claim that *B* is the grounds of *A—C* we must be prepared to back it up by showing how *A—C* is due to *A—B* and *B—C*.

The equation of middle term with principles is plausible when there is only one middle term and two premises in a proof, as in the proof

A—B—C.

Where there is more than one middle term the tendency to identify the grounds with the middle term leads Aristotle to say that there are more grounds than one and to ask which middle term is *the* ground (II.18).[5] In general, this question does not make sense. If the proof is

A—B—C—D,

the conclusion is based equally on principles *A—B*, *B—C*, and *C—D*. Both middle terms *B* and *C* are needed and neither has a claim to preference. But if *B—D* is a subject-attribute proof and *A* is a species of *B* or a particular *B*, we have an application proof, and Aristotle holds that in such a case, the reason why *A—D* is that *A—B* (and *B—D*).[6] In any case, although Aristotle switches back and forth between viewing grounds as principles and viewing them as middle terms, the two perspectives involve no fundamental difference in his view of the nature of demonstrations or scientific explanation.

Grounds of Truth or Grounds of Knowledge?

How are scientific principles grounds? Are they the grounds of our knowledge of the conclusions, so that our knowledge of the conclusions depends on our knowledge of the principles? On this view they are the *ratio cognoscendi* or *Erkenntnisgründe*. Or are they the grounds of the conclusions themselves, so that the truth of the conclusions depends on that of the principles? On this view they are the *ratio essendi* or *Realgründe*. I believe that Aristotle requires them to be both.[7]

It is clear that he considers them to be grounds of the facts stated in the conclusions of proofs. When he says that having scientific knowledge of *x* involves knowing the grounds of *x*, and knowing that the grounds of *x* *are* the grounds of *x* (I.2 71b10–12), he is *not* saying that we must know *our reasons* for claiming to know *x*, but that we must know why *x actually* holds. And when just below he says that the principles are grounds of the conclusion (b22) and justifies this claim (b30–31) by referring back to the condition of knowing the grounds, his point is that the truth of the conclusion is due to the truth of the principles, which are its grounds. Further evidence for this view is found where he says that the middle is the grounds of the conclusion,[8] not that it is the grounds of our knowledge of the conclusion.

It might be objected that principles and proofs are not grounds for our knowledge of conclusions. *APo* II.1 makes it clear that we can know that something is true (knowledge of the fact) before discovering its explanation (knowledge of the reason why), the middle of the demonstration that proves it. But if we can know provable facts without knowing their proofs, our knowledge of them is not grounded in the principles and demonstrations. This objection can be met simply by pointing out that at the stage of factual knowledge discussed at II.1–2, we do not yet have *scientific* knowledge. We are at the stage of discovery and organization, not of a finished demonstrative science.[9] There are nondemonstrative grounds for knowing[10] facts. Otherwise the view would be open to severe criticism. And yet *scientific* knowledge of a fact, which we get by having a proof of the fact (I.2 71b17–19), is grounded in knowledge of the principles on which the fact is grounded. Scientific knowledge is knowledge of the reason why, and "to have scientific knowledge of the reason why is to have scientific knowledge through the grounds" (I.6 75a35): to know the reason why x holds (to have scientific knowledge of x) is to know x through the grounds of x, through the principles on which x is based.

Knowledge of the Fact and Knowledge of the Reason Why

APo II.1's distinction between knowledge of the fact (KF) and knowledge of the reason why (KRW) is basic to Aristotle's theory. KF corresponds roughly to what we regard as knowledge (including scientific knowledge), and KRW—Aristotle's scientific knowledge—corresponds to nothing in our ordinary modern vocabulary or range of concepts. The same is doubtless true for ancient Greek. *APo* develops a special conception of *epistḗmē* and *epístasthai*, which, though based on ordinary usages of those terms, goes beyond them. It sharpens and deepens those usages—a legitimate and frequently undertaken kind of philosophical task, which may (but need not) in turn shape ordinary usage. Possibly *APo* may be regarded as a "persuasive" definition of *epistḗmē*. If *APo* clarifies a confusing concept, that is a reason to adopt the use it recommends even if doing so means altering how we apply the term.

Two Kinds of Unsatisfactory Demonstrations

APo I.13, which discusses ways we can fail to have KRW, proves to be the lengthiest treatment of the explanatory nature of scientific demonstrations and so needs to be considered in detail. The chapter begins (78a22–28) by characterizing two cases where KF and KRW differ in a single science. Next is a lengthy study of specific examples interspersed with general statements (a28–30, b3–4, b11–13, b13–15, b17–18, b20–22, b23–24, b28–30). I am inclined to follow Barnes's attempt (p. 151) to fit all the later material under the two ways described at the beginning of the chapter,[11] so that 78b13–31 treats case 1 and 78a28–b11 treats case 2.[12] The two cases are described as follows:

> **I.13 78a22–28** . . . "(1) if the deduction does not take place through immediates (since the primary grounds are not grasped, but scientific knowledge of the reason why is in virtue of the primary grounds), and (2) if it is through immediates, but is not through the grounds, but the better known of reciprocating terms."

CASE 1: NONRECIPROCATION

Case 1 is illustrated and discussed at 78b13–31. The argument that does not give KRW amounts to this:

> R1. Everything that breathes is an animal.
> No wall is an animal.
> Therefore, no wall breathes.

That is, walls do not breathe because they are not animals. But

> **I.13 78b16–23** "if this were the grounds of not breathing, it would be necessary for being an animal to be the grounds of breathing. Just as if the denial is the grounds of not belonging, the assertion is the grounds of belonging . . . so if the assertion is the grounds of not belonging, the denial is the grounds of belonging. But for things exhibited in this way, the stated condition does not obtain. For not every animal breathes."

Granting the truth of the premises, R1 gives a sufficient reason to think that no wall breathes. But it is not the most powerful argument since it does not account for why nonbreathing animals (and Aristotle thought that there are some) do not breathe. On the other hand, an argument applying to nonbreathing animals will also apply to walls. For example, if animals with lungs are the only ones that breathe, we can argue as follows:

> R2. Everything that breathes has lungs.
> No wall has lungs.
> Therefore, no wall breathes.

R2 covers nonanimals and nonbreathing animals as well, so lacking lungs is a better candidate for the grounds of walls' not breathing, by the criterion above. If the assertion ("*x* has lungs") is the grounds of belonging ("*x* breathes"), the denial is the grounds of nonbelonging.[13] Cause and effect reciprocate.[14]

Mention of "the primary grounds" (78a25) reminds us of the term for the widest possible subject, "primary universal."[15] The thought is the same; only the viewpoint is different. Regarding the primary universal, we begin with the attribute and ask what is the widest subject to which it belongs. In the present case, we begin with a subject and attribute whose relation is to be demonstrated, and ask what is the widest[16] middle term that can be used in proving it. Another example would be: Why do isosceles triangles have exterior angles equal to four right angles? The primary grounds is "rectilinear figure," not "isosceles triangle" or "triangle." We are in the situation of an application argument, looking for the primary subject or attribute by virtue of which the attribute (4R or breathing) belongs to other subjects.[17]

We also see why R1 does not take place "through immediates" (a24). If being an animal is not the primary cause of breathing, something else is. Again, suppose that this is having lungs. But then the major premise of R1 can be proved:

> Animal belongs to all things with lungs.
> Having lungs belongs to all breathing things.
> Therefore, animal belongs to all breathing things.

At least one premise of R1 is not immediate.

Is there any significance to the fact that R1 is in the second syllogistic figure? Since the alleged grounds ("animal") extends more widely than the attribute ("breathing"), the only way this can be rendered syllogistically is for it to be predicated of the attribute—which requires a second-figure syllogism (since Aristotle did not recognize the fourth figure). But then, can second-figure syllogisms *never* give scientific knowledge?

This appears to be the meaning of the opening words in the discussion of case 1:

> **I.13 78b13–15** "Also in cases where the middle is placed outside. For also in these the demonstration is of the fact and not of the reason why, for the grounds are not mentioned."

The first sentence might appear to refer specifically to cases where the middle extends more widely than the major, as in the premise "everything that breathes is an animal," so that Aristotle is objecting to this special situation, which requires a second-figure syllogism for its expression, rather than to second-figure syllogisms in general. But it is too hard to accept this interpretation in view of *APr* I.5 26b39, where the same expression ("the middle is placed outside") describes the position of the middle in second-figure syllogisms.

Against this evidence are three considerations. (1) I.14 argues that the first figure is the *most* scientific but offers no reason to think that the second-figure is *un*scientific. (2) The fact that second figure syllogisms always have denials as their conclusions is no argument against its being the correct form for some proofs, since some principles deny one thing of another (I.15). By contrast, by pointing out that third-figure syllogisms always have particular conclusions, he shows that *they* cannot be the vehicle of proofs.[18] (3) R2 is a good candidate for a proof, and (despite 78b13–15) does mention the grounds.

I therefore prefer to discount 78b13–15, and I suggest that the second sentence means only that some second-figure syllogisms produce KF and not KRW.

In these cases, as the example shows, the grounds of the conclusion is not mentioned, because the middle term that *is* mentioned is too wide.

CASE 2: RECIPROCATION IS NOT ENOUGH

Case 1 shows that the primary grounds of demonstrable facts are immediate and that the major and middle terms in proofs should reciprocate.[19] Case 2 establishes that even a middle satisfying these conditions may fail to be the grounds of the conclusion.

Aristotle's principal example here is the following pair of syllogisms:

S1. Things that do not twinkle are near.
The planets do not twinkle.
Therefore, the planets are near.

S2. Things that are near do not twinkle.
The planets are near.
Therefore, the planets do not twinkle.

S1 purports to prove that the planets are near through the middle "nontwinkling": the planets are near because they do not twinkle. S2 purports to prove that the planets do not twinkle through the middle "near": the planets do not twinkle because they are near. He supposes that all the premises (and both conclusions) are true, and that the terms "nontwinkling" and "near" reciprocate. He claims that S1 does not give KRW, but only KF,[20] because "it is not on account of not twinkling that they are near, but on account of being near that they do not twinkle" (78a37–38). This assertion is made without justification; Aristotle presumably considered it obviously true.[21] For the same reason, S2 gives KRW.

Why does Aristotle think that S2 is a successful proof and S1 not? First, "either can be shown through the other" (a39): either of the terms "nontwinkling" and "near" can be used as a middle term in a syllogistic argument with the conclusion that a subject (here, the planets) has the other term as an attribute. Second, he says that the two terms reciprocate. Third, he seems to say that both syllogisms have immediate premises. *Di' amésōn* in a26 can hardly mean anything else than that both premises in S1 are immediate. It would solve a problem[22] if it meant only that the major and middle terms are immediately related,[23] but this is not a plausible way to take the Greek.[24] S2, which produces KRW, is meant to have immediate premises.

It is obvious that the major premises of S1 and S2 are not immediate: the relation between nearness and nontwinkling can be explained. Twinkling depends on the apparent magnitude of the celestial object, atmospheric conditions, and perhaps the observer's eyes. The relation between apparent magnitude and nearness can be explained by reference to laws of optics and the absolute magnitude and brightness of the celestial object.[25] But let us suppose that the relation between nontwinkling and nearness is proved by something Aristotle would recognize as a scientific demonstration, and suppose further that the relation is reciprocal. At this point the serious difficulty arises.

For even granted that the major premises of both arguments are acceptable, it is just not possible for both minor premises to be immediate, as Aristotle says they are. For if S2 is a demonstration, then it is immediate that the planets are near, and nonimmediate that they do not twinkle. Therefore, the minor premise of S1 is not immediate, and its conclusion is. Alternatively, if both premises of S1 are immediate (as a26 seems to say), then its conclusion is not immediate, and so the minor premise of S2 is not immediate, and S2 is not a demonstration. This state of affairs is generally the case: there cannot be two competing candidates for proofs such that all their premises are immediate and yet the minor premise of each is identical to the conclusion of the other.

This is a welcome state of affairs. Without it, case 2 would be a serious obstacle to the view that any logically valid combination of immediate premises is a demonstration. If both premises in S1 *could* be immediate, we would need further criteria to decide which combinations contain the grounds of their conclusions. As it is, the view stands that if all premises are immediate, an argument *is* a demonstration.

Criteria for Grounds

Does this criticism of the formulation of case 2 mean that we can learn nothing from the example of the planets? Not at all. The point turns on whether the middle term in the proof is grounds of the conclusion, not on whether the premises are immediate. In the context of a fully organized science we will not be faced with a pair of arguments like S1 and S2. S1 will simply not appear. But in setting up a science we may be faced with rival candidates for proofs like S1 and S2, which amount to rival candidates for immediate principles: is it a basic fact about planets that they are near or that they do not twinkle? Failing to notice the relation between nearness and nontwinkling, we might think that both are immediate truths. But if we find that we can construct arguments to show both that the planets are near and that they do not twinkle, which is a principle and which a demonstrable relation? Equivalently, which of the reciprocating terms "near" and "nontwinkling" is the grounds of the other?

If we learn by induction or perception (78a33–35) that all that does not twinkle is near, the same methods will also teach us that all that is near does not twinkle. Logic cannot help here either. It can weed out badly formed propositions and badly constructed arguments, but both arguments (we may say) are sound and their premises well formed. We need a way to decide which sound argument is a demonstration, and this means a way to decide which has immediate premises, or rather, which has a minor premise that is immediate. But immediacy is a concept that can be applied only in an organized science. At the stage envisaged we may know that not both of the minor premises are immediate, but we do not know which (if either) is: we are trying to *decide* which.

The solution depends on the notion of explanatory grounds. It is impossible for each candidate to be grounds of the other. If induction, perception, and logic cannot decide which is immediate, we are to identify which is grounds of the other. The planets' nearness is the grounds of their not twinkling and not vice versa (78a37–38). But why? We need a principle for deciding, not just a clear case. (Aristotle seems to consider the present case obvious.) Similarly, in II.16 he observes that since broad-leaved and deciduous reciprocate, either can be shown through the other as middle term, but "they cannot be grounds of each other" (98b16–17),[26] yet he assumes that being broad-leaved is the grounds of deciduousness (a37–b1).[27] Also in treating lunar eclipse he assumes that eclipse and the earth's being in the middle reciprocate, and asserts "it is *obvious* that eclipsing is not the grounds of the earth's being in the middle, but that this [is the grounds] of eclipsing" (b21–22). This time he says why it is obvious. But even if the reasons he gives fail to show why it is obvious, that does not prove that it is *not* obvious. He says (a) the grounds of *x* are prior to *x* (b17), (b) the grounds of eclipsing are that the earth is in the middle, but eclipsing is not the grounds of the earth's being in the middle (b17-19), and (c) "[the earth's] being in the middle belongs in the account of eclipsing, so it is clear that that comes to be known through this and not this through that" (b22–24).

PRIORITY OF THE ACCOUNT

Statement (c) shows that Aristotle is thinking of the definition of eclipse as "a certain loss of light from the moon due to the earth's being in the middle."[28] And his description of "the earth's being in the middle" as "belonging in the account of eclipse" recalls the discussion in II.8, in which screening by the earth is called the account of eclipse, which gives KRW (93a30ff.)[29] of the attribute's belonging to the subject. On this reading of II.16, for the earth's being in the middle to belong in the account of eclipse is not just for it to be an element in the definition of eclipse, but for it to be the grounds of eclipse.[30] But this is where we came in. Our question is how to tell which of two reciprocating terms is the grounds of the other. If we already know which is "in the account" of which, we no longer wonder which of the two candidates is the real demonstration.[31]

Another approach to finding a criterion for priority in II.16 98b22–24 comes from noticing that the description of "the earth's being in the middle" as "belonging in the account of eclipse" recalls the account of the per se 1 relation: "[I call] per se all the things that belong [to something] in 'what it is' " (I.4 73a34–35). "Account" (*lógos*) replaces "what it is" (*tí esti*).[32]

On this reading, the earth's being in the middle is the grounds for eclipse because it belongs per se 1 to eclipse. We must make a few moves before this claim is reasonable. First, "eclipse" is a name for a subject-attribute connection: a certain loss of light from the moon. This connection is demonstrable via the middle term "the earth's being in the middle": the moon (on certain occasions) suffers that loss of light because (on those occasions) the earth is in the middle.

The earth's being in the middle is an element in the full "rearranged demonstration" kind of definition, which reveals the grounds of the occurrence of the relation in question.[33] This is a rather special sort of case, which does not resemble other cases of interest. It is implausible that not twinkling is in the definition of being near, or the reverse, or that deciduous is in the definition of broad-leaved, or the reverse, just as the earth's being in the middle is not likely to occur in the definition of the certain loss of light (the attribute in abstraction from its subject). Second, since the kind of relation proved in subject-attribute demonstrations is per se 1,[34] to say that the grounds must be related per se 1 to what is being proved fails to distinguish grounds from demonstrandum.

Both these problems can be solved. As to the second, the accounts of the per se 1 and per se 2 relations in I.4 are appropriate to definitional (immediate) relations, and it was necessary to extend the range of per se 1 and per se 2 beyond what was apparent from I.4 to make room for nonimmediate per se relations.[35] The present claim that the grounds of *x* is in the account of *x* is reasonably taken to mean that the grounds stands in an immediate per se relation to a term in the demonstrandum. As to the first problem, although the grounds may not be very helpful in identifying the attribute, still if it is immediately related to the attribute, it is in the attribute's fat definition. And so, given that it reciprocates with the attribute, we can prove that a subject has the attribute by showing that it possesses the grounds: if fig trees are per se broad-leaved, then they are deciduous; if planets are per se near, then they do not twinkle. The proof consists in showing that the subject possesses an attribute that is immediately related to the one being proved and that is thus in its fat definition.[36]

Elements in the definition of an attribute *C* are prior to *C* in that a proof that a subject *A* has *C* proceeds by showing that *A* has an attribute *B* that is an element in *C*'s fat definition, such that as a consequence *A* has *C*. Does this insight into the nature of proofs help our current discussion? Faced with rival candidates for proofs

A—*B*—*C* and *B*—*A*—*C*,

if we know that *B* is in the fat definition of *A*, we will prefer the first, and we will prefer the second if we already know that *A* is an element in the fat definition of *B*. But this is not enough. In the first place, this presupposes that we already know the fat definitions of *A* and *B*. But if we did, we would not be asking the question in the first place: we would know whether *A*—*C* or *B*—*C* was immediate, and the question which argument is a demonstration would not have arisen. And secondly, the immediate relation between *A* and *B* cuts both ways.[37] If *A* is immediately related to *B*, so is *B* to *A*, and so each is in the fat definition of the other.

PRIORITY OF WHAT IS BETTER KNOWN

The remaining point of attack left by II.16 is the assertion that the grounds of *x* are prior to *x* (98b17). The passage from I.13 that introduces case 2 is helpful:

I.13 78a26–30 "[The deduction] is through immediates, but is not through the grounds, but the better known of reciprocating terms. For nothing prevents the better known of reciprocating terms from sometimes being what is not the grounds [of the other], so that the 'demonstration'[a] proceeds through this."

NOTE ON TRANSLATION

a. "Demonstration" needs to be placed in scare-quotes. In fact, S1 is not a demonstration, since it does not proceed through the grounds. It is a *putative* demonstration.

Recall that principles are "better known, prior to, and grounds of the conclusion" (I.2 71b21–22),[38] and that these conditions are interrelated: "they must be prior if they are grounds" (I.2 72a31).[39] But this is not yet sufficient. We need an independent criterion to determine which of two competing claimants is prior to the other. More help comes from the association of "prior" and "better known" at I.2 71b33–72a5 and the distinction between what is prior and better known without qualification and what is so to us. The passage was discussed in a preliminary way above in Chapter 2, where I noted that "prior in relation to us" need not be taken as synonymous with "better known in relation to us," and similarly for "prior without qualification" and "better known without qualification." The point may just be that they apply to the same things. But even this connection is close enough for present purposes. We need criteria to identify what is the grounds of something else, and if Aristotle provides identical criteria for being prior and being better known, they will satisfy our need.

Since the grounds of a conclusion are prior and better known "without qualification" or "in nature," not "in relation to us," in case 2 the alleged grounds are better known "to us." But this must be treated with care. *APo* I.13 78a26–30 does not say "better known to us," and for good reason. If it said "nothing prevents the better known *to us* of reciprocating terms from sometimes being what is not the grounds of the other" there would be no need for the qualification "sometimes," since according to I.2, things that are better known to us are never better known or prior in nature, and so are never grounds of what is less well known to us. The correct interpretation is surely that what we understand more clearly is sometimes less well known in nature and sometimes better known in nature. Experts at a discipline, who have made what is better known in nature better known to themselves,[40] will not make these mistakes. For *them*, the prior and better known of reciprocating terms will be the one that is the grounds of the other. But for those who are not yet experts, sometimes the thing that is prior and better known to them will be prior and better known in nature and sometimes not. When it is not, they can be expected to make the wrong choice and mistake an argument of the fact for a demonstration of the reason why.[41] In the light of these passages we must not take the contrast between what is prior and better known "without qualification" and "to us" as implying that "we" cannot rise above the level of seeing things in just the wrong way. Rather, "to us" is introduced as a qualification to serve as a polar opposite to "without qualification," and it refers perhaps to the level of the raw beginner, at the opposite end of the scale from the expert.

This correction made, I.2 points to a way to decide which reciprocating term is grounds of the other. The question becomes, Which is without qualification prior to and better known than the other? *APo* I.2 tells us that these are "the things that are farther [from perception]" and that the things that are farth*est* [from perception] are those that are most universal." If we may infer from the latter statement that things that are farth*er* from perception are *more* universal, then we have two criteria for priority, which are asserted to coincide. But do they?

In one relevant way Aristotle believes they do. In his account of concept formation (II.19),[42] he says that we begin by perceiving individuals of a kind (for example, Callias), then the first universal (human) comes to rest in our mind and then more and more general universals (animal, etc.) (100a15–b3). This gives a good sense to the idea that things that are more universal (more general) are more remote from perception.[43] And this approach is relevant to proofs: triangle is more universal than isosceles triangle, and triangle is the grounds for isosceles triangle—2R: isosceles triangles have 2R because they are triangles.[44] But this application is not enough to solve our present problem, since it does not touch cases where the two terms in question reciprocate, for neither term is more general than the other.

If universality fails as a criterion, can anything be made of remoteness from perception? We might suppose it is to be applied as follows. It is obvious that the planets do not twinkle: we can see it. But not so that they are near. It is also obvious that the moon suffers eclipse, but not that the earth is in the middle, and that the moon waxes in a certain way, but not that it is spherical.[45] Perhaps, then, being remoter from perception means being less capable of being observed, and the criterion amounts to the thesis that facts that are more capable of being observed are accounted for by facts that are less capable of being observed. This would be a powerful and interesting theory. We would need criteria for being more capable of being observed, but it might be easier to find them than criteria for so abstract a characteristic as being prior in nature. There is even a kind of plausibility about such an interpretation, if it is taken as a claim that observed facts can be accounted for by general laws. (The particular/universal contrast is relevant here too, since what is observed is a particular instance or occurrence.) We seem to be on the verge of the covering-law model of scientific explanation and the observational/theoretical distinction.[46]

Nevertheless, this interpretation cannot tell the whole story, and from Aristotle's point of view it does not even tell an important part. For the view of *APo* is that the conclusions of scientific demonstrations are universal, and they hold "always and everywhere." They apply only incidentally to particulars, at a certain time and place.[47] Particular cases are explained by reference to universal per se facts, and one of the functions of application arguments is to provide such explanations. This is an important part of science, but *APo* is little concerned with trivial applications of universal scientific facts to particular cases. Its main interest is in explicating the relations of those universal facts, which occurs on the level of universals, and even there application arguments play a subordinate role in the structure of sciences.[48] More importantly, it is doubtful that Aristotle would insist that the "remoteness from perception" criterion fits all the cases of grounds discussed in *APo*. Is being broad-leaved less easily observed than being deciduous?[49] In geometry, it is presumably more easily observed that a figure is a triangle than that it has 2R, and certainly more easily observed that a figure is a circle than that it cannot be squared. Further, Aristotle would surely hold that a piece of iron glows red because it is hot and not that it is hot because it is red, yet being red and being hot are equally easily observed.

Neither of the criteria Aristotle offers succeeds in determining what is prior and better known in nature, so as to determine which of two reciprocating terms is the grounds of the other. To judge by recent attempts to do the same,[50] it is no easy job. No wonder, therefore, that Aristotle fails to complete it successfully. (We may credit him with an interesting first attempt.) But even if he fails to provide clear criteria, we need not abandon the concepts for which the criteria were intended. There remain the theses that reciprocating terms that belong to the same subject cannot each be the grounds of the other's belonging to that subject, and that demonstrations prove their conclusions to hold through the grounds on which those facts actually depend. This view of the nature of demonstration and the belief that the relation "is the grounds of" is asymmetric can stand even without clear criteria for determining what is the grounds of what.[51]

Aristotle's view amounts to the claim that principles are naturally prior to and better known than other scientific facts, which are proved from, depend on, and are grounded in them. To ask which of two rival arguments is a demonstration amounts to asking which of the minor premises is a principle. The question which term is the grounds of the other amounts to asking which one's relation to the subject is a principle on which the other's relation to the subject depends. Moreover, since demonstrations prove an attribute to belong to a subject through the grounds on account of which it actually belongs, Aristotle is committed to a natural order of priority and intelligibility, a ranking of facts in a science in terms of being naturally prior and better known, which the science makes manifest in its proofs. The science explicates the network of relations among the subjects and attributes that constitute its subject genus.[52] "For Aristotle, an axiomatic system (i.e., a demonstrative science) is not just a preferred ordering of humanly constructed knowledge, but a mapping of the structure of the real."[53]

Priority, Primacy, and Ungrounded Grounds

But principles are not just prior (*prótera*) to the conclusions, they are primary (*prôta*, I.2 71b21). Since what is primary is indemonstrable (b26–27), a principle (72a6–7), and immediate (a8),[54] and since no proposition is prior to an immediate one (a8), it follows that nothing is prior to what is primary, so that "primary" is explicitly the superlative of the comparative "prior." But since the priority of the principles is closely related to their being better known (sc., in nature) than and grounds of the conclusions, their superlatively prior character should also be reflected in these conditions too. Now it is no problem to apply it to the former: we may simply say that the principles are *best* known in nature of all the propositions in a science. And surely Aristotle would agree. The application to the condition that the principles be grounds of the conclusions is not so straightforward, but that is what interests us here. If *x*'s being the grounds of *y* entails that *x* is prior in nature to *y*, it follows that if nothing is prior to *y*, nothing is the grounds of *y*: the principles are ungrounded. This does not mean that there is no reason to accept the principles, only that within the formal structure of the relevant demonstrative science, there are no further principles on which the principles are grounded.

It is well known that Aristotle believes that principles are indemonstrable and no other propositions within a science are prior to them (I.3 72b18–20, b26–28). Is he aware that his principles must be ultimate as regards groundedness as well?

Evidence is not hard to find. First, there is the final sentence of I.6, which concludes an argument (75a28–37) that the principles of proofs are per se propositions. Even if the conclusion of an argument is per se, unless the premises are per se the conclusion is not known to be per se; further, the argument will not give the reason why, will not proceed through the grounds. "Therefore it is necessary both for the middle to belong to the third (sc., the minor term in the argument) because of itself and for the first (sc., the major term) to belong to the middle [because of itself]" (a35–37). The specification "because of itself" (*di' hautó*) is significant. The point of saying that in principles the attribute belongs to the subject "because of itself" is to contrast them with provable propositions, where the relation holds because of something else, the middle or principles used in the proofs. This description of principles is exactly what is needed to make the point in question: the principles are not further grounded in other propositions; they are grounded in themselves.

Similarly in the account of hypotheses and postulates in I.10:

> **I.10 76b23–24** "What must be because of itself and must seem [to be because of itself] is not a hypothesis or a postulate."

I argued[55] that this sentence contrasts what is provable (hypotheses and postulates are here demonstrable propositions) with what is not, the principles, by characterizing a principle as "what must be because of itself," as opposed to the demonstrable conclusions, which presumably are because of the principles.[56]

The final piece of evidence comes from the discussion of the possibility of proving the principles of each science from a set of higher (more ultimate) principles, where Aristotle makes a claim that holds generally whenever there is a demonstration:

> **I.9 76a18–22** "The person who knows from more ultimate grounds knows better, for he knows from things that are prior when he knows from ungrounded grounds. And so, if he knows better and best, that scientific knowledge would be better and best."

The shift from comparative to superlative is abrupt, but the passage obviously describes principles as "ungrounded grounds": knowledge based on principles is based on the ultimate grounds, which have nothing prior to them, and so are ungrounded. We might say not *further* grounded, since I.6 and I.10 indicate that they are *self*-grounded. But the point is the same.

CASE 2A: ANOTHER CASE OF NON-EXPERT KNOWLEDGE?

The sentence bridging I.13's treatment of case 2 and case 1 presents a puzzle:

> **I.13 78b11–13** "In cases (2A) where the middles do not reciprocate and the one that is not grounds is better known, the fact is shown, but not the reason why."

This is a footnote to the preceding discussion of case 2, where KF and not KRW results from taking the better known of reciprocating terms as middle when it is not the grounds.[57] In case 2A, since they do not reciprocate it is not true that "either can be shown through the other" (78a39). We cannot form syllogisms related as S1 and S2 (p. 216) are related. Only the proof (corresponding to S2) can be formed. But then how can it be said that the fact is shown, but not the reason why? Since the syllogism corresponding to S1 cannot be formed, how can the fact be shown at all?

Aristotle may be guilty of a blunder, but he may have in mind a non-expert to whom what "is not the grounds is better known." He knows that the planets are near better than that they have heliocentric orbits, whereas their having heliocentric orbits is the grounds of their being near. Since not everything near goes round the sun (e.g., the moon), but everything that goes round the sun is near, the following syllogism (corresponding to S2) yields KRW because the premises are grounds of the conclusion.

S3. Everything that goes round the sun is near.
The planets go round the sun.
Therefore, the planets are near.

But the following syllogism (which corresponds to S1) has a false premise, and so does not even yield KF.

S4. Everything that is near goes round the sun.
The planets are near.
Therefore, the planets go round the sun.

It is incorrect to form S4 and then transform it into S3 (as we can move from S1 to S2). The non-expert knows that the planets are near but not why, and so has simple KF. If someone shows him S3, he will not accept it as a proof since its conclusion is better known (to him) than its premises. Thus, it cannot give *him* KRW even though it reflects the knowledge of the expert. From his non-expert point of view, S3 proceeds in the wrong order, and so does S2. From his perspective, S3 has the same "backwards" look to it that S1 has to the expert. The point is that a person who does not (yet) know the principles better than other propositions in a science cannot have KRW. Even if offered a demonstration, he cannot grasp it as a demonstration, and so cannot understand it as showing anything other than the fact.[58] On this reading, case 2A involves a false premise and therefore, although related to case 2, is not properly a case of KF within a science.

I.13 Summary

Can Aristotle reasonably think he has explained all the ways that KF and KRW differ in the same science? This should amount to an examination of all the ways demonstrations may fail. One way he does not consider is the cases where the premises are not all true,[59] but such arguments are not "in" the science at all. In fact, they are positively unscientific, as he has just said in the previous chapter (I.12 77b16–27).[60] He also fails to consider invalid arguments[61] and arguments with premises that are true but not per se, but none of these arguments are "in" the science either.

After these ways, the obvious way for a demonstration to fail is for it to depend on true scientific propositions that are not principles. Case 1 covers this kind of failure, and although the discussion at 78b13–31 covers special cases that arise only in the second syllogistic figure, the general statement of case 1 at 78a24–26 (quoted earlier in the present chapter) also covers first-figure cases, where, for example, in proving *A—D* we use only one middle term (*B*) instead of two (*B*,*C*), so the premise *B—D* is not immediate.[62]

Case 2 seems to require more: that we must put convertible principles the right way round. But all it actually requires is that premises be principles, and case 2 applies where we are not yet certain which propositions are principles. Thus case 2 is concerned not with blunders that can be made once the principles are identified, but with errors that may arise while a science is being organized. Aristotle can reasonably suppose that he has covered the possible ways to go wrong in proofs within an organized science and also what is likely to be the main source of trouble in forming proofs in the organizational stages.[63]

The remainder of I.13 (78b32–79a16) considers pairs of disciplines one of which can have only KF and the other of which has KRW, frequently without KF (a4). The passage contains a number of problems deserving of detailed treatment, but since it adds nothing useful for present concerns I omit discussion.[64]

To judge from I.13, KF results only when a proof misfires or one discipline relies on another for its principles. But this is a mistaken notion to adopt in view of II.1–2 and II.8, where KF is said to be gained in the process of discovery, before we find a proof that provides KRW. First we discover *that* it is so, and then we proceed to discover *why*. This doctrine is not at all in conflict with I.13, whose message is that scientific knowledge of *x*, which is KRW, depends on having a demonstration of *x*, which depends on immediate principles and reveals the grounds of *x*. Without such a proof, we do not have KRW, and so have (only) KF. In the context of II.1–2 and II.8, this would correspond to discovering the fact, attempting to discover the demonstration of it, but failing to discover the proof. What we think is the proof is flawed, so we do not get KRW and we remain with KF.

Scientific Explanations and the Four Causes

After exploring the natures of grounds and scientific explanation and their connections with other aspects of demonstrative science, it will be interesting to see how well the notions of grounds and explanation correspond to certain other concepts, so as to gain a wider perspective on these ideas. We may begin by comparing the way principles are grounds (*aítion*) of conclusions with the ways Aristotle identifies of one thing's being *aítion* of another in his well-known doctrine of the "four causes."[65] First let us eliminate the "final cause," which is described as

> ***Phys*.II.3 194b32–35** "that on account of which, as health [is the final cause] of going for a walk: why is he going for a walk? we say 'in order to get healthy,' and when we have said this we think we have given the cause."

I can think of no plausible way to regard principles as "that on account of which" the conclusions hold true. Equally great problems arise if we try to see them as the "efficient cause" of the conclusions. Aristotle defines this kind of cause as

> ***Phys*.II.3 194b29–32** "that from which comes the first beginning of change or rest, as the one who planned is the cause, and the father [is the cause] of the child, and generally that which makes [is the cause] of what is made and that which produces change of what undergoes change."

But conclusions are not produced or changed or at rest, nor do principles cause any change or rest in them. Since principles and conclusions are eternally true, they do not change at all, nor is it accurate to say that principles bring it about that the conclusions are true. They account for their truth, but there is no suggestion that conclusions would have been different if the principles had been otherwise; they just cannot be otherwise, and Aristotle does not take this hypothetical approach to the matter.

The "material cause" seems equally unpromising, since it is characterized as

> ***Phys*.II.3 194b24** "that out of which as a constituent a thing comes to be."

This does not appear to describe the relation between principles and conclusions, since principles are not found in conclusions as bronze is found in a statue and silver in a bowl (Aristotle's two examples at b24–25). But another passage brings more kinds of relations under the heading of "material cause,"[66] one of which is that between "hypotheses" and conclusions: "and the hypotheses are causes of the conclusion as 'that out of which' " (*Met.* Δ.2 1013b20–21). If Ross[67] is right to take "hypotheses" as premises, we still do not have the relation we need, since the relation between principles and conclusions of proofs is not the same as that between the premises and conclusion of *any* syllogism. But Aristotle says "hypotheses," not "premises," and uses the word in the previous chapter for principles of proofs. In identifying uses of the term *archḗ* (*APo*'s word for scientific principle—the plural is *archaí*), he says, "that from which the thing first becomes known is also called the *archḗ* of the thing; for example, the hypotheses [are *archaí*] of demonstrations" (*Met.* Δ.1 1013a14–16). Since he is *defining archḗ*, he cannot say that the *principles* of demonstrations are *archaí* because his word for "principle" *is archḗ*. Thus he uses a different word for the same idea. Ross's view that "hypotheses" here means simply premises cannot be right. This makes it very likely that the word bears the same sense, "principles of demonstrations," in the next chapter. Confirmation is provided by the words immediately following: "Also causes are said in as many ways, for all causes are *archaí*" (a16–17). If discussing principles prompts him to speak of causes in Δ.1, and if in Δ.2 he then mentions, among his examples of causes, hypotheses as causes of conclusions, his thought is pretty clearly running along similar lines in both cases. It therefore seems that he claims that the principles of a demonstration are the "material cause" of the conclusion.

It can be objected that the *Physics* also calls hypotheses material causes of conclusions (195a18), where there is no reason for thinking that "hypotheses" refers specifically to scientific principles or "conclusions" to conclusions of proofs, and that *Metaphysics* Δ.2 is "almost word for word identical with *Physics* 194b23–195b21" and probably "belongs originally to the *Physics*."[68] Objectors will then say that the "causal" relation found in demonstrations is not said to be that of material cause.

But even if Aristotle does mean that the principles of a proof are the material cause of the conclusion, it is uncertain what the claim amounts to. It extends the notion of material cause to cover more ground than it did originally,[69] but it is not clear what the point is except to find some place in the four "causes" for this kind of *aítion*, which is so prominent elsewhere. For the premises of proofs have nothing to do with the philosophical projects in which the notion of matter is used. They do not persist in the conclusions, are not the underlying substratum of change, or principles of individuation, or unknowable in themselves, and so on. Moreover, *Physics* II.9 200a15–30 shows that the relation of premises to conclusion "is the converse of the relation of a material cause to that whose cause it is. The premises necessitate and are not necessitated by the conclusion; the material cause is necessitated by and does not necessitate that whose *aítion* it is" (Ross, p. 639). So, even if Aristotle did think that principles are the material cause of conclusions, it is unclear why he thought so or what he thought that identifying them in that way would achieve. Explicating principles in terms of the material cause is a dead end.

The remaining possibility is that the principles are "formal causes" of the conclusion. This type of cause is defined as

> ***Phys.*II.3 194b27–29** "the account of the essence and the genera of this (e.g., the [ratio] 2:1 [is the formal cause] of the octave, and generally number) and the parts in the account"

and alternately as

> ***Met.*A.3 983a27–28** "the substance and the essence."

This is the most promising candidate yet, since subject-attribute proofs depend on the essences of the subject and the attribute (as do the other two kinds of proofs in somewhat different ways). A subject-attribute proof demonstrates that a subject *A* has attribute *B* per se, by virtue of its nature or essence: it shows that *B* belongs to *A* as a consequence of one or more of the elements in *A*'s essence. The proof can be sketched as follows:

A — essence of *A* — *B*

The proof may show that *B* belongs to *A* by showing that *A*'s essence entails that *A* satisfies the defining conditions of *B*; that is, *A* has as attributes essential attributes of *B*. In this case, the proof is outlined as follows:

A — essence of *A* — essence of *B* — *B*

It is therefore true to say that such proofs proceed via the essence of the subject and attribute. The reason why *B* belongs to *A* is that *A* and *B* have the essences that they do. If the middle term is the "cause" and the middle term is the essence,[70] we have a straightforward reason to say that the kind of "causal" relation characteristic of demonstrations is the formal cause.

The same type of reasoning also holds for application arguments. There the form is:

specific subject A — generic subject B — attribute C,[71]

where B—C is immediate or is established by a normal subject-attribute proof, and B is a genus of A, so that A is related per se 1 to B. A is a species or another subclass of B and is defined as such. The remarks in the previous paragraph apply to the proof of B—C. A—B is also an essential relation. Here as before the conclusion A—C depends on the essences of A and C.

Existence proofs also proceed via the essence. Derivative subject B is proved to exist by showing that something (A) whose existence follows from that of the primitive subjects has as per se attributes the essence of B. The proof goes as follows:

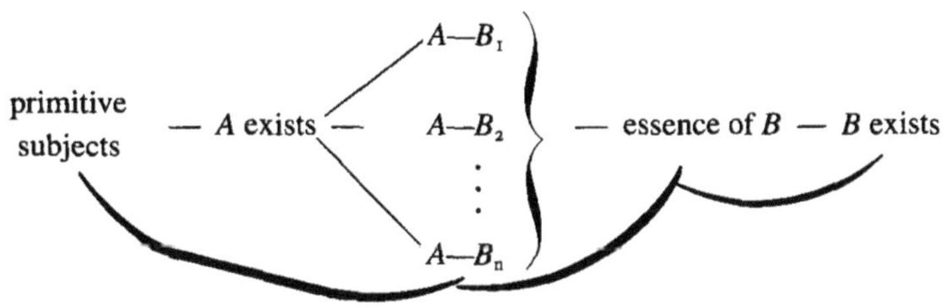

B's essence is the middle and so the "cause."[72]

In all three kinds of proofs the essence (formal cause) is the kind of "cause" in play. But although it is in play, it is not quite the kind of "cause" required. In subject-attribute proofs, the essences involved are the essences of the subject (minor term of the conclusion) and attribute (major term). This holds for the other kinds of proofs, *mutatis mutandis*. The relation here is between one term in the proof and another. But the relation we are investigating is between facts, not terms; between principles and conclusion, or between the principles of a science and its provable facts. And we must keep in mind that claims that the middle (term) is the "cause" are equivalent to claims that the indemonstrable facts that contain the middle term are the "cause" of the conclusion. It is therefore too simple to say that the kind of cause relevant to demonstrative science is the formal cause. That type of cause is relevant in the ways we have seen, but it is not the same relation as the one characteristic of scientific explanations.

This result also applies to II.11, which purports to show that all four "causes" can serve as middles of demonstrations, so that the scientific knowledge that results may be based on any one of the four.[73] It may happen that final, efficient, or material factors enter into a thing's definition—a chair's purpose may be part of the definition of chair. In such cases, the essence contains nonformal elements. But it remains that if such an element is used as a middle term in a proof that the thing has some attribute, it does so by virtue of being part of the thing's essence. The nature of the explanation is the same, only the type of explanans is different. It follows that II.11 does not help solve our present problem.

In general, we should not expect any of the four "causes" to apply here, since they are introduced in the *Physics* in a context alien to demonstrative science. Aristotle says that he must investigate the kinds and number of causes "that have to do with coming into being and passing away and all natural change" (*Phys*. II.3 194b21–22). But these kinds of phenomena are not all applicable to scientific facts. The examples given to illustrate the four "causes" show an interest in accounting for things (especially natural objects, but many examples mention artifacts), not facts. Also, if the four "causes" are to be understood as generative factors,[74] they will have nothing to do with the eternal (and therefore ungenerated) relations characteristic of science. It is best to refuse to identify the relation between scientific principles and conclusions with any of the four "causes," while recognizing that the relation *depends* on the formal cause and perhaps the material cause as well.

Aristotelian Grounds and Humean Causes

It is no more helpful to try to understand the "causal" relation between principles and conclusions in Humean terms, for Hume and his followers are concerned with events, not subject-attribute relations.[75] Moreover, they are interested in particular events, not universals, and in their contingent, not necessary, effects. The kind of things that are cause and effect are vastly different from principles and conclusions of Aristotelian demonstrations, and the nature of the causal relation is vastly different from the relation between principles and conclusions. It is for this reason that I have shunned the translation "cause" for *aitía* and *aítion* throughout.

Aristotelian Explanations and the Deductive-Nomological Theory

The way a demonstration explains its conclusion bears certain relations to the "covering law" or "deductive-nomological" view of scientific explanation. The latter view applies primarily to particular events locatable in time and space, and the explanation refers to initial conditions, which are also locatable in time and space.[76] The emphasis on events rather than subject-attribute connections is un-Aristotelian, but more importantly the concern to explain particulars is the reverse of Aristotle's emphasis on universals, with particulars being explained only incidentally. Also relevant is the fact that this view of explanation violates Aristotle's prohibition on reciprocating attributes' explaining one another. (One of the best examples in the literature on this topic is a pendulum whose period can be deduced from its length or vice versa.) The covering law theory is not grounded in a view of a natural ordering of priority and intelligibility, and so even though its explanations may resemble Aristotelian demonstrations in important respects, the conception of explanation underlying them is different.[77]

It is more useful to compare Aristotelian proofs with modern views on explaining scientific laws, since each proposition of an Aristotelian science can be thought of as a scientific law. (The deductive-nomological view can describe the application of those propositions to individual cases, an aspect of science largely neglected in *APo*.) An enterprise similar to that of demonstrative science is carried on today in certain areas of science: "In the more advanced physical sciences, at any rate, the paramount concern is with the explanation of laws and, in consequence, with the systematic interrelations of laws."[78] The deductive-nomological view can be extended to cover explanations of scientific laws. Nagel asserts that explanations of scientific laws have the following features: (a) "all explanations of laws seem to be of the deductive type"; (b) "all the premises are universal statements"; (c) "there is more than one premise, each of which is essential in the derivation of the explicandum"; (d) "the premises, taken singly or conjointly, do not follow logically from the explicandum"; and (e) "the premises of satisfactory explanations do appear to be more general than the explicanda."[79] I shall restrict myself to indicating briefly how well these agree with Aristotle's requirements.

Aristotle explicitly requires (a) and (b). He also requires (c), since irrelevant premises are not grounds of conclusions. He would reject (e). The demand that conclusions be as general as possible—that the attribute be proved to belong to as wide a subject as possible, and wherever possible to a subject with which it is coextensive—presupposes that premises need be no more general than conclusions. When such proofs employ syllogisms in Barbara, all the premises are as general as the conclusion *and no more so*. He would probably reject (d) too. The discussion of case 2, in which the middle term reciprocates with the attribute of the conclusion, suggests that he thinks there can be two propositions each of which follows from the other. Of course, neither proposition follows from the other one taken alone. Nor does either follow from the other in conjunction with the same additional assumption, since the major premises are different (cf. S1 and S2 on p. 216 above). On the other hand, if we suppose that the two major premises are both consequences of a single law stating a biconditional (a celestial object is near if and only if it does not twinkle), then each minor premise can be deduced from the other in conjunction with this law. In any case, if it turns out that Aristotle agrees with Nagel it will be because of technicalities in the syllogistic, not because of Nagel's reasons. It results that the nature of explanation of scientific laws envisaged by Nagel is not identical with Aristotelian "real" explanation in order of natural intelligibility.[80]

Demonstration, Teaching, and Ideal Explanations

At this point it is worthwhile saying that despite important similarities between Aristotelian demonstrations and modern views of scientific explanation, Aristotelian principles are not exactly explanations of the conclusions. I have preferred to translate *aitía* and *aítion* as "grounds" or "explanatory grounds," partly to avoid the inappropriate translation "cause," and partly to avoid the word "explanation," since even though demonstrations are explanations, they are explanations of a special kind, and it seemed a good idea to find a somewhat different term. Moreover, "explanation" is more squarely in the realm of language than is desirable. An explanation is a series of propositions whose truth accounts for a fact. But Aristotle's premises and conclusions are simultaneously or indifferently facts and propositions. He moves back and forth between the two ways of looking at them, sometimes saying that the principles and conclusions must be true (so that they are propositions) and sometimes treating them materially (e.g., "the conclusion belongs per se to a genus," I.7 75a40–41). "Grounds" and "explanatory grounds" can be used on both levels, since we may say that the premises (propositions) are the grounds of a conclusion (another proposition) and that certain facts are grounds of others. "Explanatory grounds" is clumsier, but conveys the idea that premises are grounds of conclusions by being the grounds (both as facts and as propositions) on which demonstrations are based, since demonstrations provide a certain sort of explanation of their conclusions and show that one proposition follows from others and also that one fact is necessitated by others.

I shall conclude this chapter with a few words about the context in which such explanations are appropriate. It is important to keep in mind that "explanation" is a relative term. Explaining is basically making clear to someone something he does not understand. It follows that what needs explanation is not understood, so when a person understands something, it cannot be explained to him. Moreover, in general when different people do not understand the same thing, different explanations may be appropriate, depending on their knowledge and other factors. Likewise, what it takes to explain something to a single person will be different at different times. Aristotle recognizes these facts:

> ***Top.*VI.4 141b15–19** "Absolutely, then, it is better to try to make what is posterior known through what is prior, inasmuch as such a way of procedure is more scientific. Of course, in dealing with persons who cannot recognize things through terms of that kind, it may perhaps be necessary to frame the expression through terms that are intelligible to them."

> ***Top.*VI.4 141b34–142a6** "Moreover, those who say that such definitions, viz. those which proceed from what is intelligible to this, that, or the other person, are really and truly definitions, will have to say that there are several definitions of one and the same thing. . . . Moreover, to the same people different things are more intelligible at different times . . . so that those who hold that a definition ought to be rendered through what is more intelligible to particular individuals would not have to render the same definition at all times even to the same person." (Trans. Pickard-Cambridge)

What he says here about definitions applies equally well to explanation. His solution is to call for a single definition:

> ***Top.* VI.4 142a7–11** "The right way to define is not through terms of that kind, but through what is absolutely more intelligible: for only in this way could the definition come always to be the same. Perhaps, also, what is absolutely intelligible is what is intelligible, not to all, but to those who are in a sound state of understanding, just as what is absolutely healthy is what is healthy to those in a sound state of body."

The analogous claim for scientific explanations would be that demonstrations are directed at experts, people with a sound understanding of the subject.[81] We are experts when "the same thing is more intelligible both to us and absolutely" (*Top.* VI.4 141b24–25), when we have managed "to start from what is more knowable to oneself and make what is knowable by nature knowable to oneself" (*Met.* Z.3 1029b7–8). For experts, demonstrations make possible the appropriate understanding of all the facts in a science by proceeding from principles, the most intelligible facts, to conclusions, which are less intelligible.

But for the non-expert this is not the case. Insofar as what is less intelligible absolutely is more intelligible to a person, a demonstration does *not* explain its conclusion. But this means that proofs are not appropriate vehicles for teaching. Teaching-learning situations demand presentations appropriate to the students' level, and students are not experts. For Aristotle the goal of learning is to reach the point where proofs based on principles that are most intelligible in nature reflect our own understanding of the subject, but such proofs are not the necessary or the sufficient means for reaching that point.

On the other hand, experts do not learn from demonstrations since they already know. They not only know the principles better than the provable propositions, but also know them *as* principles of those propositions, which is to say they already know the relations of logical consequence and real dependence that proofs display. Proofs only reaffirm what they already know. It follows that proofs are not the best way to teach a subject to anyone. By the time we reach the state when they *would* be the appropriate way to learn, we have nothing left to learn.

It is not hard to think of circumstances where teaching *is* being done through demonstrations. University courses in mathematics and theoretical physics are the most obvious examples. But even here it is not clear that teaching is done precisely according to the rules of Aristotelian demonstrative science. Frequently, steps in the reasoning are not explained, either because they are "obvious" or because they are "beyond the scope of this course," which is to say that the demands of formal rigor are not always satisfied. Perhaps more importantly, in presenting an axiomatized version of a subject, teachers frequently make no effort to go through the stage where the facts are known (KF) prior to being arranged in a deductive system. The consequence is that although students learn how some facts follow from others, they may not have a very sturdy grasp on the truth of what they are proving.[82]

Demonstration is not primarily a teaching device but a kind of ideal explanation. Experts do not learn from demonstrations, but demonstrations reflect their understanding of the subject. This understanding is the goal toward which non-experts strive and against which they are measured. It is as if a demonstration of a fact is the correct answer to the question "why does the fact hold?" and we are graded by how closely our understanding of why it holds approximates the demonstration. The value of the demonstration in teaching non-experts is normative, since it shows them how they should understand the discipline, what facts they should find most intelligible, and the way they should view less intelligible facts as dependent on the most intelligible ones. To become experts they must first get factual knowledge and then learn how the facts are related. A formal presentation of the subject will help in both, perhaps not mainly by stating the facts or even by stating what their relations are,[83] but by presenting the subject in such a way that learners can see that the relations are what they are, and also why they must be that way. At that point they have become experts.

Our Knowledge of the Principles

CHAPTER XVIII

APo SAYS A good deal about conditions principles must satisfy, about the types of principles required and their roles in proofs. To complement this material it should tell how the principles are discovered, since determining them is the indispensable first step in constructing such a science. Such an account would also quell the suspicion that either there are no principles that satisfy the theory's requirements, or if there are they cannot be identified. These problems are especially acute if *APo* is not derived from contemporary mathematical practice. If geometry were a demonstrative science, Aristotle could say, "to reach the principles of any science, copy the geometers—look to your own subject matter and follow their example." But his conception of demonstrative science is a goal arguably not attained even in Greek mathematics,[1] and *APo* may contain a program for improving geometry as well as organizing other branches of knowledge. In any case, even a recommendation to copy geometry is in important ways inappropriate for sciences whose subject matter is nonabstract, nonmathematical, and less precise than geometry.

APo's treatment of this crucial topic is disappointing—only occasional brief and sketchy passages, which are far from adding up to a complete account. Still, when they are supplemented by discussions in other works, generosity can find a coherent view that covers most of the ground. The purpose of this final chapter is to assemble this material and consider the account that emerges. This account is Aristotelian, and if it plugs the hole *APo* leaves open, there can be no objection to using it to complete what *APo* says. Still, if it seems curious that this account is not found in *APo* but has to be extracted from such diverse works as *Topics*, *Prior Analytics*, *Metaphysics*, *De Anima*, *Nicomachean Ethics*, and *De Partibus Animalium*, the following consideration may help alleviate it.

Many of Aristotle's works treat a single topic from a single standpoint.

There is no general work on biology, but a series of treatises of limited scope. Of the most general treatises, *De Partibus Animalium* examines the causes of the parts of animals, and *Historia Animalium* treats the differences among animals, the characteristics that belong to them all, and the causes of these. Similarly, *Physics* treats not the wide range of material included under the heading "things concerned with nature," but only the principles of that subject, and this is all it professes to do.[2] Although this interpretation requires further discussion even for the works mentioned and would probably not hold for every one of Aristotle's works,[3] still it can throw light on his selection of topics. He frequently fails to treat topics of primary importance to the subject and whose importance he should have recognized. In some cases, the reason is that the topic in question is off the point of his discussion. For example, although he is deeply interested in biological classification,[4] the biological works contain no systematic classification of animals into genera and species since the avowed purpose of those works is not to classify animals, but to discuss and account for their parts, their generation, their movement, and so on.[5]

The same can be argued for *APo*. Since the theory it sets out makes scientific principles so important, we naturally wonder how they become known and we expect Aristotle to satisfy our curiosity. But our expectations, although relevant to the subject matter, take us beyond the author's intention. For *APo* discusses the nature of demonstrative sciences, bodies of knowledge *already arranged* as systems of demonstrations that derive conclusions from basic principles.[6] Thus conceived, the work need not consider how to discover facts and how to identify principles. These are related but separate issues.

In fact, *APo* does contain material relevant to these issues, mainly in I.18, I.31, and II.19. But I.18 and I.31 do not purport to be about these questions, and treat them only incidentally. Although I.18 is the most informative passage in *APo* I on induction and the transition from particulars to universals, its purpose is to show that if any of the senses were lacking, some scientific knowledge would be lacking also. The chapter is connected with I.16 and I.17, which take up other kinds of ignorance. (*APo* I.18 presumably treats "ignorance in virtue of a negation," referred to at I.16 79b23.) *APo* I.31 argues that scientific knowledge cannot be obtained directly via sensation (87b28), but requires demonstration. Moreover, II.19 introduces the topic of how principles come to be known as if it had not been covered earlier.[7] Nevertheless, each of these chapters contributes to the topic at hand, and other parts of *APo* II fill in some of the gaps. The task before us, then, is to examine these passages, to determine Aristotle's views on the kind of knowledge we have of scientific principles and how we obtain it, to see whether they are coherent, to determine the gaps, and to supplement the gaps where possible from Aristotle's other writings.

Before taking up these passages, however, we should look at others that maintain that even practitioners of a science should not trouble about principles, but simply accept them. Aristotle frequently says that scientists must assume (*lambánein*) principles (esp. I.10 76a31ff.) or posit (*títhesthai*) them (e.g., I.2 72a21–22, I.10 76b12) or know them in advance (*proginṓskein*) (I.1 71a11, I.2

72a28, a35), without any consideration of how they gain that knowledge. More strongly, he asserts that scientists are not required to account for the principles (I.12 77b5–6). But if it is not the expert's job to justify the principles, whose job is it? Or, if no justification is possible, how can we know the principles better and be more confident of them than the conclusions (I.2 72a25–b4)? Knowledge and confidence seem alarmingly close to blind faith.

These problems can be solved by considering the contexts in which the troublesome claims occur. Whenever Aristotle talks of assuming or positing principles or knowing them in advance, he is contrasting them with conclusions: in a formally organized science conclusions are proved from principles, and principles are not proved, but posited. It does not follow that they are arbitrary or that there is no rational, systematic way to discover them; such matters just lie outside the scope of the discussion.

Also, the point of I.12 is not that scientists have no idea why the principles of their sciences are what they are, but that it is inappropriate to ask them questions about the principles while they are proving theorems. The subject of I.12 is not how principles are known, but what questions fall under a science (77b6–9), and what I.12 says makes good sense in the context of a body of knowledge already organized as a demonstrative science: qua geometer, a person knows the truths of geometry in the appropriate ways and can state its principles and prove its conclusions. A geometrical account of a conclusion is a proof, and there is no such account of principles.

Perception and Science

On the other hand, *APo* occasionally steps outside the context of completed demonstrative sciences and offers glimpses of how to gain knowledge of principles. Induction (*epagōgē*) should play a major role in the procedure, since every argument is deductive or inductive (*APr* II.23 68b13–14) and inductive arguments reveal the universal through the fact that the particular is obvious (I.1 71a8–9). Since scientific principles and conclusions are universal (I.8 75b21ff.), and since conclusions are reached by deduction from principles but not vice versa, Aristotle should hold that we learn principles by induction.

This view is explicit in I.18, a brief chapter that argues that the absence of any of the senses would result in the lack of some scientific knowledge. The main thrust of the chapter is clear despite some obscurity in the connection of thought: demonstration proceeds from universals, we can reach universals only by induction, and perception is prerequisite for induction since induction proceeds from particulars[8] and perception, not scientific knowledge, has particulars as its proper objects.

Knowledge of principles is ultimately based on perceiving particular occurrences of phenomena. Induction elevates us from the level of the perishable world, the here and now to which the truths of science apply only incidentally

(I.8 75b24–30), to the universal, from the "this" (*tóde*) to the "such" (*toiónde*).[9] And the universal level is where the principles are found.

This sketchy account says nothing about how induction works, nothing about the ontological status of universals or the psychological events associated with induction. Also, although principles and conclusions are both universal, it says nothing about how to identify which of the universals gained through induction are principles. It does not even address the need to separate the universal connections that fall under a science from those that do not. Notice in Aristotle's defense, however, that his failure to provide this information here is simply explained by the fact that the subject of I.18 does not require it.

APo I.31, which refutes the view that scientific knowledge can result from perception alone, stresses that the object of demonstrative science is universal whereas that of perception is particular, and bases its arguments on this and related facts. Although not containing the word "induction" (*epagōgḗ*), it refers to induction and offers further information:

> **I.31 88a2–5** "As a result of observing this [connection] occurring many times, after hunting out the universal we would have a demonstration; for the universal is clear from several particulars."

Although this passage offers no precise account of how induction takes place and seems to ignore the distinction between reaching universals and organizing them into a system of principles and conclusions, it reinforces I.18's view of induction and suggests that repeated observations and careful thinking are needed to reach universals. But some cases are simpler:

> **I.31 88a12–14** "If we saw some things, we would not investigate [further for the universal that explains the connection], not on the grounds that we had knowledge by means of seeing, but on the grounds that we had the universal as a result of seeing."[10]

In such cases we gain understanding

> **I.31 88a16–17** "by seeing each case individually but at the same time grasping that it is the same in all cases."

Some universal connections are easier to grasp than others. In no case is the grasping mere perception, but in some it follows closely on the heels of perception. An example shows that if we perceived more finely, some current puzzles would have obvious solutions (87b39–88a17). Whether or not we grasp the universal simultaneously with perceiving a particular case,[11] the important point is that the move from particulars to universals is not a mechanical operation requiring a certain quantity of perceptions as input. The resulting grasp (*noêsai*) that the connection holds in all cases, not just the ones at hand, sometimes requires a search and sometimes is apparent as soon as we see a single case.

Moreover, as I.34 points out, some people have a knack for quickly grasping (89b12) connections that explain phenomena. This knack, "sharpness" (*anchínoia*), is defined as the ability to hit the middle in an imperceptible time, and is characterized in terms of perception:

> **I.34 89b14–15** "On seeing the extremes, the person recognizes all the middle terms that explain them."

APo I.34 confirms that induction is not a mechanical or even an automatic process, but a matter of grasping universals, which may happen quicker or slower depending on the universal and the person in question.

APo I.31 and I.34 share an interest in scientific explanations. It is not merely a matter of moving from a particular event to a description of it that substitutes universal for singular terms. When Aristotle says that scientific knowledge is a matter of recognizing the universal (I.31 87b38–39), he goes on to equate knowledge of the universal with knowledge of the grounds and knowledge of the reason why (87b40–88a2). "After hunting out the universal we would have a demonstration" (88a3–4) means that the universal we find will be the scientific grounds indicated in the demonstration. Also note the emphasis in I.34 on "the middle," "why," and "because," and the identification of grounds with middle terms (I.34 89b11–15).

The preceding discussion has introduced three conceptions of universals: universal concepts (horse), universal connections (all triangles have 2R), and universals as grounds of conclusions. These conceptions are interrelated: a universal connection is a connection between an attribute and a universal concept, and the universal ground of a conclusion is a universal connection. In talking about universals as grounds of scientific knowledge, Aristotle is referring not to a separate type of universal, but to an important role of universal connections in demonstrative sciences.

The three conceptions point to three requirements for an account of how we come to have knowledge of principles. It must explain (1) how we move from individuals (Socrates) to universal concepts (man), (2) how we move from individual connections (my hand is now blocking my sight of the sun) to universal connections (an opaque object placed between a source of light and a surface prevents the light from that source from reaching the surface), and (3) how we identify which universal connections are the grounds of which. Clearly (3) is quite different from (1) or (2). Although I.18 bears only on (1) and (2), I.31 and I.34 raise the issue of (3). Admittedly they do not call attention to how different (3) is from (1) and (2), but this apparent failure can be attributed to the topics of the chapters. In treating the inadequacy of perception to constitute scientific knowledge, I.31 displays different reasons why awareness of particulars, which is all perception is, fails to meet the conditions of scientific knowledge. One reason is that the latter has universals (both concepts and connections), not particulars, as its object. Another is that scientific explanation works at the level of universals even when the explananda are particulars. These points are naturally expressed in terms of the distinction between particular and universal and the roles of universals in science.

The Path to the Principles

The picture so far is the following. Scientific knowledge, which is on the level of universals and involves the ability to explain facts by means of demonstrations, is not identical with acts of perceiving, which inform us that individual instances of connections are currently taking place, and which do not operate on the level of universals or account for what is being perceived. However, scientific knowledge stems ultimately from perception. If we observe enough occurrences of a connection (and in some cases one occurrence may be enough), we can come to have universal and causal knowledge of it.

The question how we move from perceived individual cases to the level of universals is taken up in the final chapter of *APo*, which is introduced as an epilogue. Aristotle announces that he has completed his discussion of deduction, demonstration, and demonstrative science, and will take up two questions concerning the principles: how they come to be known and what the state that comes to know them is (II.19 99b17–18).[12]

The answer to the first rejects the views that knowledge of principles is in us innately and that it comes from some other preexisting knowledge. Therefore we possess a capacity (*dúnamis*) for gaining knowledge of them.[13] After I.18 and I.31, it is no surprise that this "innate capacity for discriminating" (*dúnamin súmphuton kritikēn*) is identified as perception (99b35).

The account of the path from perception to the principles is sketchy and has been taken in various ways. It falls into two parts (100a3–14 and a14–b5), of which the first identifies a number of stages in rapid fashion. I shall argue below that there are five: (1) perception, (2) memory, (3) "experience" (*empeiría*), (4) the stage where we have reached the level of universals, and (5) the stage where we grasp the principles.[14] The first part of this account reads like a normal empiricist version of the origin of knowledge, and I do not propose to discuss the nature of memory or say more about perception. The first difficulty comes in determining the nature of "experience." The parallel passage in *Metaphysics* A.1 offers assistance.

> **II.19 100a4–6** "Experience arises from frequent memory of the same thing, since memories that are many in number constitute a single experience."

> ***Met.* A.1 980b28–981a1** "Experience arises in people from memory, for many memories of the same thing result in the capacity of a single experience."

The *Metaphysics* passage goes on to point out that experience resembles science and art,[15] and, like *APo*, to say that science and art arise from experience. Unlike

APo it identifies two differences between the two stages: (1) experience stays on the level of particulars, whereas science and art are on the level of universals (981a15–16), and (2) people at the stage of science and art can explain what they know or do, whereas people on the level of experience cannot (a28–30). These differences are valuable clues to Aristotle's view of the nature of experience and for understanding *APo* II.19. The first is stated together with an example:

> ***Met.* A.1 981a5–17** "Art arises when from many notions of experience there comes a single universal grasp concerning similar things. For it is a matter of experience to have a grasp that this thing helped Callias when he had this illness, and also Socrates and so for many individuals. But it is a matter of art to grasp that it helped all people of a certain kind—marked off in a class—who had this illness, for example, phlegmatic or bilious people burning with fever. For as far as performance goes, experience appears no different from art, but in fact we see people of experience who are even more successful than those who have a theoretical knowledge of the subject without experience. The reason is that experience is knowledge of the particulars, whereas art is knowledge of the universals, and all actions and comings-into-being are concerned with the particular."

Our question is precisely how experience differs from memory and from art and science. Where does it fit between them? At first sight the past tenses of the verbs in the passage might suggest that experience is hardly different from memory, which should also contain information such as "this helped Callias when he had this disease," but since experience also applies knowledge about the past to present cases, it is an advance on memory. We may now wonder how it can take place without reaching the level of universals. For the fact of recognizing that Callias, Socrates, and so on were cured by the remedy in question and that therefore it should be applied to another individual with the same illness seems to presuppose that those cases are "marked off in a single class"; this is true whether the class in question is the class of people with the particular illness or a certain kind of people with the illness (as if the treatment for phlegmatic people is different from that for others). But since marking the individuals off in a single class is characteristic of art, the distinction between experience and art is in danger of collapse.

The solution, I think, comes from two directions. The first has been set out by Ross.

> After having described [experience], however, as produced by many memories of the *same* object, Aristotle proceeds to describe it as embracing a memory about Callias and a memory about Socrates. These are not the same object, but only instances of the same universal; say, "phlegmatic persons suffering from fever." An animal, or a man possessing only *empeiría*, acts on such memories, and is

> unconsciously affected by the identical element in the different objects. . . . Experience . . . is a coagulation of memories; what is active in present consciousness in virtue of experience has not been experienced together. Therefore (a) as embodying the data of unconsciously selected awarenesses it foreshadows a universal; but (b) as not conscious of what in the past is relevant, and why, it is not aware of it as universal.[16]

People with experience can recognize one case as like others, but not say why or in what respect. They have not *consciously* marked off the individuals into a class, cannot "put their finger" on the common element (the class concept). They cannot explain their actions since they lack articulate knowledge of the concepts needed for an adequate explanation. For this reason too they are poor teachers (981b7–9). Aristotle gives no credit to the way a master teaches a trade to an apprentice, the teaching consisting in on-the-job training, which leads to experience.[17]

The other approach to the problem is suggested by the assertion (981a12ff., quoted above) that a person with experience tends to do a better job than someone who knows universals but lacks experience, the familiar picture of the novice with plenty of book learning. The mention of such cases shows that the final stage, or at least knowledge of universals and the ability to give explanations, can be attained without experience. For this not to fly in the face of Aristotle's account it must be taken as follows. In the history of human learning, relevant experience must precede achieved scientific knowledge, but once the latter is reached it can be taught to others without their having to go through the preliminary steps. On the other hand, for practical applications there is no substitute for experience, which can be attained only by familiarity with individual cases. Sophomoric individuals who have knowledge without experience can mark off different kinds of cases. They know remedies for phlegmatic people suffering from fever and for bilious people with the same illness. They may be unable to recognize a phlegmatic or bilious person, but the distinctions will be clear in their minds. People with experience, however, are good at spotting the kind of case at hand, whether or not they can articulate (in technical language or otherwise) the classes and differentiae involved.

The two approaches to the problem of the relation between experience and knowledge of universals show that articulate universal knowledge is neither a necessary nor a sufficient condition of experience. They also indicate the intellectual sophistication present at the level of experience. The second difference between experience and the more developed states agrees with this view. According to it, people with mere experience cannot properly explain what they know: they know the fact (*hóti*), but not the reason why (*dihóti*), as Aristotle says in language familiar from *APo*.[18] The *Metaphysics* passage contrasts proper scientific (*dihóti*) knowledge not with scientific knowledge manqué, but with a kind of knowledge that in its nature cannot pretend to be scientific knowledge. It stays on the level of particulars, and so does not have explanations at all. Behind this charge are the views that explanation works at the level of universals, and that explanations cover particulars only because the particulars are instances of the universals that are explained. Perception is the most authoritative knowledge of individuals, but does not give us the "why" of anything; it tells us that fire is hot, but not why.[19] This feature of perception also holds for experience since in neither case has knowledge of the universal been attained.

It is clear, however, that even if the knowledge associated with the ability to give an explanation requires knowledge of universals, the two types of knowledge are not the same. It is conceivable that we may have the latter without the former—we may grasp the appropriate universals without understanding how some are explained by means of others. Scientific knowledge requires both kinds (when we have them both, we *have* scientific knowledge), and Aristotle owes us an account of both.

Did he even recognize that they are different? There is a small amount of evidence that he did. After saying that memory arises from perception and experience arises from memory, he says:

> **II.19 100a6–8** "From experience, *or* from the whole universal at rest in the soul, the one beside the many, which is one identical thing in all of them (the many), [arises] the principle of art and science."

The issue in interpreting this passage is how to take the conjunction "or" (*ḗ*). Is it (a) disjunctive (the principle of science comes *either* from experience *or* from the universal in the soul), (b) explicative (it comes from experience, *that is to say*, from the universal in the soul), or (c) progressive (it comes from experience, *or rather* from the universal in the soul, which is the next stage after experience)? The first can be eliminated straight off. Aristotle is not offering alternative accounts of the origin of principles, and worse, talk of the universal's being in our soul apart from the perception-memory-experience route would defeat the anti-Platonic account he is presenting. The second is adopted by most translators and commentators,[20] but it contradicts the *Metaphysics* view that experience remains on the level of particulars. The parallels between *Metaphysics* A.1 and *APo* II.19 are so close that we should be loath to find any significant disparity in the conceptions of experience present in the two chapters. Accordingly I prefer the third interpretation of "or," on which "the universal at rest in the soul" is a stage intermediate between experience and scientific knowledge.[21]

Finally comes the stage where we know principles (100a8).[22] This stage is an advance on the level of knowing universals.[23] The nature of the advance can be understood if we keep in mind that "principle" is a relative term: a principle is a principle *of something*. To grasp something as a principle is to understand how the things of which it is a principle depend on it.[24] In the case of demonstrative science, to grasp something as a principle is to grasp it *as* true, immediate, primary, and prior to, grounds of, and more naturally intelligible than the connections for which it is a principle (cf. I.2 71b20–22).[25] At this point we have scientific knowledge of the entire subject matter of a science, since recognizing principles involves recognizing that they are principles and why they are principles, and also recognizing that the universals that are not principles are dependent on the principles and how they are dependent. Thus, knowing principles entails appropriate knowledge of nonprinciples, and this amounts to possessing the whole demonstrative science.

Aristotle mentions five levels of awareness, from the direct and unreflective awareness of perception to the highly structured awareness of a body of knowledge that we have upon knowing the principles. Each succeeding stage brings a significantly different level of awareness, and the resulting picture serves Aristotle's purpose well. He then claims that his account satisfies the initial conditions (cf. 99b25ff.) of the chapter:

> **II.19 100a10–11** "So the states (memory, etc.) are not innate[26] in a determinate form and come not from higher[27] states, but from perception."

Next he offers a simile:

> **II.19 100a12–14** "As happens when a rout has occurred in a battle and one man has stopped, another stops and another, until it reaches the original position[28]—the soul is such as to be able to have this happen."

This simile has favorably struck some commentators[29] and has given rise to much speculation over the circumstances imagined in the simile and its precise application to the matter at hand. However, like some other famous similes of Aristotle's,[30] this one does nothing to make the ideas clearer.

Grasping the Universals

Aristotle's answer to the question how the principles come to be known is of limited value. In effect he says that we come to have this state as a result of having had a series of other states. He says nothing about how we get from one state to the next and little about what it signifies to have these states. Nor does he argue for this theory of the acquisition of knowledge. In my opinion, he thought the account, or at least the first part of it, would be generally accepted: "From perception arises memory, *as we say*, and from frequent memory of the same thing happening arises experience, *for* many memories in number are one experience." "As we say" refers to the preceding passage, which states without proof that in some animals there comes to be a "remaining of the perception." The plausibility of this view must rest on opinions Aristotle expects his readers to share. Likewise, "for many memories are one experience" will be an argument or explanation (as "for" suggests) only if it is an appeal to ordinary views. As it stands, it only repeats the initial claim about the origin of experience.

Even if Aristotle can expect general agreement so far, the agreement should be qualified, since the account can only work for perceptual memory, and he does not argue that this is the only kind, or the basic kind on which all other kinds depend, or the kind relevant to learning scientific principles. Moreover, even if the appeal to ordinary views carries him as far as the level of experience, it is unlikely to take him any farther. Perception, memory, and experience are familiar conceptions, but the conceptions of universals and of scientific principles are technical philosophical ideas. Aristotle owes us a clearer explanation of what is involved in coming to know universals and principles. He acknowledges as much and pays part of this debt in the following lines:

> **II.19 100a14–b3** "Let us say again what was just said, but not said clearly. When one of the indiscriminables[a] has stood still,[b] the first universal is in the soul. (For to be sure it is the particular that is perceived, but perception is of the universal, e.g., of man and not of Callias man.) Again, a stand is made in these until the universals without parts[c] stand still, for example, first such and such an animal, and then animal, and also similarly here."

NOTES ON TRANSLATION

a. "Indiscriminables" refers to least general universals, *infimae species*; cf. II.13 97a37–39, b31.

b. Notice the frequent use of the verb "stand," which echoes the metaphor of the routed soldiers stopping in their flight, and to some extent continues it.

c. "The universals without parts" are presumably the most general universals. These will be "without parts" in that they cannot be further analyzed into genus and differentia.[31]

The picture is fairly clear and well adapted to a system of classification by genus and species. The claim is that by noticing a number of relevantly similar individuals we come to have the concept for an *infima species*. When we have done likewise for all[32] the species falling under a genus, we may consider them together and get the generic concept. This same process can be continued stage by stage until we reach the most general universals.[33]

The passage presents four main difficulties of interpretation. (1) It fails to call attention to the difference between the first and subsequent steps in this process, although the move from individuals to universals is logically different from that from less general to more general universals. This oversight or error has been seen elsewhere (e.g., I.24), and I will not say more about it. However, the comment in parentheses can be read as a hint of Aristotle's views on this issue.[34] (2) It claims to clarify what has been said before,[35] but can it be seen as clarifying all that is said in 100a3–14? (3) It discusses the acquisition of increasingly general universals, but what relation does this topic have to the subject under discussion, how the principles of sciences come to be known? (4) What is the process of "coming to have" the concepts?

(2) The question here is what Aristotle refers to as "what was just said." *APo* II.19 100a3–14 is so sketchy that we would like a fuller account of the whole process from perception to principles, but the passage simply does not do that. On the other hand, if he thought he could base some of his account on ordinary views about perception, memory, and experience, the part needing additional clarification would be the final two stages. In fact, the present passage amplifies the account of the first of these two stages, the advance from experience to knowledge of universals, pointing out that there are universals of different levels of generality and claiming that we come to know universals in order of increasing generality. "What was just said" refers to 100a6–7.

(3) The examples (b2–3) prove that a14–b3 deals with the acquisition of universal concepts, whereas what is needed is an account of how we come to know connections of subject and attribute, since the principles and conclusions of scientific demonstrations are connections, not concepts. On the other hand, 100a3–14 can be read as treating either universal concepts or universal connections. Ross and Barnes prefer to see universal concepts throughout both passages, but consistency is achieved at the price of taking the entire discussion as misdirected.

I prefer to understand the passage as answering the initial question of the chapter, how principles come to be known. This is the best interpretation of the perception-memory-etc. account, since we can perceive individual instances of a connection as well as individual instances of universal concepts (e.g., I.31 88a1), and the same goes for memory. The example of experience in *Metaphysics* A.1 is of a connection, not a concept, and "universal" is used of connections in *APo* (e.g., I.8). Finally, the principles of sciences must be connections, not concepts.

If 100a14–b3 is taken in the way just suggested, it is by no means irrelevant to the initial question. There are close relations between knowledge of universal concepts and knowledge of universal connections. Connections are connections between terms, and in scientific proofs all the terms must be universal.[36] Universal concepts of different levels of generality have their place in sciences, as the oft-repeated example about isosceles triangles versus triangles having 2R makes clear.[37] It is therefore useful to make the point that we come to have knowledge of universal concepts at these different levels.

Moreover, the same holds for knowledge of connections of different levels of generality, and the same example applies. In fact, if we read 100a14–b3 in terms of universal connections, its doctrine of how we come to know connections of increasing degrees of generality is related to the method sketched in I.5 for telling when we have a universal connection (74a36–b4). The attribute 2R belongs to the isosceles bronze triangle, and it belongs when abstraction of "being bronze" and "being isosceles" is made, but not when "figure" or "limit" are abstracted. First we notice that there is a connection between a subject and an attribute, and then we discover that there are also connections between wider subjects and that attribute. Thus the fact that the examples are examples of concepts, not connections, no longer needs to be taken as a sign of confusion. For knowledge of universal connections presupposes knowledge of universal concepts; moreover, it is reasonable for Aristotle to believe that we gain knowledge of universal concepts and connections in similar ways.

The interpretation offered makes 99b35–100b3 relevant to the stated purpose of II.19 and makes 100a14–b3 relevant to the preceding passage, which it claims to explicate. *APo* II.19 100a14–b3 explains only part of what needs to be explained in the preceding passage and says nothing about moving from the stage of knowing universals to the stage of knowing the principles, which will involve the quite different processes of discovering logical relations among the connections and determining which are most intelligible and explain the rest.

(4) To understand what is involved in "coming to have" knowledge of universals, we must be clear what is at issue. The process sounds automatic and unreflective: something that we have the capacity to do (99b35) and that is the sort of thing that can happen in our mind (100a13–14). But the text is also compatible with the view that progress from one stage to the next requires systematic intellectual effort. Still, Aristotle does not explain how the mind transcends the level of particulars to reach universals: he does not explain what if any mental effort is required, or specify what change occurs in us when we come to grasp universals, or say what it is to have a universal "at rest in the soul." These would be grave omissions in a general theory of universals or of epistemological psychology; however, Aristotle does not intend to present such a theory, but to say how scientific principles come to be known (*gnōrimoi*, 99b18).

This question is to be understood in terms of the use of *gnōrimon* elsewhere in *APo*. Recall the opposition between things "prior and more *gnōrimon* without qualification or in nature" and things "prior and more *gnōrimon* to us" (I.2 71b33–72a5). Scientific principles are most *gnōrimon* in nature and so *least gnōrimon* to us. *APo* II.19 gives a sense to the claim that the things more *gnōrimon* to us are "nearer perception" and those more *gnōrimon* in nature are "farther from perception" (I.2 72a1–3). Similarly for the statement that the latter are most universal and the former are particulars (I.2 72a4–5). What Aristotle wants, then, is an account of how we pass from what is more *gnōrimon* to us to what is more *gnōrimon* in nature,[38] or of how what is *gnōrimon* in nature comes to be *gnōrimon* to us.[39]

Gnōrimon and its forms cannot be translated throughout as "known" without awkwardness: what is by nature less known may be better known than what is by nature better known.[40] "Knowable" and "intelligible" are better, but require explanation. The most helpful approach is offered by the discussion of definition in *Topics* VI.4,[41] which requires definitions, like demonstrations, to proceed from what is prior and more *gnōrimon*—the only way to avoid a plurality of definitions of the same thing (141a26–b2). It continues as follows:

> ***Top.* VI.4 141b5–142a10** "What is prior is without qualification more *gnōrimon* than what is posterior, as point is to line, line to surface, and surface to solid. . . . But sometimes it falls out in the op-

> posite way for us. For what is solid is most directly perceived, and the surface is more so than the line, and the line than the point. For most people know such things first, for it is characteristic of ordinary intelligence to learn these, and of precise and fine intellect to learn the others. Now speaking unqualifiedly, it is better to try to get to know what is posterior through what is prior, since this way is more scientific. However, for those who cannot get to know something in this way, it is perhaps necessary to frame the account [*ló-gos*] in terms of what is *gnṓrimon* to them. Examples of such definitions are the ones defining a point as the extremity of a line. . . . But we must not fail to notice that such definitions cannot express the essence of the definiendum, unless the same thing happens to be more *gnṓrimon* both to us and without qualification. . . . The view that definitions in terms of what is more *gnṓrimon* to each person are really definitions is open to the objection that there are many definitions of the same thing. For different things turn out to be more *gnṓrimon* to different people, and not the same to all, so that different definitions would be given in terms of what is more *gnṓrimon* to each. Moreover, different things are more *gnṓrimon* to the same person at different times—perceptible things at the beginning, and the opposite as they get more precise, so that on this view the same definition will not always be given even to each person! It is clear, then, that this is not the way to define, but it should be done via what is without qualification more *gnṓrimon,* for only this way will there always turn out to be one and the same definition. Perhaps further, that which is without qualification *gnṓrimon* is not that which is *gnṓrimon* to all, but what is *gnṓrimon* to those properly disposed in their intelligence."

The greatest value of this passage is the refinement it gives to the notion of what is *gnṓrimon* to us. The "we" here is not all human beings but "the many" (b12), those who are not "properly disposed in their intelligence" (a10), who do not, or do not *yet*, have a "precise and fine intelligence" (b13), in short, who are not experts. Aristotle does not deny that what these people can grasp is somehow *gnṓrimon*, namely, relatively to them. To this extent, what can be known can be known by individuals, and the claim of ordinary folk to knowledge is not rejected, though it is distinguished from the ideal. Further, what is without qualification *gnṓrimon* is not beyond human reach, but is what is *gnṓrimon* to the expert. Moreover, experts are made, not born.[42]

This analysis supports translating *gnṓrimon* as "intelligible," since different things are intelligible to different people and to the same person at different times, and we can speak of what is intelligible without qualification as a standard for certain concepts, such as definition, proof, and explanation. The latter will be what experts agree to be satisfactory in relevant ways.[43]

As to the question how what is more *gnôrimon* without qualification comes to be more *gnôrimon* to us, part of the answer is now at hand: it comes to be so as we come to be experts, as our intelligence comes to be correctly disposed. Connecting *Topics* 142a2–4 with *APo* II.19 100a3ff., we see that these changes are a matter of moving from the condition where perceptible particulars are the most intelligible to us through the stages of memory, experience, and thought at the level of universals, and finally reaching thought that reveals the truth of things through their relation to the principles. This process consists in enlarging and enriching our awareness from the level where we are limited to the immediate apprehension of individuals by perception to the highest stage, where we see individuals as unimportant except as instances of scientifically explainable universal truths.

Aristotle speaks of this process not as a shadowy way of constructing universals out of the residue left by particulars,[44] but in terms of the verb *gnōrízein*, "become acquainted or familiar with," "come to know," "recognize," and the related adjective *gnôrimon,* "familiar" or "intelligible."[45] *APo* II.19 begins by asking how the principles come to be known and ends its discussion of this question by saying "we must come to know the primary things by induction" (100b4).[46] *APo* I.31 proves that perception cannot provide scientific knowledge by saying that perception must have particulars as its objects, but scientific knowledge is a matter of coming to know the universal (87b37–39). Although inadequate as a full description of scientific knowledge, this claim throws light on the approach of II.19. The universal is present in the particulars (100a7–8), and our task is to recognize it, to become familiar with it. Aristotle identifies the cognitive dispositions a person has during the process of becoming familiar, and quite reasonably supposes that he has given an answer to the question how we become familiar with them. By that question he merely intends to ask what stages we pass through in coming to know the principles. He does not need to explain what mental work is needed to do this, what change occurs in us when we come to grasp a universal, or what it means psychologically or physiologically for us to have a universal "in the soul." He can be content with indicating that there is a point at which we recognize what the many particulars we have perceived have in common, and instead of giving an account of the mechanism by which we do this, he can say simply that we can do so because of the way our soul is constituted (100a13–14).

APo II.19 100a14ff. should be read in this light. It tells more about the stage where the universal comes to be at rest in the soul, but consistently with the rest of the passage does not discuss psychological events. The earlier statement needs amplification because what it mentions as a single stage really covers several substages, since there are universals of different levels of generality and according to Aristotle they are learned in sequence and do not all equally proceed from experience.

The stretch of text beginning at 100a14 concludes with this summary statement:

II.19 100b3–5 "So it is clear that we must come to know the primary things by induction; for also that is how perception plants the universal in us."

It is easiest to take "that is how" (b5) to refer to induction (b4)[47] and "induction" as the term for the process described at a15ff.[48] This is some kind of argument, as "for" shows. The main issue is how to take "primary things" (b4), an expression frequently used for scientific principles.[49] If it has this sense here,[50] the argument goes: perception instills universal concepts by induction, as we have seen; so it is clear by analogy that we come to know principles by induction. On this reading Aristotle is confused as to what he has shown. The previous passage said nothing about principles, only something about universals, and no reason is given to think we reach principles and universals the same way. This interpretation can be maintained if 100b3–5 summarizes not just a14ff., but everything from 99b35, which does take us as far as the level of principles. This view is implausible, however, since 100b3–5, with its references to induction and the universal, is closely related to 100a15ff. but not to the rest of the section.

Alternatively we may take "primary things" as the most general universals in a science (as at I.10 76a32).[51] The passage then asserts that perception instills the most specific universals in us by induction, and so by extension (or, as we have just seen) we come to know the most general universals by induction. On this view, the passage summarizes 100a14–b3.

Induction

To decide between these views we must determine what Aristotle means by induction. Is it precisely the process that leads from perception of individuals to recognition of universals through the intermediate stages memory and experience? Is it something more general that includes this process among others? Or is it something that presupposes the capacities to perceive, recall, and so on, but differs from the exercise of these capacities?[52]

The words *epagōgḗ*, *epágein*, and *epaktikós* have a variety of uses in Aristotle, some technical and some not.[53] The technical, dialectical, and logical uses vary. Ross's survey and analysis[54] deserve mention, but more helpful for now is Engberg-Pedersen's schema of uses of *epágein* and *epagōgḗ*[55] In his view, "the root idea of Aristotelian *epagōgḗ* in its full, technical sense" is "something like 'attending to particular cases with the consequence that insight into some universal point is acquired' or 'acquiring insight into some universal point as a consequence of attending to particular cases.' "[56] (Note that "particular cases" can include individuals, as opposed to universals, and specific, as opposed to general, terms. The "particular"/universal distinction also covers items that can appear as subjects and attributes in predications, and also connections of subjects and attributes that are expressed in subject-predicate propositions.) This "root idea" is not found in all cases where the words are used in logical contexts,[57] but this or something like it is the use relevant to *APo*.

Epagōgḗ is prominent in a variety of contexts. *Topics* I.12 identifies it and deduction (*sullogismós*) as the two forms of dialectical reasoning and describes it as the more convincing and clear of the two, more intelligible from the point of view of perception, and appropriate to use in dialectical discussions with the many (105a16ff.). Its use in dialectical situations is suggested in *Topics* VIII.2 157a34ff.: a number of cases are produced to force one's opponent to grant a general claim or produce a counterexample, and except in unusual circumstances, he may not offer the case in dispute as a counterexample.[58] Likewise, *epagōgḗ* is said to play an important part in teaching and learning,[59] whether done cooperatively or alone.

Epagōgḗ does not require a large number of cases, just enough for the individual in question to recognize the common principle or element in the cases given. The number needed will depend on the aptitude of the individual and the value of the cases; in some situations a single case will suffice.[60]

The induction described at *Topics* VIII.2 157a34ff. is a form of argument. Another kind of argument associated with induction is found in *APr* II.23. This is so-called "perfect" induction.[61] Aristotle's example is as follows:

> ***APr* II.23 68b19–29** "Let *A* stand for 'long-lived,' *B* for 'bileless,' *C* for particular [species of] long-lived animals, such as man, horse, and mule. Now *A* belongs to all *C*, since every *C* is long-lived. Also *B* belongs to every *C*. So if *C* and *B* are convertible and *B* is not wider in extension, *A* must belong to *B*. . . . But we must grasp that *C* is composed of all the particulars, since the *epagōgḗ* proceeds through *all* of them."

This argument has little in common with the other. Both are called *epagōgḗ* and both yield a general conclusion, here, that all bileless species of animals are long-lived. But here it is essential to know that all cases (in this example, species) have been covered, a requirement not generally needed for the former kind of *epagōgḗ*.[62]

Epagōgḗ need not be restricted to the two kinds of argument so far identified. Hamlyn suggests that the inductive arguments for whose use Aristotle praises Socrates[63] are like the one in *Meno* where Socrates leads a slave to understand a geometrical relation in a particular diagram and *ipso facto* to see that it holds for all similar cases.[64] Although Aristotle does not specify exactly what features of which Socratic arguments are inductive, there is no reason to deny that this type of argument is covered by the "root idea" of Aristotelian *epagōgḗ*.

There may be other ways to advance to general or universal knowledge. Aristotle's assertion that all convictions result from deduction or induction[65] leaves the field wide open. Since the one proceeds to the universal or general and the other proceeds from it,[66] *any* way of proceeding from the particular or specific to the universal or general should count as *epagōgḗ*.

In my view, *APo* portrays *epagōgḗ* differently from (though perhaps not incompatibly with) *epagōgḗ* by argument. *Epagōgḗ* has no place in achieved demonstrative sciences. Its use in dialectical argument is irrelevant,[67] as is its use in teaching. Demonstration is a form of teaching, but is quite different from *epagōgḗ*.[68] But *epagōgḗ* does play a role in reaching scientific principles, and in this role it is connected with perception.[69]

The following is a sketch of its role. Coming to know scientific principles involves two stages of thought: acquiring universal concepts and connections, and organizing them into a deductive system. *Epagōgḗ* has to do with only the first stage. It consists in passing from the level of particulars to that of universals, of whatever generality, which may or may not happen as the result of argument. It is a matter of coming to see a particular as a particular of a given kind, coming to see Callias as a man.[70] We *come to see* the universal in things. We do not in general see it right off, although that happens in some cases.

That *epagōgḗ* has to do with grasping the universal is obvious from such passages as I.18 81b2 and II.19 100b4–5.[71] On the reading advanced above, 100b3–5 summarizes a15ff.; the latter in turn expands part of the previous discussion (a6–7), which moves too quickly from the level of experience to that at which universals come to be at rest in the soul. The expanded treatment speaks of the occasion when the first (most specific) universal comes to be in the soul, and further speaks of becoming acquainted with more and more general universals. Perception brings us to have the universal "in our soul" by *epagōgḗ* (b5). This situation is described at a15–16: "when one of the indiscriminables has stood still, the first universal is in the soul," a reference to a3–6, which identifies the stages from perceiving particulars to grasping universals. It also picks up the doctrine of I.18 on the relation between perception, *epagōgḗ*, and the grasp of universals. But the process does not stop here. *Epagōgḗ* not only gets us to the most specific universals, but also advances to the most general. So far the passage does not contain an argument, only an analogy: we learn the most specific universals from perception by means of *epagōgḗ*; similarly we learn the most general universals by means of *epagōgḗ*. I suggest that the argument depends on a tacit premise that all movement from *kath' hékaston* to *kathólou* (which covers movement both from particulars to universals and from specific to general) is due to *epagōgḗ*. The argument goes: (1) the move from perceiving particulars to grasping the most specific universal is due to *epagōgḗ*; (2) the move from perceiving particulars to grasping the most specific universal is a move from *kath' hékaston* to *kathólou*; (3) all moves from *kath' hékaston* to *kathólou* are due to the same thing; (4) therefore, all moves from *kath' hékaston* to *kathólou* are due to *epagōgḗ* (from 1, 2, 3); (5) the move from grasping the most specific universal to grasping the most general universal is a move from *kath' hékaston* to *kathólou*; (6) therefore, the move from grasping the most specific universal to grasping the most general universal is due to *epagōgḗ* (from 4, 5).

This reconstruction reflects the parallel Aristotle sees between moving from particulars to universals and moving from specific to general. It also accounts for his using the parallel as a basis to infer that both processes are due to the same thing.[72] Therefore, *APo* requires nothing more of *epagōgē* than its normal function of getting us to know universals (the more general) from knowing particulars (the more specific). The passage does not support the view that *epagōgē* gives us knowledge of principles.[73]

Although *epagōgē* can occur through argument, *APo* stresses not argument but perception. The crucial passage is this:

> **II.19 100a16–b1** "For he perceives the particular, but the perception (or, his perception) is of the universal, for example, man, but not Callias [the] man."[74]

This passage raises several problems. It seems to maintain that whenever we perceive something, we necessarily perceive it as an instance of the universal it falls under. But in the first place this claim seems plainly false. We frequently fail to classify correctly things we perceive. Second, it seems to give too big a role to perception: if perception does so much, what need is there for memory and experience? Third, it makes a distinction between what is perceived and what the perception is of—a tenuous distinction.[75] Finally, it appears flatly to contradict a passage that says that perception is *not* of the universal:

> **I.31 87b39–88a2** "If we were on the moon and saw the earth acting as a screen, we would not know the cause of the eclipse. For we would perceive that it is now eclipsing, but not in general why. *For the perception was not of the universal.*"

According to Ross,[76] 100a16–b1 explains how "the 'standing still' of an individual thing before the memory is at the same time the first grasping of the universal; this is made easier to understand by the fact that even at an earlier stage—that of perception—the awareness of an individual is at the same time awareness of a universal present in the individual; we perceive an individual thing, but what we perceive in it is a set of qualities each of which can belong to other individual things." Ross claims that what is involved in seeing Callias and identifying him as a man is the same as what is involved in seeing Callias and identifying him as pale. Both pale and man can belong to other individuals and so are universals. And even at the earliest stage (perception) we are aware of perceptible qualities.

Some reject this view of perceptible qualities. According to Owen,[77] we see a *particular* pallor, though other individuals can have this same particular shade too. (In this respect colors are different from substances.) And so, how from seeing a particular pallor we attain the notion of pale remains just as much a problem as how we attain the notion of "human" from seeing an individual human.

A further objection may be granted by those who do not accept Owen's view of individuals. For Aristotle, we grasp perceptible qualities differently from other aspects of things. "Perception is affected by what has color or flavor or sound, but not by the way in which each of them is called [Callias, man] but as such and such [pale]" (*De An.* II.12 424a21–24): we perceive color and so on directly, but not so Callias or man. The former are proper objects of the senses, the latter incidental objects.[78] The distinction between what we do and do not perceive directly is central to Aristotle's theory of perception and fundamental to the empiricist epistemology of *APo* II.19. He says nothing about how we become aware of incidental objects, but this process is obviously different from our direct perception of proper objects.

Two passages from *De Anima* suggest a way to bridge the gap between perceiving proper objects and grasping universals:

> ***De An.* III.8 432a4–5** "The objects of the intellect are among (or in) the sensible forms."

> ***De An.* III.4 429a15–18** "[The part of the soul by which it knows and understands] must be receptive of the form . . . and as the faculty of sense is to sensible objects, so must the intellect be related to intelligible objects." (Trans. Hicks)

If objects of the intellect *are* sensible forms, and if the intellect stands in the same relation to its objects as the senses do to theirs, then in apprehending sensible forms we *are* apprehending intelligibles. Ross's account can be sustained if (1) sensible forms are the same as proper objects of the senses, (2) objects of the intellect are (or include) universals treated by science, and (3) 432a4–5 means that the objects of the intellect are among (are a subset of) the sensible forms.

(1) The notion of "sensible form" is basic to the view that it is not the things perceived that are in the soul, but their forms: "the rock is not in the soul, but only its form" (*De An.* III.8 431b29), and "the sense organ receives its object minus its matter" (*De An.* III.2 425b23–24).[79] Perceived forms are sensible forms,[80] "the forms of sensible things, the qualities in them which constitute them what they are, the red of the red thing, the sound of the resonant thing, the flavor of the flavored thing."[81] Thus sensible forms appear to be the proper objects of perception.

(2) *De Anima* agrees with *APo* that the objects of scientific knowledge are universals, and these are in some way in the soul:

> ***De An.* II.5 417b22–25** "Actual sensation has particulars as its objects, whereas scientific knowledge has universals. These are somehow in the very soul. Therefore it is in a person's power to think [*noêsai*] whenever he wishes, but not to perceive, since that which is perceived must be present."

This passage makes a contrast between the objects of perception and of science, and connects science with thinking (*noêsai*). It is reasonable to conclude that the objects of thought (*tà noētá*) are, or include, universals treated by science.

(3) The claim that the objects of the intellect are in the sensible forms could mean that they are *among* perceptible forms, a subclass of them (so Hamlyn), or *inherent* in them, existing in them (Simplicius, Hicks, Rodier, Ross). The first interpretation promises a simple solution to our problem: universals with which science deals are intelligible forms and so perceptible forms grasped by perception, like other perceptible forms. But this interpretation conflicts with the notion of "perceptible form" for reasons given above (pp. 253–54); nor does it fit the following lines as well as the second interpretation does, since the objects of the intellect include "abstractions and all the states and attributes of perceptible things" (*De An.* III.8 432a5–6) and it is hard to see how to count these things as perceptible qualities.

De Anima III.8 432a4–5 is therefore best taken as saying that objects of the intellect are *inherent in* perceptible forms, but how we recover intellectual forms from sensible forms remains obscure. Aristotle is silent on this subject elsewhere as well.[82] Since the way we come to know universals described at *APo* II.19 100a15ff. must not in general be assumed to be the same as the way we grasp perceptible qualities, Ross's account is unsatisfactory.

Another approach comes from viewing *epagōgē* in terms of coming to have a special kind of perceiving: coming to see a particular as a particular of a certain kind. *APo* II.19 100a16–b1 refers to the stage where *epagōgē*. has taken place, where we have reached the first (most specific) universal, where we perceive Callias no longer as a mere particular but as a particular of a certain sort: we now spot the universal element in particulars. At this point we can direct our attention to the universal element, in which case our perception is of human, not of Callias or of Callias the human. (We can also direct our attention to the particular in question, so that our perception is of Callias or Callias the human.)[83]

This interpretation also resolves the apparent conflict with *APo* I.31 87b39–88a2. *APo* I.31 contrasts perception with scientific knowledge on the grounds that we perceive particulars ("the this and here and now," 87b30), whereas scientific knowledge has universals as objects, and that perception does not explain things, whereas scientific knowledge does:

> **I.31 87b39–88a5** "For this reason, even if we were on the moon and saw the earth acting as a screen, we would not know the cause of the eclipse. For we would perceive that there is an eclipse now, but not generally why; for perception was not of the universal. However, as a result of observing this [connection] occurring many times, after hunting out the universal we would have a demonstration; for the universal is clear from several particulars."

> **I.31 88a12–17** "If we saw some things we would not investigate [further for the universal that explains the connection]—not on the grounds that we had knowledge *by means of* seeing, but on the grounds that we had the universal *as a result of* seeing; for example, if we saw the glass perforated and the light going through, it would

> also be clear *why* it burns—by seeing each case individually, but grasping at the same time that it is the same in all cases."

These passages do not treat proper objects of perception, but are concerned with connections rather than objects. What is seen is the individual eclipse; acquaintance with the earth, moon, and so on is taken for granted. But these considerations make no important difference to the account, which makes it clear that we perceive individuals, not universals. In general, knowledge of universals comes from considering individuals. No matter whether it takes one or many observations, seeing the phenomenon, though necessary, is not a sufficient condition for grasping the universal. Our minds work in such a way that the natural and expected result of perceiving something enough times is that we grasp the universal under which it falls: we see it *as* an *A*. But this result is due both to the acts of perceiving and to other activity of our minds.[84] Although Aristotle never provides the explanation and justification this view needs, it is intelligible as it stands.

What is important for our purposes is that the person imagined to be on the moon and seeing the earth acting as a screen has not studied enough cases to know the universal. When he sees the eclipse he sees the particular, but not as an instance of the appropriate universal. Thus, his perception is not of the universal. Upon seeing enough cases he will be in the situation described at II.19 100a16–b1, and his perception will be of the universal. The conflict between I.31 and II.19 is resolved by recognizing that the "perceivers" are at different stages.

Perception provides the material for induction. It presents particulars to our consciousness in such a way that our minds can operate on them and come to recognize the universals in them. This is the meaning of 100b5: "This [induction] is the way in which perception implants the universal." Aristotle fails to give a detailed account of induction, if that means a full description of the cognitive and psychological processes involved in moving from perceiving individuals to perceiving them as individuals of a type. But on the present interpretation, induction plays exactly the role it should in *APo*, enabling us to take an important step toward grasping scientific principles.

To summarize. *Epagōgē* is the way we come to spot individuals as individuals of a kind. Equivalently, it is the way we become aware of universals in particulars. It can happen by argument and in other ways. The way *APo* II.19 sketches consists not in argument but in acquiring expertise at spotting relevant general features in things. *Epagōgē* is not simply habituation that occurs automatically as cases are piled up, but the development of a kind of insight into individual cases, an ability to grasp universals present in them. It occurs in humans as a measure of their rationality, but is by no means a mechanical or inevitable result; it may happen quickly or slowly or not at all, depending on the person and the nature of the cases. Teachers can help by providing good examples and pointing out relevant features, and such teaching may proceed not only by presenting case studies but also by basing arguments on them. But learning need not take place in this way; indeed it cannot when a science is being learned and systematized for the first time. Thus *epagōgē* has a central role in getting scientific knowledge for the first time as well as in a context of teaching and learning.

Noûs

The second question of II.19 is what state (*héxis*) comes to know the principles (99b18). *APo* II.19 100b5–17 identifies the state as *noûs*. Traditionally, *noûs* was rendered "intuition" or "intuitive reason," and was seen as a rationalist element in an otherwise empiricist account, a combination that provoked varying reactions.[85] Recent work has rejected the rationalist view of *noûs* and seen the passage as telling a thoroughly empiricist story.[86] In general I favor the empiricist line, although I find that that line of interpretation has been overstated.

Most importantly, *noûs* is a state, and so is on a par with memory, experience, and the state characterized by the universal at rest in the soul.[87] *APo* II.19 100b5ff. supplies a name to this state, already described as the *archē* of art and science.[88]

Archē does not here mean indemonstrable principles. They are the *objects* of *noûs*, to be sure. But in calling *noûs* the *archē* of science Aristotle is asserting the priority of having *noûs* over possessing scientific knowledge. Roughly speaking, scientific knowledge depends on *noûs*: if you want to get it you first need *noûs*, since the objects of *noûs* are the principles of the propositions that are the objects of *epistēmē*.[89]

APo II.19 100b5ff. argues that *noûs* is the state that knows the principles. The passage pays excessive attention to terminology and lacks philosophical content.[90] It says nothing significant about *noûs* but appears to pull the term out of a hat.[91] It also contains some unjustified and unelucidated premises. Still, its point is clear: we cannot know principles the same way we know other facts, which we know as a result of proofs based on the principles. There is no proof of principles. If *epistēmē* is demonstrative knowledge of conclusions of scientific proofs, undemonstrated knowledge of principles must be something else. Aristotle gives reasons (the unjustified premises) to identify this knowledge as *noûs*.

Are scientific principles the only objects of *noûs*, or can *noûs* also apprehend provable scientific facts? This query is prompted by the first part of II.19. We know universal connections before organizing them into a finished demonstrative science. We then have knowledge of the connections, although not demonstrative knowledge. What would Aristotle call this knowledge? After organizing the connections we will have demonstrative knowledge of some of them, but not of the principles. The grounds of our knowledge of the former will then be different from what it is now. But not so for the principles. As a result, Lesher argues in an oft-cited article,[92] if our knowledge of them *after* the science is organized is *noûs*, our knowledge of them *before* it is organized must also be *noûs*. But at that point our knowledge of them is no different in kind from our knowledge of the other facts. Accordingly, *noûs* is simply the knowledge resulting from induction. It follows that whereas we have *noûs* of principles, we have both *noûs and epistēmē* of demonstrable connections. This view entails that our grasp of principles is perhaps weaker and certainly no stronger than our grasp of the conclusions, a result contrary to Aristotle's view (I.2 72a25ff.) that we have more confidence in principles than in conclusions based on them.

The key to these issues is *APo*'s use of *noeîn* and *nóēsis*. Lesher cites passages where *noeîn* and *nóēsis* have objects that are not principles (most clearly I.31 88a4–7, a14–17, and I.34 89b11ff.).[93] I have shown that in these chapters induction is the means by which we grasp or come to know the universal in particulars. Aristotle uses *gnōrízein* and related words for "coming to know," and *noeîn* and *nóēsis* for "grasp."[94] Further, he distinguishes the stage of grasping universals (whether concepts or connections) from that of possessing demonstratively organized scientific knowledge. Since at the former of these stages we grasp by induction connections that turn out to be conclusions in the organized science, and also truths that turn out to be principles, we can speak of *noeîn* and *nóēsis* of both principles and conclusions. So far Lesher and I agree.

We part company over *noûs*. Not a single passage of *APo* is naturally or plausibly read as saying that there is *noûs* of anything else than principles. To be sure, when *noûs* is called the *archē epistēmēs* (I.33 88b36) and the unit of demonstration and scientific knowledge (I.23 85a1), Aristotle does not say that principles are its *only* objects. But without positive evidence it would be unwise to say they are not, as it would be for someone, on being told that horses have four legs and a tail, to say that he had heard nothing to disprove his theory that *some* horses have two legs and no tail. Unless *noûs* and *nóēsis* are identical, there is no basis for the claim that there is *noûs* of both indemonstrable and demonstrable connections. The occurrences of the word in *APo* I (the two just mentioned and that in I.33 89b8, which contributes nothing to the issue) do not suggest that *noûs* has as objects anything other than principles. In *APo* II the word occurs only in Chapter 19, where again it is called the *archē epistēmēs* (100b15; cf. a8)[95] and described as the state whose objects are principles (b12).

Although *noeîn* and *nóēsis* apply more generally, the only objects of *noûs* are principles. In support of this view we have not only the repeated insistence that this is the role of *noûs*, but also II.19's concern with precise terminology, which should make us wary of applying *noûs* in places not specifically authorized by that discussion. Also it is not quite right to say that *noûs* is identical with *nóēsis* qua restricted to principles. *Nóēsis* is generally used in *APo* for the grasp of universals, the state corresponding to the stage of seeing individuals as individuals of certain kinds. On the other hand, *noûs* is the state where we recognize principles *as principles*, which requires not only that we spot principles as universals instantiated in individual cases, but that we understand other universals as dependent on them. Thus, identifying the *archē epistēmēs* as a separate and further level is a signal that *noûs*, the grasp of principles, is a different, higher level than *nóēsis*, the grasp of universals. All the more reason to give it a different name and to call it by one of the names associated in the philosophical tradition with the highest kind of intellectual apprehension.[96]

This is not to say that *noûs* is an infallible faculty. It is always true in the way *epistēmē* is always true (100b6–8): if we have *noûs* or *epistēmē* of *p*, then necessarily *p* is true. But this does not mean that *noûs* and *epistēmē* are self-advertising. It does not follow that we know when we have them, or that there is any way to be sure we really do have them. The evidence does not exclude a stronger interpretation, but I prefer this weaker one, which does not run afoul of the standard objections to an infallible faculty of intuition. I also prefer to read the description of *noûs* as more precise (*akribésteron*) and truer (*alēthésteron*) than *epistēmē* (100b8ff.) along the same lines.[97]

This way of distinguishing *noûs* from *nóēsis* avoids the need to hold that our grasp of principles is the same as our grasp of scientific facts before we organize them. *Noûs* is different from *nóēsis* not because we have stronger grounds for knowing its objects but because we know them in a different way, the way appropriate to scientific principles. And *noûs* is better than *nóēsis* because it is more difficult to attain and because it reflects our status as experts, in which the naturally most intelligible facts are most intelligible to us, so that we put more confidence in them than in the other facts of the science.

Dialectic, Expertise, and the Principles

APo II.19 identifies the intellectual states that constitute the path to knowledge of principles but says practically nothing of the process by which we go from one state to the next. From perception to memory, from memory to experience, from there to knowledge of universals, and from this to knowledge of principles as principles—each of these steps needs explanation. I have argued that Aristotle is not bound to provide such explanations, given the questions he asks and the way he interprets them. Since the first three steps are general problems of his psychology and epistemology and not peculiar to the theory of demonstrative science, I will say nothing more about them. However, the last step is intimately connected with the theory. Coming to know principles as principles requires discovering which facts are most intelligible and which facts follow from which, in short, organizing the facts into a demonstrative science. Since this business will be largely different in different sciences, it is not surprising that Aristotle fails to offer detailed instructions. It is clear to the expert astronomer that the planets' nearness is more intelligible than their not twinkling, but it is not in general clear how to show that one fact is more intelligible than another. Aristotle might reply that it is a matter of coming to see things the right way, the way experts do. True, perhaps, but unhelpful. On the other hand, he does discuss some techniques of organization.

Dialectic is mentioned only once in *APo* (I.11 77a29–35),[98] but what we are told about it is significant: (1) it is common to all sciences; (2) it has no determinate subject matter, no single subject genus; (3) it does not demonstrate, but asks questions.[99] Further information on these and other features of dialectic is found in *Topics* and *Sophistici Elenchi*.[100] I will not take up the nature and methods of dialectic except as needed to show its roles in demonstrative sciences. These are two, both related to discovering principles. First, dialectic examines received opinions (*éndoxa*) concerning the subject matter of sciences with a view to determining their truth or falsity. Second, it draws consequences of propositions of a science, to put them into relation with one another as a preliminary step toward determining the principles (the final choice being reserved for the expert), and it tests the suitability of proposed principles.

The use of dialectic to discover truth is not infrequently mentioned in *Topics*, although its role in debate is more prominent. Aristotle distinguishes the philosopher's use of dialectic from the dialectician's on the grounds that the former investigates on his own and has no need for expertise in framing a strategy of questioning or putting questions to an opponent.[101] The point is simply that many debating techniques are useful in discovering the truth. They are not cheap tricks[102] but ways of testing for soundness, consistency, presuppositions, implications, form, relevance, and so on, of propositions and arguments.[103]

A prominent passage identifies one of dialectic's functions:

> ***Top.* I.2 101a36–b4** "It is also useful with a view to the principles of each science. For it is impossible to say anything about them on the basis of the proper starting points of the science in question, since the starting points are primary to everything, but it is necessary to base our account of these on the received opinions on each subject. And this is peculiar to or anyway most appropriate to dialectic; for since it is a technique of criticism, it contains the path to the principles of all disciplines."

This key statement needs filling out from other sources. First, the notion of received opinion.[104] Dialectic uses premises that are received opinions, or as Aristotle says, the views of all or most people or of the wise.[105] In the case of a science, dialectic starts with received opinions on its subject matter and tests them by systematically exploring questions and problems they raise in order to determine their truth or falsity. *EN* recommends and follows this method:

> ***EN* VII.1 1145b2–7** "We must, as in all other cases, set the different views before us and, after first discussing the difficulties, go on to prove, if possible, the truth of all the received opinions about these affections, or at least the greatest number and the most authoritative. For if the difficulties are solved and the received opinions are left standing, it will have been proved sufficiently."

This account appears to conflict with *APo*, for there it seemed as though we spot universals after observing enough individual cases, and here it seems as if we need dialectical investigation. Also, the starting point of the investigation appears different in the two accounts: perception of individual cases in *APo*, and received opinions in *Topics* and *EN*. But it is premature to write off *Topics* as superseded by *APo* or irrelevant: the two accounts, far from conflicting, complement one another.

The examination of received opinions is prominent in many Aristotelian works, but often the nature of the subject matter does not demand such careful treatment. When the nature of the subjects, attributes, and connections is obvious on inspection or clear through repeated observation, the *APo* account will suffice. But in many important cases it is not such clear sailing. When observed phenomena are interpreted differently, dialectic is needed to settle the disputes and determine the truth.[106]

A common feature of the *APo* and *Topics* accounts is that both are related to induction, which *Topics* declares to be one of the two principal kinds of dialectical reasoning (I.12), defining it as "the approach from the particulars to the universals," and which *APo* calls the means by which we recognize universals on perceiving a number of particular instances. The two accounts can be seen as describing the same sort of thing—reaching the universal, which is done differently in different cases.

Finally, received opinions are digests and interpretations of observations. The fact that they can conflict and be incorrect reflects the possibility of incorrectly interpreting what we perceive. Supplementing *APo* with the *Topics* account leaves less room to accuse Aristotle of too naive an empiricism. He recognizes that it is not always a straightforward matter to elicit the proper universals from observed particulars. This is true not only for common conceptions about a subject and the opinions of the wise, but also for reports of perception.[107]

So far, dialectic proves useful in reaching scientific truths by serving as a systematic method to analyze conflicting opinions resulting from observation. However, there is a further step in determining principles, namely, systematizing the truths once they have been recognized, seeing which truths follow from which and which truths are principles. In my view, dialectic is responsible for a great deal, though not all, of this work.

The clearest statement of the procedure comes from *APr*:

> ***APr* I.30 46a3–30** "It is necessary to consider the attributes that belong to both [terms between which we want to prove a connection] and to provide as many of these as possible, and to examine these by means of the three terms (the different figures of syllogism) identifying the ways in which they form valid arguments and the ways in which they form invalid ones. . . . We have spoken generally of the principles of syllogisms . . . so that we only need to look to a smaller number of points and certain definite ones, and we must make our selection of premises concerning each particular thing. . . . Most of the premises in each subject are proper to the science, and so it is the function of experience to provide us with the principles of each. . . . And so, if the attributes belonging to each thing have been grasped, it is our job to reveal the demonstrations readily. For if nothing is omitted in the investigation of the at-

> tributes that really belong to the things, we will be able to discover the demonstration of everything that can be proved and to prove it, and on the other hand we will be able to make clear where there is no demonstration. In general, it has been pretty well stated how to select the premises; but we have gone through it in detail in our treatise on dialectic."

This passage focuses on two separate activities: identifying facts and arranging them to form proofs. It emphasizes that both are found in every branch of knowledge and claims to have made the second more efficient by showing how to select from the whole number of scientific facts those that can be relevant to the connection to be proved. If we are looking for a proof of a universal affirmative connection (*AaB*), the syllogistic reveals that the proof must proceed by one or more syllogisms in Barbara, so that the only candidates we need consider for premises are other universal affirmative connections. If we have a list of all connections within the science, we can look for a *C* such that *AaC* and *CaB* are on that list. If we find such a *C*, we have the premises for a sound argument to the desired conclusion. This is what "selecting the premises" means. The method's generality suggests that it belongs to dialectic, and the final sentence confirms that it is. Since the final words refer to virtually all of *Topics*,[108] *APr* I.30 is a general sketch of how to form proofs, with an appeal to *Topics* to fill in details. The importance of dialectic in science could hardly be more perspicuous.

This seems right as far as it goes. But Aristotle realizes that a proof of *AaB* is not simply a sound argument with *AaB* as a conclusion. Even the passage being considered may refer to this point in asserting that experience[109] provides the principles of each subject. Thus, he recognizes the need for a third activity, identifying the principles, which takes place after the facts are recognized and before they are formed into proofs. If we have a complete list of connections in a science and if the principles have been identified, then the procedure to find proofs is more complex but more satisfactory. First look for a *C* such that *AaC* and *CaB* are connections and also principles. If such a *C* exists, we have a demonstration of *AaB*. If there is no such *C*, look for a *D*, such that *AaD* and *DaB* are connections, and then follow a similar procedure to find proofs of *AaD* and *DaB*. Aristotle calls this procedure "condensing the middle" (*APo* I.14 79a30f., I.23 84b31ff.). It is not clear that he intends this more complex procedure in *APr* I.30, despite the mention of principles.[110]

On the other hand, it is significant that it is dialectic's job to arrange facts into demonstrations. *Topics* goes farther. First and quite generally it tells us that dialectic is useful in discovering principles: they cannot be investigated on the basis of *themselves*, and so must be treated on the basis of received opinions, which is primarily the job of dialectic, "for since it is a technique of criticism, it contains the path to the principles of all disciplines" (*Top*. I.2 101a36–b4).

Scientists correctly use the tools of dialectic to investigate principles. The techniques of demonstration are inapplicable,[111] and dialectic is appropriate since it is critical.[112] Like Socrates, it examines proposed principles by trying to refute their claim to be principles. Moreover, it proceeds by asking questions.[113] Like Socrates, it puts proposed principles to the question, rather than proving that they are correct or incorrect. Since the questions should be answerable "yes" or "no," scientists must *propose* principles for investigation. They may not ask "What is a human?" or "In how many ways is 'the good' used?" but "Is 'the good' used in this way or that?" or "Is a human a two footed animal?"[114] If the answers are not apparent, they can be pursued by appropriate lines of questioning (*tópoi*) as identified and discussed in *Topics*. Scientists must ask and answer them and must therefore be aware of these considerations.[115]

Still, dialectic shows what kinds of questions to ask and how to examine the answers. The lines of questioning introduce relevant considerations, but do not determine either the actual questions or the answers. The considerations raised in *Topics* are useful in treating *any* science, but precisely because these considerations are general, dialecticians as such are not responsible for answering questions they ask within the scope of each science.[116] Even more clearly, the dialectician is not responsible for the received opinions. These are the views of all or of the many or the wise.[117] Again, the scientist is in control of this technical information. Thus, in investigating principles, scientists propose a proposition as a possible principle, then use dialectical considerations to raise questions about it and test it, and finally supply answers from their specialized knowledge.

We have seen how this approach works in investigating truth and falsity. It also bridges the gap discussed in *APr* I.30 between using dialectic to construct proofs and discovering principles. Recall the specimen arguments of *APo* I.13:

S1A. Things that do not twinkle are near.
S1B. The planets do not twinkle.
S1C. Therefore, the planets are near.

S2A. Things that are near do not twinkle.
S2B. The planets are near.
S2C. Therefore, the planets do not twinkle.

These two syllogisms can be constructed out of four connections, since S1B = S2C and S1C = S2B. Accordingly, if S1A, S1B, S1C and S2A were all found among the truths of astronomy, the methods of *APr* I.30 would license both syllogisms. However, only the first is a demonstration. The premises of proofs are prior to, more intelligible than, and grounds of the conclusions, and these relations are asymmetric. To construct proofs, therefore, we must be able not only to form valid syllogisms but also to identify which connections are prior to, more intelligible than, and grounds of others. Dialectic suffices for the former task, but the latter requires expert knowledge of the subject.

By deploying dialectic alongside expert knowledge, scientists can systematically identify candidates for proofs and determine the relative priority and intelligibility of the connections the science treats. The primary and most intelligible connections will turn out to be premises in proofs of other connections, and themselves indemonstrable. These will be the principles.

The foregoing discussion identified two roles for dialectic in demonstrative science. It is useful in determining facts insofar as they are not apparent from observation but require examination of received opinions. This function is associated with coming to grasp universals. Its second role has to do with organizing the science into a system of demonstrations based on indemonstrable principles. In this role it plays a part in the move to the highest level of apprehension discussed in II.19. In both roles dialectic cooperates with technical knowledge. It constitutes a battery of techniques for criticizing, analyzing, and testing proposed truths and principles. These techniques can lead a candidate to be rejected, but passing the tests of dialectic is only a necessary, not a sufficient, condition for being a truth or a principle. In general, when dialectic leaves more than one candidate in the field, the technical expertise of the specialist in the science will determine the final choice.

Aristotle says little about how specialists make this choice, which is no doubt one of the most difficult and controversial stages in organizing a science. The parts of *APo* devoted to the topic of advancing from knowledge of the fact (KF) to knowledge of the reason why (KRW) have a bearing on the present question.

Recall that once we have KF we go on to seek KRW. This amounts to asking whether the fact can be proved, whether there are terms that can serve as middles in a demonstration of the fact, and next asking what the proof is, what the middle terms are.[118] Although these questions do not require a special investigation if the middle is obvious to perception (II.2 90a26–30), the distinction between KF and KRW holds nonetheless.

To advance from KF to KRW we must grasp the fact in a relevant, nonincidental way. We must know "something of the thing itself" (II.8 93a22) or "something of its essence" (a29), which gives a starting point in investigating the middle. To judge by the examples in II.8 and the account in II.10, in investigating a subject-attribute relation, we must know it as a relation between that subject and that attribute (we must know that thunder is a certain noise [attribute] in clouds [subject]). We then proceed to look for a middle term that satisfies both formal criteria and criteria of priority, explanatoriness, and intelligibility. The techniques of dialectic determine whether a candidate satisfies the formal criteria.

But formal criteria may fail to decide. Three formally acceptable candidates for the explanatory middle term of lunar eclipse are screening (of the earth), turning (of the moon), and quenching (II.8 93b5–6). Of these, we are simply told that the middle is the "account" of the major term, "for eclipse *is* screening by the earth" (b6–7). I have argued that "account" means no more than grounds in the sense appropriate to scientific proofs, and that Aristotle is wrong to think that the middle is the grounds of the major since "eclipse" is not the major but the name for the connection between the minor ("moon") and what is actually the major ("a certain loss of light").[119] Still, Aristotle's point is that to locate a middle we must identify candidates and select the one that is the grounds. It is left to the scientist to locate candidates that satisfy the formal criteria and to investigate which is the middle. This is the place for scientific research, and *APo* says precisely nothing about it.

Suppose we decide that screening by the earth is the middle that accounts for eclipse. We next examine the relations between subject and middle and between middle and attribute to see whether either can be further accounted for. The way this requirement is stated (b12–14) suggests that terms that can account for one or the other of these relations are candidates rejected in the former investigation. None of the candidates in the example will serve, but we can imagine how it might work. Suppose that we are considering the relation *A—D* and that of the formally satisfactory candidates *B*, *C*, *X*, and *Y* for the middle, we choose *C*, so that the argument goes

A—C—D.

We then ask whether *A—C* or *C—D* can be explained. Suppose we find that *B* and *X* are formally satisfactory middles of *A—C*, but *Y* is not, and of the two, suppose *B* is the grounds of *A—C*. Then we have

A—B—C—D.

We continue this process until there are no further explanations. Then we have a proof of the original subject-attribute relation and the indemonstrable principles on which it is based.

APo I.13's treatment of candidates for proofs of the planets' nontwinkling and their nearness agrees. It too stresses that proofs proceed via middle terms that are grounds of their conclusions. The example is distinctive in showing how the conclusion of an argument may be an indemonstrable principle, so that special care is needed to ensure that we have a proof, not an argument that goes in the wrong direction. In constructing a proof, we must locate the grounds of the conclusion, in fact the primary grounds (78b3–4): the premises must be immediate, and so must have no further grounds. It seems unproblematic for Aristotle that the planets' nearness is the grounds of their not twinkling and not vice versa, so he may think that if more than one candidate is formally acceptable, it will be obvious which is the grounds of which. In some cases, no doubt, it *will be* obvious. In others, research will settle the issue. But there may be cases where it is not clear which of two or more true propositions each derivable from the others is the grounds of which.

Aristotle's advice not to assume that the better known is the grounds recalls the distinction between better known to us and better known in nature. What is better known in nature is always the grounds and what is better known to us never is. But since people can make what is better known in nature better known to themselves,[120] the advice can be directed neither to experts, for whom what is better known in nature and what is better known to them coincide, nor to novices, for whom they are antithetical, but to those in between. For such people sometimes they coincide, but sometimes they do not, and a warning not to trust intuitions about what is better known is appropriate. Aristotle urges such people to be sure that what they choose as middles actually are the grounds of the conclusions. Admittedly, his criteria for determining which of two facts is better known in nature do not succeed.[121] Still, this does not mean that there are *no* adequate criteria, and without proof that there are none, that the conception of a natural order of priority, intelligibility, and explanation is inapplicable, Aristotle can reasonably follow his instincts (his belief that nearness can explain nontwinkling and not vice versa) and base his theory of science and scientific explanation on that conception, so that identifying workable criteria remains a problem for investigation.

In looking for unexplainable facts we are looking for principles, which are true, primary, immediate, more intelligible than, prior to, and grounds of the other facts in the science. The final three conditions are the key to identifying them. The procedure indicated in II.8 yields a relative ordering: *A—C* and *C—D* are prior to, more intelligible than, and grounds of *A—D*, and *A—B* and *B—C* are of *A—C*. Going through the procedure for all the facts of the science will give us demonstrations of all the provable facts. The remainder are principles, which, being indemonstrable, are primary, ungrounded, and most intelligible.

Success in identifying principles will depend on expertise in a subject and on facility with the techniques of dialectic. The arrangement of facts one generation or school finds best may be rejected by another. Aristotle's view of scientific principles and how they are grasped leaves room for progress and for disagreement. It even allows different arrangements of a subject matter to be equally good from the point of view of logic, though it does *not* allow for more than one to be correct. Sciences can be arranged into deductive systems, Aristotle holds, and in them certain truths are primary and more intelligible than the rest. As our expertise increases, we are better able to recognize these principles. But expertise depends on observation, and since at a given time the relevant facts may not all be in, we can only hope to organize the material at hand in as satisfactory a way as possible. In general, it will be exceedingly uncommon that experts in a field will agree that they have all the relevant information. (Even if they agree, they may be wrong.) Accordingly, it is safest to admit that one's body of facts may be incomplete, and safest to admit that one's principles may someday be replaced by more ultimate truths.

Aristotle unsurprisingly says little on this topic,[122] since he is mainly concerned with achieved sciences. He supposes that the facts are in and have been organized, so that the principles are identified and the proofs constructed, and he describes such a system. It is not his purpose to discuss problems in setting up such a system, and perhaps they could not be described in much detail. The interest would reside in particular cases. In fact, his theoretical works consider theories of his predecessors (the experts of their day) and criticize the principles they proposed. Whether he believed that the principles he proposed himself might one day be shown to fail is best considered in another context. But the answer has no bearing on the present discussion.

The Kinds of Principles

A central concern of this study has dropped out of sight in the present chapter—whether the methods described for reaching principles yield the types of principles identified in *APo*: axioms, definitions, and existence claims. In view of earlier discussion, I will concentrate on definitions and address the others here only briefly. Axioms or common principles may well turn out to be unprovable facts in the province of an appropriately general science:[123] first philosophy for axioms like the law of noncontradiction, and the general science of quantity for axioms like EQUALS. Primitive existence claims may be determinable in the way discussed above. *Some* existence claims will be needed to introduce the basic subjects used to prove the existence of others, and the expert will decide, after examining candidate existence proofs produced according to the method of *APr* I.30, which existence claims are prior to others and which are primary. The principal difficulty is whether the primary, immediate (nonexistential) connections identified as principles turn out to be definitions or parts of definitions. (By "part of a definition" I mean this: if *A* is defined to be *BC* [or *BC* . . . *N*], then "All *A*'s are *B*" and "All *A*'s are *C*" are parts of the definition of *A*.)

Again the best hope of a solution to this problem is II.8's procedure for discovering immediate connections. Suppose this procedure has been carried out completely, so that we know which connections are immediate and which are not. For any subject *S* in a science's subject genus, a number of attributes A_1, A_2, . . . A_n are immediately related to *S*. The obvious suggestion is to call "*S* is A_1 & A_2 & . . . & A_n" the definition of *S*. A_1, A_2, . . . A_n are the essential attributes of *S*;[124] the other attributes follow from these (together with other principles). There is no guarantee that "*S* is A_1 & A_2 & . . . & A_n" is a definition by genus plus differentia, but it is a "fat definition."[125] A similar account holds for definitions of attributes. Recall that immediate connections go in two directions, so that *A* can be an element in *B*'s fat definition and vice versa.[126] Thus the doctrine of fat definitions allows for a simple solution. The method for locating the immediate facts does yield definitional connections of subjects and attributes, and these can simply be collected and formed (by logical conjunction) into scientific definitions.

The preceding account must be defended against the charge of begging the question. *APo* II.8's procedure for finding grounds for a demonstrable connection presupposes that we know that the connection holds (93a26–27). Further, Aristotle believes that analogous claims hold regarding subjects: we must know a subject exists before investigating its essence (II.1–2). But if we do not know what the subject *is*, that is, what its essence is, how can we know that it exists? If we know that something exists, how do we know that it is the subject in question?

These questions can be answered in a way Aristotle would find congenial. I shall answer by describing how to set up a demonstrative science. The extent to which the description is plausible will reflect the plausibility of the enterprise of organizing bodies of knowledge in demonstrative form.

Assembling a Science

By the time we are in a position to think of constituting a discipline as a demonstrative science, we must have a certain amount of factual knowledge about a set of entities (subjects), and some reason for treating these entities and facts together. For instance they may all have to do with numerical properties or celestial objects. We must have a workable if tentative idea of what the entities and phenomena are. We are prepared to admit more phenomena and entities and revise our understanding of the ones of which we are already aware. We may even jettison some of the entities and phenomena, either because further investigation reveals them to be spurious, or because as we systematize our knowledge they prove unrelated to the rest. We next marshal the facts systematically, arranging the subjects into genera and species and establishing which attributes belong to which subjects. We are now able to form application arguments. An attribute belonging to a genus also belongs to all the species of the genus, and does so because it belongs to the genus.[127]

Our goal in organizing knowledge is to account for as many facts as possible by showing them to be grounded in other facts. In effect we ask why each fact holds and we look among the other facts for possible explanations. We use the dialectical tools described in *APr* I.30 to find candidates that satisfy formal requirements. If there is more than one candidate, we eliminate those that do not explain the fact at hand. If no other known fact explains it, we must regard it (tentatively) as basic. If we do find other facts that account for it, we then ask whether there are explanations of these other facts, and continue until we reach facts we cannot explain further.

The same holds for entities. We begin with a more or less accurate idea of what a given subject or attribute is. At the least we know how to identify it reasonably well. We then ask why it has these identifying attributes and try to account for them in the way described. Some may prove to be consequences of others, or of facts about the entity not initially used to identify it. Further, some of the initial identifying attributes may be found either not to hold or to hold accidentally (e.g., that the sun is the largest celestial object). In these ways we determine the entity's scientific essence. We also ask why each entity exists and try to account for its existence by showing that its existence follows from and is grounded in that of others. We do the same for the others until we reach the basic entities on whose existence that of all the rest depends.

We can now organize the material into a first approximation of a demonstrative science. For each subject and attribute we gather the unprovable connections and constitute the scientific essence. We identify common principles that are used in the proofs of our science but apply more widely. We identify the primitive subjects and posit their existence. We now have a set of principles that we cannot prove and from which we can derive the remaining facts.

This is a first approximation of a demonstrative science because although it is in demonstrative form, it is not yet clear that the arguments are demonstrations. Even assuming that the arguments are valid, that all the propositions are true, and that the premises are naturally prior to (and more intelligible than) the conclusions, it is still not clear that the premises are immediate. Although our current state of knowledge does not let us prove any of them, further research may produce facts that explain those now thought immediate. It follows that we should keep asking why even after our material is in demonstrative form. In fact, it is always appropriate to ask why, since further investigation (research) may always yield further facts, of which some may be provable and others not. We are never in a position to know that we have reached unprovable facts or found all provable facts, and therefore never in a position to know that we have a finished demonstrative science.

Research and the Good Life

APo says little about the need for continuing research and the impossibility of reaching the position of knowing that you have scientific knowledge. He sketches the nature of a finished demonstrative science, and the picture he offers us has as much value as an ideal toward which to direct one's efforts as it would have if fully organized demonstrative sciences were easy to achieve. Also, the concept of experts, for whom what is most intelligible in nature is most intelligible to them, has value as an ideal even if it means that we can never know whether we, or others, are experts.

Ideals characteristically provide patterns for life. The ideal of expert knowledge demands a life spent looking for more facts and better explanations. Only so is there hope of becoming an expert. It is no more than hinted in *APo* that like a Socratic quest for knowledge these mutually dependent activities have no end. On the other hand, we may suppose that Aristotle's own scientific research taught him the need for a lifelong quest for knowledge. Also pertinent is the contrast between intelligibility to us and in nature. Aristotle recognizes how hard it is to come to know what is most intelligible in nature, the first principles, which are grounds for everything else. In introducing the study of metaphysics he makes some observations that apply to demonstrative sciences in general:[128]

> ***Met.* A.2 982a21–b4, b28–983a5** "Now of these characteristics, that of knowing all things must belong to him who has in the highest degree universal knowledge; for he knows in a sense all the instances that fall under the universal. And these things, the most universal, are on the whole the hardest for men to know; for they are farthest from the senses. And the most exact of the sciences are those which deal with first principles. . . . But the science which in-

> vestigates causes is also *instructive*, in a higher degree, for the people who instruct us are those who tell the causes of each thing. And understanding and knowledge pursued for their own sake are found most in the knowledge of that which is most knowable . . . and the first principles and causes are most knowable; for by reason of these, and from these, all other things come to be known, and not these by means of things subordinate to them. . . . Hence also the possession of it might be justly regarded as beyond human power; for in many ways human nature is in bondage, so that according to Simonides 'God alone can have this privilege,' and it is unfitting that man should not be content to seek the knowledge that is suited to him. . . . But the divine power cannot be jealous . . . nor should any other science be thought more honorable than one of this sort. For the most divine science is also most honorable." (Trans. Ross)

The knowledge that is divine is universal knowledge of first principles and causes. Since these matters are remote from what is familiar to the senses, knowledge of them is hardest to acquire. But that knowledge is the most honorable and insofar as we can, we must bend our efforts toward acquiring it.

EN acknowledges the difficulty of reaching scientific knowledge. *APo*'s distinction between scientific knowledge (*epistḗmē*) and *noûs* recurs in *EN* VI.3 and VI.6, and in *EN* VI.7 the state of the scientist, which combines *noûs* and *epistḗmē*, is called *sophía*, "wisdom." The final chapters of the work ask what the best life is for a human being and offer a barrage of reasons in favor of a life dedicated to wisdom. In fact, they tend to speak of a life devoted to theoretical wisdom (*theōría*),[129] which is the characteristic activity of the wise person (*sophós*).[130] Occasionally Aristotle identifies life in accordance with *noûs* as the most fortunate life (X.7 1178a6–8), but he draws no distinctions between this life and that of wisdom and theoretical wisdom.

What are the activities of such a life? Not actions (*práxeis*), which are the province of practical wisdom (*phrónēsis*), but theoretical activity. When *theōría* and *theōrētikòs bíos* are translated "contemplation" and "contemplative life," Aristotle's ideal life seems not to be a life of research and discovery, but one of reflecting on knowledge we have. But the contrast of theoretical with practical and productive wisdom is made in terms of their results. Productive knowledge aims at "contriving and considering how something may come into being" (*EN* VI.4 1140a10–12, trans. Ross); practical knowledge has action as its end (*Met.* α.1 993b21); and theoretical knowledge has truth (*Met.* α.1 993b20–21; *EN* VI.2 1139a27–29). It is best to reach the truth, to attain wisdom, since "those who know will pass their time more pleasantly than those who inquire" (*EN* X.7 1177a26–27, trans. Ross). But if we have not reached the divine state of certainty about principles and proofs, the best thing is to aim for it as a goal, to attempt to discover the truth and the real grounds of things. Only by pursuing this knowledge can we approach the ideal. Speaking of the divine state of achieved wisdom, Aristotle eloquently calls for people to devote themselves to its pursuit:

***EN* X.7 1177b26–1178a2** "But such a life would be too high for a man; for it is not insofar as he is man that he will live so, but insofar as something divine is present in him. . . . If reason (*noûs*) is divine, then, in comparison with man, the life according to it is divine in comparison with human life. But we must not follow those who advise us, being men, to think of human things, and, being mortal, of mortal things, but must, so far as we can, make ourselves immortal, and strain every nerve to live in accordance with the best thing in us; for even if it be small in bulk, much more does it in power and worth surpass everything." (Trans. Ross)

NOTES

CHAPTER I

1. I shall restrict myself to the Presocratic "scientific" tradition, which (whatever unity it has and whether or not it is correctly isolated from the Hippocratic medical tradition) Aristotle viewed as his earliest Greek scientific heritage. Some early Hippocratic treatises, notably *Ancient Medicine*, would be worth pursuing in connection with the issues I shall be treating. The picture presented in this chapter would doubtless be more complex if Hippocratic material were brought in. On this general topic, see G. Lloyd 1979, chaps. 1, 3, and G. Lloyd 1978, pp. 37–51.

2. DK frag. 1, trans. Guthrie.

3. DK frag. 17, lines 6–8.

4. DK frags. 26, 25, trans. Guthrie.

5. From DK frag. 12, trans. Guthrie.

6. Interest in the cosmos occupies different places in the thought of different Presocratics. It is a main interest of Anaximander and Democritus, but subordinate to metaphysical issues for Parmenides and probably to religious concerns for Empedocles and the Pythagoreans. Yet despite their disparate interests all found it important to speculate about natural phenomena.

7. DK frag. 100.

8. DK 21A33.

9. Kuhn 1970, pp. 18f., 37.

10. Guthrie 1971–81, 1:51.

11. This is a possible but not a necessary implication of D.L. IX.18. Cf. Guthrie 1971–81, 1:364.

12. For their low reputation, see DK 28A16, 28A18.

13. My comment pertains to both of the somewhat different versions in Aristotle, *De MXG*, chaps. 5–6, and Sextus Empiricus, *Adversus Mathematicos* VII.65–87.

14. For estimates of Empedocles' verses see DK 31A21–27. The Romans thought them fine poetry; the Greeks, mere verse.

15. Burkert 1972.

16. However, see below, p. 18.

17. The catalog was made by Thrasyllus and is preserved in D.L. IX.46–49.

18. D.L. I.15.

19. G. Lloyd 1979, pp. 88–98, shows that public lectures and debates were common in medicine and on scientific topics related to medicine. He supposes that the same holds for all sciences. The following material documents this thesis for the Presocratics.

20. I do not hesitate to speak of the Eleatics because the circumstances that affected how they presented their works were the same as those that affected scientists. I doubt the public would have regarded a presentation of Parmenides as different in kind from one made by a scientist.

21. Here I accept Plato's remarks (*Parmenides* 128c–d) on the purpose of Zeno's book, with, e.g., Guthrie 1971–81, 2:88.

22. Plato, *Parmenides* 128d–e.

23. Ibid., 127c–d.

24. Plutarch, *Pericles* 4.

25. Aristotle frag. 65 Rose (taken from D.L. VIII.57).

26. Plutarch, *Pericles* 4.

27. DK 29A15.

28. Athenaeus, *Deipnosophistae* XIV. 620d; D.L. VIII.66.

29. Cf., e.g., Plato, *Hippias Major* 285c.

30. Plato, *Hippias Minor* 363d.

31. We may doubt the details of the story about Herodotus, but I find the general picture plausible.

32. D.L. II.10 (same as DK 59A1); also Pliny, *Natural History* II.149 (same as DK 59A11), and Plutarch, *Lysander* 12 (same as DK 59A12).

33. Philostratus, *Vita Apollonii* I.2 (same as DK 59A6); D.L. II.10 (same as DK 59A1).

34. Aelian, *Varia Historia* X.7 (same as DK 41A9).

35. Plato, *Hippias Minor* 368b.

36. Plato, *Phaedrus* 270a.

37. Despite Lysias's assertion that one of Heracles' purposes in instituting the Olympic festival was an exhibition of wisdom, there is no evidence of formal artistic or intellectual contests. See Drees 1968, p. 62.

38. Plato, *Phaedrus* 270a.

39. Trans. Guthrie 1971–81, 2:68.

40. Cf. Porphyry, *Vita Pythagorae* 18–19 (same as DK 14A8a), and the story of the ascendancy of Pythagoras and his followers in Croton.

41. See DK frags. 1, 34, 50.

42. As Lloyd has argued (G. Lloyd 1979, pp. 97f.).

43. Kuhn 1970, pp. 177f., 181.

44. This passage and the following come from Kuhn 1970, pp. 13–17.

45. This paragraph is based on Kuhn 1970, pp. 18ff., 36ff., 163f.

46. As recounted by Simplicius, who quotes Eudemus extensively (*In Phys.* 60.22–68.32 Diels).

47. Heath 1921, 1:202; cf. ibid., p. 328.

48. Heath 1921, 1:201.

49. Eudemus, taken from Proclus, *Comm.* 66.7–8.

50. G. Lloyd 1979, p. 112.

51. Heath 1921, 1:90; cf. ibid., pp. 215f.

52. Szabó has argued at length for this thesis (Szabó 1969 and 1978; also previous articles).

53. The quotations and other information in this paragraph come from Proclus, *Comm.* pp. 66–67.

54. In general, see Heath 1921, 1:202ff., 316–53.

55. Aristotle discusses, for example, *álogon*, *kéklasthai*, and *neúein* (*APo* I.10 76b9).

56. Recall in this connection Proclus's interpretation of the word "elements," according to which the elements are those theorems an understanding of which leads to knowledge of the rest. Proclus, *Comm.* 71.24–72.13; also Aristotle, *Met.* Δ.3 1014a35–b3.

57. Heath 1921, 1:321.

58. This distinction must have been an ancient one in the crafts, including medicine.

59. Protagoras, frag. 7; cf. Plato, *Protagoras* 318d–e.

60. Plato, *Republic* VII.530–31.

61. *APo* I.7, 9, 13.

62. This work, which appears in the corpus of Euclid's writings, is more likely to be based on a work of Euclid's than actually by him: Heath 1921, 1:444f.

63. Plato, *Republic* VII.529–30.

64. *APo* I.7, 9, 13.

65. Ibid.

66. For a discussion of these disciplines and their relation to pure mathematics, see McKirahan 1978.

67. The proof is set out and discussed in McKirahan 1978.

68. Esp. Solmsen 1975.

CHAPTER II

1. See Stewart 1892, 1:1–3, for discussion of this kind of opening sentence.

2. For argument by example see *APr* II.24, and for enthymeme, see *APr* II.27. Recall also *APr* II.23 68b13–14: "For all beliefs are based on either deduction or induction."

3. For translation of this passage, see Chapter 3 below.

4. This passage is translated and discussed in the section "Applications to Individuals" in Chapter 14 below.

5. Burnyeat 1981, p. 105, translates *epistḗmē* as "understanding," but allows (p. 108) that "English could supply alternative ways of registering Aristotle's contrast between *epístasthai* and *gignṓskein* . . . : for instance, suitably introduced and circumscribed, a contrast between a richer and a more ordinary concept of knowledge, for the first of which the label 'scientific knowledge' might do after all."

6. Barnes, p. 97.

7. This issue is discussed in the section "Grounds of Truth or Grounds of Knowledge?" in Chapter 17 below.

8. "Without qualification [*haplôs*]" (71b9) is perhaps a signal of this.

9. Aristotle uses other words having to do with knowledge in *APo*, notably the verbs *eidénai* ("know"), *gignṓskein* [also *ginṓskein*] ("know," "be aware of"), *gnōrí-*

zein ("become familiar with," "come to know," "recognize"), *xuniénai* ("understand"), and *noeîn* ("grasp"), the nouns *gnôsis* ("knowledge"), *noûs*, and *nóēsis* ("grasp"—discussed in the section "*Noûs*" in Chapter 18 below), and the adjective *gnṓrimos* ("intelligible")—discussed in the section "Grasping the Universals" in Chapter 18 below). Barnes, who chooses different words to translate many of these terms, discusses Aristotle's use of these and other words on p. 90. I think that *epistḗmē*, *epístasthai*, *noeîn*, *noûs*, and *nóēsis* are the only words given special senses in *APo*, and I am inclined to think that Aristotle succeeds in using them consistently in these senses. Although it may be true that *eidénai* is sometimes tacitly used as a synonym for *epístasthai*, in some places at least it contrasts implicitly with *epístasthai* (e.g., II.1 89b23–35).

10. Ross, p. 507.

11. Aristotle uses *ginṓskein* here instead of *epístasthai* to avoid circularity.

12. Grammatically, "that this cannot be other than it is" may depend on either "we think" or "we think we know." For a defense of the second interpretation, which I adopt, see Barnes, p. 97; also cf. I.33 89a6–8.

13. It will also turn out that *A* knows the grounds of *x*'s being necessarily true. See I.6 74b26–32, translated and discussed at the beginning of Chapter 7 below.

14. See below, Chapters 7, 17.

15. He puts off the question whether there is another way. The answer is at first affirmative—there is scientific knowledge of immediate, indemonstrable principles (I.3 72b18–24)—but his authoritative account (II.19) distinguishes between demonstrated scientific knowledge (*epistḗmē*) and undemonstrated knowledge of principles (*noûs*), thus affording a negative answer.

16. Cf. I.3 72b18–22 (translated and discussed later in this chapter), where Aristotle rebuts opponents of demonstrated scientific knowledge (cf. 72b5–6) simply by denying without argument one of their claims.

17. Or "more intelligible than." The interpretation will be discussed in the sections "Principles Are Better Known, Prior, and Grounds" later in the present chapter and "Priority of What Is Better Known" in Chapter 17 below.

18. Aristotle recognizes this at I.10 76b10.

19. Mure's translation, "the premises of demonstrated knowledge must be true, etc.," is wrong.

20. The Greek is *ek*.

21. The Greek avoids using a noun for "things" by employing the neuter plural.

22. The word appears at 71b23, but is not defined until 72a5–7.

23. His caution is perhaps even irritating; cf. Barnes, p. 98.

24. Barnes, p. 99.

25. The Greek would presumably be *émmesos*—cf. Philoponus, *In APo* 47.15 (Wallies)—but the word is not Aristotelian.

26. This interpretation of "immediate" differs from that of Barnes (p. 99) in requiring that what is lacking is a term that would be the middle term in a *proof* of the immediate proposition, rather than merely a *syllogism* that has the immediate proposition as a conclusion.

27. On the other hand, *contra* Barnes, I would note that *prótasis* is sometimes correctly translated as "premise"—e.g., in *APr* I.25, where the *protáseis* of a syllogism are contrasted with its conclusion (*sumpérasma*).

28. *APo* I.2 71b26–29 has "indemonstrable" rather than "immediate." For the equivalence (in Aristotle's syllogistic logic) of these terms, see the discussion of 71b26–29 earlier in the present chapter.

29. "Appropriate" (*oikeîai*) principles are not the same as "proper" (*ídiai*) principles. The common principles used in the proofs of a science are appropriate to that science, or to the proofs in which they are used and to the conclusions of those proofs, even though they are appropriate to other sciences as well.

30. The alternative translation "a principle is an immediate premise of a demonstration" makes this point just as well.

31. These arguments are presented clearly and discussed by Barnes, pp. 106–8.

32. The assumption is not made just by the first position, as Barnes states (p. 108).

33. See Ross 1949, Index Verborum, s.v. ὁρισμός.

34. This is presumably Ross's intention, but it requires an unusual meaning of *hóroi*.

35. Barnes, p. 109 with references.

36. See Barnes, pp. 109–12, for analysis.

37. Cf. *APr* II.2 53b7–8, b11–15.

38. Cf. *APr* II.2 53b26–II.4 57b17.

39. For a different interpretation see Barnes, pp. 98–99.

40. Recall Socrates' complaint in *Phaedo* 97–99, esp. 99a–b.

41. Cf. Owen 1986d.

42. See the section "Priority of What Is Better Known" in Chapter 17 below.

43. Barnes's "more familiar," while acceptable for what is *gnōrimṓteron* to us, sounds intolerable to my ears for what is *gnōrimṓteron* in nature.

44. Cf. *Top*. VI.4 141b9–14.

45. It is characteristic of Aristotle to refer to the principles not relatively ("better known to the expert") but absolutely ("better known in nature or without qualification") (cf. *Top*. VI.4 142a9–11). See Evans 1977, chap. 3, esp. pp. 68–73.

46. He speaks this way in other places too, esp. I.24.

47. Ross's attempt to explain this away (pp. 54f.) is weak.

48. Aristotle does not use *epístasthai* here, merely *ginṓskein*.

49. Any other view would run afoul of I.1 71a1, a6, where *dià proginōskoménōn* is equivalent to *ek prohuparchoúsēs gnṓseōs*. Cf. also 71a11, a12 (*prohupolambánein*).

50. See below, Chapter 17, esp. the section "Aristotelian explanations and the Deductive-Nomological Theory."

51. Cf. II.19 99b20–21.

52. These kinds of priority are worth comparing with the varieties of priority discussed in *Cat*. 12.

53. The same problem arises for the view that *noûs* is more *true* (*alēthésteros*) than *epistḗmē* (II.19 100b11–12).

54. Strycker (1968 and 1969) discusses Aristotle's use of *mâllon* in other texts and says that it can mean "more" or "rather," neither of which suits the present passages.

55. See Barnes, pp. 104–6, for analysis of Aristotle's arguments.

CHAPTER III

1. Widely differing accounts of the principles can be reached depending on what passages are taken as basic. The literature is vast. In addition to Ross and Barnes, see also Frede 1974, Fritz 1955, Gomez-Lobo 1976–77 and 1978, Heath 1949, Hintikka 1972 and 1974, Landor 1981, H. Lee 1935, Mignucci 1975, Mueller forthcoming, Upton 1985. Space limitations compel me to be very sparing with discussion of this secondary material.

2. This kind of proof is discussed below in Chapter 15.

3. Cf. *Met.* Γ.3 1005a20: "the things called axioms in mathematics."

4. See the section "The Composition of the Subject Genus" in Chapter 4 below.

5. Those wanting to take "principles" in its normal meaning of "indemonstrable premises of demonstrations" will be inclined to take *hóti ésti* (76a31, a33) in the veridical sense: "that they are the case."

6. See I.1 71a11–17, translated and discussed in Chapter 3 below, and I.28 87a38–39, translated and discussed in Chapter 4 below.

7. *De Int.* 6.17a32–33.

8. Bonitz (s.v. οἷον) also gives both meanings for the expression *hoîon légō*. He cites *APo* II.13 97b15 and *De Sensu* 7 448a23 (a misprint for 448b23) in support of "e.g." and (in addition to the present passage) *APo* II.13 97b5 in support of "i.e." But 97b5 does not contain the expression. (Is this a misprint for 97b15, where the expression clearly means "e.g."?) And since the present passage is disputed, the only parallels favor "e.g." I do not regard this evidence as conclusive, however. The fact that *hoîon* frequently means "i.e." in Aristotle is adequate warrant for supposing that it could have that force in the expression *hoîon légō*.

9. This matter is discussed further in Chapter 9 below.

10. Cf. II.7 92b8–11.

11. This account should be welcome to Barnes (p. 136), who sees that this interpretation is needed, but is unable to justify it. His interpretation of 76b24 as "necessarily seems to be the case" instead of "necessarily seems to be because of itself" forces him to see Aristotle as referring to axioms, not principles in general.

12. Barnes 1975b, p. 77. See below, pp. 231–34.

13. See my discussion in the section "Principles Are Better Known, Prior, and Grounds" in Chapter 2 above.

14. Cf. I.3 72b24, translated and discussed in the section "Principles are Primary and Immediate" in Chapter 2 above.

15. *Contra* Barnes, p. 137, who acknowledges reasons for saying that they are different and yet says that they are the same.

16. Cf. H. Lee 1935, p. 116 n. 2, and Heath 1949, pp. 55–56.

17. See the sections "Geometrical Diagrams" and "Existence vs. Constructibility" in Chapter 12 below.

18. See my discussion of I.10 76a31–36 earlier in the present chapter.

19. See my discussion of I.32 88b23 in Chapter 4 below.

20. The question whether each axiom must be used in every science will be taken up at the beginning of Chapter 6 below.

21. The subordinate sciences, which form an important exception to this rule, will be treated below in Chapter 5.

CHAPTER IV

1. Cf. *perì hó* (I.10 76b22).

2. I use the word "things" loosely, to cover things as diverse as the objects of mathematical and of natural sciences. For Aristotle, all of these "things" are reducible to concrete particular substances that become the subjects of different sciences in different ways. The point is nicely made by Zabarella, *De Tribus Praecognitis*, chap. 2, in Zabarella 1966, pp. 502e–503a. See below, p. 63 and Chapter 8.

3. Axioms are discussed below in Chapter 6.

4. The subalternate sciences are discussed below in Chapter 5.

5. *APo* I.6 75a28–31 and 74b11–12 (where *hoútōs* refers to *kath' hautá*, 74b6–7). The latter passage will be discussed at the beginning of Chapter 7 below.

6. Ross, p. 532; Barnes, p. 129.

7. Cf. *Met.* Γ.2 1004a9–10, 1005a11–13.

8. See I.2 72a37–39 and a30–32, translated and discussed at the end of Chapter 2 above.

9. If *B* is the "cause" of *C* and *A* the "cause" of *B* and *A* has no "cause," then knowledge of *C* that depends on knowledge of *B* depends on a "caused cause" and knowledge of *C* depending on knowledge of *A* depends on an "uncaused cause." The principles of sciences are "uncaused causes," but here, *ex hypothesi*, the first-order principles *are* "caused," and the ultimate principles are "uncaused causes."

10. I take it that Aristotle is characterizing the kind of relations to the subject and predicate of the conclusion that a relevant principle could conceivably have.

11. Cf. Bonitz, 152a16–21.

12. Cf. I.28 87b1–2.

13. This assumption is false (at least in geometry), despite its superficial plausibility.

14. *Pace* Ross, p. 604.

15. The Greek has a present indicative apodosis following a protasis with *ei* + optative. Cf. Goodwin 1890, sec. 500.

16. Commentators (Zabarella 1582, comm. ad context. 192, in Zabarella 1966, pp. 1013c–1014b; followed by Mure, ad loc.; Ross, p. 604) have found here an otherwise unknown view of Aristotle's, that each genus actually has one primary principle, and have attempted to say what such a principle would be. But since the view is not Aristotle's, these guesses are futile.

17. A suitably loose sense of "belong to" is intended here.

18. Some were discussed above on p. 49.

19. The context from 76b11 makes it clear that the genus is meant. Cf., note b, on I.10 76b21–22 in Chapter 2 above.

20. See notes b and c, on I.10 76b11–15 in Chapter 2 above.

21. I doubt that the wording can be taken as asserting it. For a modern, "all the things composed of the primary things" would (trivially) include the primary things, but not, I think, for Aristotle.

22. *Pace* Barnes, pp. 182–83, *Met.* Δ.25 1023b23 does not warrant this interpretation.

23. "As such" (*hêi autó*) is equivalent to "per se." Cf. my discussion of I.4 73b28–29 in Chapter 7 below.

24. Postulates 1–3 of Euclid's *Elements* serve as the existence claims for these primitive subjects, from which all the other subjects are derived (e.g., squares; cf. bk. I, prop. 46). The relation between Aristotle's basic existence claims and Euclid's postulates 1–3 is discussed in the section "Differences between Aristotle and Euclid" in Chapter 11 below.

25. This is Barnes's view, p. 128.

26. Joachim 1922, pp. xxii–xxiv, gives a similar interpretation.

27. *APo* I.10 76b17–18 (quoted above as evidence for (1)) does not say that arithmetic hypothesizes that number exists, only that it is not equally clear that number exists

and that cold and hot exist. Arithmetic might make clear that number exists by hypothesizing that units exist (cf. b3–6).

28. Chapter 8, below.

29. The value of this approach can be seen more easily in modern geometry, which has more than one kind of spatial magnitude. Euclidean space can be contrasted with Riemannian space, say, and the properties of figures depend on the kind of space in which they are located—i.e., on which kind of figures they are (Euclidean or Riemannian ones). Conclusions proved about figures in one of those spaces will contribute to knowledge of that space. Interestingly, for some purposes these spaces are defined in terms of per se properties of their subjects. Thus, metric differential geometry can define Euclidean space as a space where, *inter alia*, the sum of angles in a triangle is equal to two right angles.

30. Joachim 1922, p. xxii; Ross, p. 597; Barnes, p. 120; Mignucci 1975, pp. 575f.

31. See *Top*. IV for tests for the genus/species relation.

32. See discussion of I.7 75a38–b20 earlier in the present chapter.

33. See below, Chapter 8.

34. See Bonitz, 151a57ff.

35. For example at *Met*. E.1 1026a26–27.

36. For objections to this view, see Zabarella, *De Tribus Praecognitis*, chap. 14, in Zabarella 1966, pp. 527–30.

37. *Met*. Γ.2 1004a4–5, trans. Ross.

38. *Met*. Γ.2 1004a5–6, trans. Ross.

39. *Met*. Γ.2 1003b19–20 need only mean that every genus of things (we may exclude the categories here) falls under one science and one perception. His example, "articulate sounds," is not the only kind of thing that we hear.

40. Despite arguments that categories are not *summa genera*, Aristotle sometimes treats them as such. See Bonitz, 152a16–21.

41. It is possible that "universal mathematics" is meant as the study of the general attributes of all quantities. Cf. *Met*. E.1 1026a26–27.

42. Cf. *Met*. Γ.2 1004a6–9: the varieties of mathematics correspond to *génē toû óntos*, whatever Aristotle may mean by saying that the branches of mathematics are arranged in a serial order. (It is not clear that he means the ordering arithmetic, plane geometry, solid geometry, which corresponds to the complexity of the subjects studied. Cf. I.27 87a34–37 and *Top*. VI.4 141b5–8.)

43. This evidence refutes the commonly held view that the subjects of a science are substances, so that the distinction between subjects and attributes is the same as that between substances and accidents. See also S. Mansion 1976, p. 155: "les sciences ont le plus souvent pour objet des accidents, tels la quantité et le mouvement, qui n'ont qu'une essence relative."

44. Cf. *Met*. Γ.2 1003b22–1004a2, 1004a9–b17, esp. b10–17.

45. Cf. *APo* I.7 75b17–20.

46. Again the common principles and subordinate sciences are exceptions.

47. There is no indication that Aristotle thought that metaphysics could do this work. See Ross, pp. 537, 543; and Barnes, pp. 133, 143.

48. Plato, *Republic* VI.511.

49. Ross, p. 64; Barnes, p. 186.

50. References to Speusippus can also be seen. Cf. Cherniss 1944, pp. 73–75 n. 55.

51. For the same idea, see Kuhn 1970, pp. 15f., 163f.

CHAPTER V

1. For an approach of this sort, see McKirahan 1978. For the most part in what follows I shall be stating the relevant conclusions of that discussion. The reader is referred to the article for a thorough treatment, including an examination of relevant scientific texts.

2. Also cf. I.9 76a9–10 with I.7 75b10–11.

3. I use this word to distinguish aspects of the existence of visual lines in this world from their abstract geometrical properties.

4. Aristotle fails to recognize explicitly the need for such application rules. On the other hand, he says nothing contrary to the present interpretation.

5. Aristotle mentions mechanics as well as optics (I.9 76a24), but cf. I.13 78b37.

6. Cf. pseudo-Aristotle, *Mechanica*.

7. This idea is presented in the section "Axioms and Subordinate Sciences" in Chapter 6 below.

8. These are only the leading branches. There are also semigroups, which have a place between sets and groups, and Abelian groups, which fit between groups and rings, not to mention special areas of algebra that would introduce new dimensions of complexity into the picture.

9. See the analysis of the relation between arithmetic and mathematical (as opposed to acoustical) harmonics (cf. I.13 79a1–2) in McKirahan 1978, pp. 206–15. The relation between a more and a less general branch of algebra is the same.

CHAPTER VI

1. Mignucci 1975, pp. 143f., following Zabarella, holds that the *axiômata* here are premises, not common principles. He claims that since I.7 is concerned with genus crossing in proofs, "the things from which the proof [proceeds] can be the same" (75b2–3) means that in some cases (the subordinate sciences, b14–17) the same *premises* can occur in different sciences; any reference to axioms is irrelevant. But since I.7 75a39–b2 is so closely parallel to I.10 76b9–15 (cf. b22), where *axiômata* is given as a synonym for "common principles," the point must be that although the same *axioms* (i.e., common principles) may be used in different sciences, still the impossibility of genus crossing is guaranteed by the difference in genera (cf. 75b3ff.)

2. Cf. 75a39ff. The slight variation in the lists is probably due to Aristotle's immediate purposes in these contexts.

3. *APo* I.10 76b11ff. is translated and discussed in Chapter 3 above.

4. The passage is discussed in the appendix to this chapter.

5. *APo* I.10 76b20–21, I.11 77a26–27, a30–31, I.32 88a36–b3, discussed in Chapter 4 above.

6. Ross's well-balanced treatment (pp. 55–57, 65, 511, 531f.) is a notable exception, although I disagree with it on some counts.

7. In view of this fact, Philoponus's suggestion that some axioms apply to only one science (*In An. Post.* 34.10–13 [Wallies]) cannot stand. Philoponus places undue weight on the axioms' self-evidence. *APo* I.2 72a16–17 may look like a definition of "axiom," but it is axioms' common character, not their self-evidence, that is stressed elsewhere. Note also that *koiná* is interchangeable with *axiômata* in the lists of ingredients of demonstrations and sciences (I.7 75a41–42, I.10 76b14, I.11 77a27).

8. *Met.* K.4 1061b20–21 remarks that EQUALS applies to all quantities, but does not say that it applies *only* to quantities. Theophrastus recognized that EQUALS applies only to quantities and that other axioms apply to absolutely everything (Themistius, *In Post. An.* 7.3–5). I owe this point and this reference to Ian Mueller. In *APo* Aristotle is unaware of this restriction or, what is more important, of the implications of this restriction for the assimilation of the two kinds of common principles. Cf. I.11 77a26ff. In *Met.* Γ, the universal applicability of the axioms makes them the subject matter of first philosophy, the study of being qua being (Γ.3 1005a27–29). Although it recognizes the use of axioms in mathematics (1005a20), *Met.* Γ discusses only LNC and LEM.

9. Cf. *Met.* Γ.3 1005b15.

10. *APo* I.11 77a10–25 recognizes that this is so for LNC and LEM (as does *Met.* Γ.4–7), but stops short of noting that such principles are for that reason different from principles like EQUALS. Shortly below, both LEM and EQUALS appear as examples of common principles (77a30–31).

11. This deduction is syllogistically invalid, strictly speaking, since the premises and conclusion do not predicate "one thing of one thing" (*APr* I.1 24a16–17), but the important thing for our purposes is that Aristotle regarded it as valid (*APr* I.24 41b13–22; cf. b6–7).The question of how strictly Aristotle observed the formal requirements of the syllogistic is discussed in the section "Proofs and Syllogisms" in Chapter 12 below.

12. The meaning of these prepositions is discussed in the section "The Alleged Regulative Role of Axioms" later in the present chapter.

13. In *Met.* Γ, LNC and LEM are deductive principles (e.g., Γ.3 1005b7, b9). LNC is called "by nature a principle for all the other axioms" (1005b33–34), but it is not clear what this claim means. Since LNC is not an explicit premise in the arguments for LEM (Γ.7), the point may be that LNC underlies LEM as it underlies all other propositions. Since LNC cannot be demonstrated (Γ.4 1006a5–11), Aristotle proceeds to "demonstrate in the manner of a refutation" (1006a15), which differs from demonstration proper by beginning from a significant assertion made by someone (especially a person who denies LNC), instead of ultimate premises (1006a11–27). He apparently has the same view of his arguments for LEM (cf. Γ.7 1012a21–24 with Γ.4 1006a24–25).

14. The terms "categorial" and "transcendental" are introduced by commentators; nothing in *APo* corresponds to them. The former term is misleading if taken to suggest that such principles will in general hold for all the sciences whose genera fall within a given category. In fact EQUALS does hold for the quantitative sciences known to Aristotle (though it would not hold in branches of modern mathematics where the concept of subtraction is not defined), and he says that equal is a property (*ídion*) of quantity (*Cat.* 6 6a26; *Met.* Δ.15 1021a12). But nothing in Aristotle's theory shows that he thought that all such principles hold with precisely that degree of generality.

15. This role of LNC and LEM is not stressed in *Met.* Γ. *Sullogistikaì archaí* (Γ.3 1005b7) is reasonably translated by Kirwan (1971) as "principles of [sc., deductive—see Kirwan's commentary ad loc.] reasoning." *Met.* B.2 996b26 is to be taken similarly.

16. Mure, ad loc., notices this. See also Ross, ad loc.

17. Cf. Zabarella 1966, pp. 786aff.; Pacius 1966, p. 299.

18. Cf. *Met.* Γ.3 1005a23–26.

19. Despite I.1 71a14, translated and discussed in Chapter 3 above.

20. Euclid, *Elements*, bk. I, prop. 1, uses it in this way.

21. Zabarella 1966, pp. 786c–788a.

22. *Cat.* 6 5a38.

23. This is also the view of Averroes, whom Zabarella (1966, p. 787d) says he is following, and of Philoponus, *In An. Post.* 123.24–26 (Wallies).

24. Pacius 1966, p. 299, classifies *kat' analogían* alongside of "equivocal" and "univocal."

25. See Bonitz, 152a16–21.

26. This was discussed in the section "Interpretations of the Subject Genus" in Chapter 4 above.

27. *Met.* E.1 1026a26–27, K.7 1064b8–9, M.2 1077a9–10, b17–22.

28. I shall return to this subject in the section "Axioms and Subordinate Sciences" later in the present chapter.

29. For the term, see above, pp. 70–71f.

30. For this proposal, see note b on I.10 76b11–15 in Chapter 3 above.

31. Mueller (forthcoming) uses this passage as the starting point for discussing the principles, so that the axioms are the only premises used in demonstration. The resulting interpretation is radically different from mine.

32. Barnes (p. 135) supports this reading against Ross (pp. 56, 531, 602), who thinks that this and the following passage point to the use of axioms as principles of reasoning.

33. LNC is also presupposed in *reductio* arguments, but Aristotle does not regard it as a premise. For the use of LEM as a premise, see Barnes, ad loc., p. 142.

34. This idea is discussed at the end of Chapter 5 above.

35. See I.7 75a38–b20, translated in Chapter 4 and discussed in Chapter 5 above.

CHAPTER VII

1. See the section "The Role of Existence Claims" in Chapter 10 below.

2. For this distinction see note e on I.10 76b3–10 in Chapter 3 above and discussion of II.7 92b4–7 in Chapter 10 below.

3. Cf. also 94a9–12, II.3 91a1, and references in Bonitz, 524b48–61.

4. Aristotle's claim is actually stronger: if we know x, we know that x cannot be otherwise.

5. *Pace* Barnes 1981, p. 26.

6. Necessity has nothing to do with existence claims; see Chapter 10 below.

7. See discussion of I.2 71b20–22 in the section "Six Requirements for Scientific Principles" in Chapter 2 above.

8. For the move from "cannot be otherwise" to "necessary," cf. *De Int.* 13 22b6–7. For that from "scientific knowledge" to "demonstrative knowledge," cf. *APo* I.2 71b16–17.

9. Barnes, p. 112.

10. This point has nothing to do with Aristotle's mistaken belief that a necessary conclusion may follow when only the major premise is necessary (*APr* I.9 30a15–23). The things that are at stake are the grounds of necessary conclusions and our knowledge of them, not simply premises from which they follow. In the situation envisaged, "it is possible for it [the middle] not to be" (74b31): B belongs contingently to C, and perhaps also A belongs contingently to B. And Aristotle holds that if either premise is contingent, the conclusion cannot be known to be necessary.

11. *Cat.* 2 1a24–25 specifies that the "in" relation intended is not the relation between parts and whole.

12. This is to be construed broadly, as including such relations as that of lines to triangles, in which the parts are proper parts, and such relations as that of points to lines and of moments to time, in which strictly speaking the parts are not component parts. Cf. *Phys*. VI.I.

13. Barnes's comment (p. 113) regiments this relation as the predication relation, which is incompatible with the point-line example, but has the advantage of making the relation directly relevant to syllogistic reasoning.

14. Proclus, *Comm*. 97.7. This definition seemingly was known to Aristotle (*De An*. I.4 409a4–5).

15. See, for example, *Phys*. VI.1 231a24–25. Cf. Bonitz, 162a13–15.

16. "*Sano modo*," Zabarella 1966, p. 702a.

17. *APo* I.4 73b30–32; cf. b33–34, b38, b40, and for an extreme abbreviation, I.24 85b11.

18. This extends a suggestion of Ross, p. 519.

19. The plural verb is unusual with neuter plural subject, but it can hardly be construed as a participle agreeing with *autoîs*.

20. The dative of *hósois* is required by the construction "to be predicated of x in its definition" (*enhupárchein tinì en tôi lógōi tôi tí esti dēloûnti*)—cf. note c on I.4 73a34–37 earlier in the present chapter.

21. This is doubtless the basis for the similar claim that serves as premise (4) of the argument at I.6 74b5–12, discussed above, pp. 83–84.

22. *Contra* Ross (p. 522), who takes *toîs katēgorouménois* as subjects, citing *APr* I.32 47b1, so that the first things mentioned are predicates belonging to their subjects (per se 1) and the subsequent things are predicates being belonged in by their subjects (per se 2). The point of 47b1ff. is not so much that *katēgoreîsthai* is used of the subject, but that Aristotle can use words in the opposite voice: there, *katēgoreîsthai* in the active as well as passive, here *enhupárchein* in the passive as well as active. (This is the only occurrence of a passive form of the word listed by either Bonitz or LSJ.) In both places Aristotle is making the point that one thing can be predicated of another and another of it.

23. See Ross, p. 522.

24. For the notion of opposition involved, see *Cat*. 10 11b38– 12a20, 12b26–13a3, 13b13–35.

25. *Kampúlon*, the word translated "bent," covers both curved and zigzag lines.

26. This is true whether we take "oblong" to mean a number of the form m(m + 1) (cf. Heath 1921, 1:82) or any nonsquare composite number (Plato, *Theaetetus* 148a).

27. Cf. note b on I.10 76b23–34 in Chapter 3 above.

28. *A* "belongs" in any of the ways permitted according to the discussion in the section "Per Se 1" earlier in the present chapter.

29. At I.22 84a24 we are to understand *hupárchein*, but this will be Aristotelian shorthand for *enhupárchein*. Cf. 84a20.

30. Euclid, *Elements* VII, def. 11.

31. "Primary" distinguishes the essential description of prime from any derivative descriptions that may state necessary and sufficient conditions for being prime.

32. Euclid, *Elements* I, def. 23.

33. Ibid., VII, def. 9: literally, "that [number] which is measured by an even number according to an odd number" (*artiákis dè perissós estin ho hupò artíou arithmoû metroúmenos katà perissòn arithmón*). The position of the article indicates that "even-times-odd" is predicate—not subject ("even-times-odd *number*"), as it is commonly translated.

34. This result is proved by Euclid, *Elements* IX, prop. 28.

35. Cf. Euclid, *Elements* I, defs. 20, 22.

36. The same question arises for compounds of differentia and subject, like "odd number," and the proposed solution applies to them as well.

37. *APo* I.10 76b3–4, b13. These passages are translated and discussed in Chapter 3 above. Also I.7 75a42–b2. This kind of demonstration is discussed in chapters 12 and 13 below. For other kinds of demonstrations and their relation to this primary kind, see below, Chapters 14, 15.

38. This discussion assumes that the subclass is not empty. In fact, the theory of science requires us to prove that (nonprimitive) subjects exist before we prove that they have further attributes. This move avoids the embarrassment of attributes like "even-and-odd," which might seem to belong per se 2 to "number," but which actually fails to specify a non-empty subclass of numbers.

39. "Has been proved" is more precise than "can be proved" because of the requirement that we cannot know the per se facts about a subject until we know that it exists (II.7 92b4–8).

40. There is no need to add *leukón* with Ross, since what is wanted is not another improper predication, but another improper subject.

41. For discussion of the two types of predication, see the section "Subjects and Attributes" in Chapter 8 below, and Barnes, pp. 116–18.

42. Ross understands it in this way, p. 519.

43. The same construction is used again at I.5 74a37, b3.

44. I shall indicate a scientific predication or "connection" between subject *A* and attribute *B* by writing "*A—B*."

45. I argue for this view in the section "In Every Case, Per Se, Universal, and Necessary" later in the present chapter.

46. Cf. 73a35, a38–39 for these very examples.

47. Cf. I.2 72a4–5, quoted and discussed in Chapter 2 above. "Prior" (*próteron*) means "more general," so "primary" (*prôton*) (its superlative) will mean "most general." Cf. this use of "prior" at 73b39: triangle is prior to isosceles.

48. The importance of this procedure of stripping off or subtraction for understanding Aristotle's notion of "abstraction" is the subject of Cleary 1985.

49. Aristotle has "figure" and "boundary" after "triangle." I suppose figure is one step more general than rectilinear figure. The relevance of "boundary" is easily seen by referring to Euclid, *Elements* I, def. 14; cf. defs. 6, 13.

50. Commentators frequently say that Aristotle requires that premises of demonstrations state relations that hold universally (in the sense defined in I.4). This is false and deeply misleading about the nature of scientific demonstrations. The only way for premises of demonstrations to be "universal" would stem from genus-plus-differentia definitions where the final differentia is related per se 1 to and is coextensive with the definiendum. (Cf. *Met.* Z.12.) But such premises will not get us very far in constructing demonstrations.

51. Cf. the many occurrences of the words "proof" (*apódeixis*) and "prove" (*apodeíknumi*, *deíknumi*) between 73b33 and 74b4.

52. Per se 3 and per se 4 relations are obviously not relevant.

53. Tiles 1983 argues that it is not.

54. The statement that straight and point belong to line "as line" (*hêi grammêi*) is clearly an adaptation of the expression "as itself" (*hêi autó*) to the particular case, as is *hêi trígōnon* in the other example.

55. For the parallel to be exact it should be "a certain loss of light from the moon."

56. *APo* II.8 93a29; also "something of the thing itself" (*ti autoû toû prâgmatos*, 93a22).

57. Below in Chapter 16 we will see that Aristotle is not entitled to much of the apparent force of this argument. It turns out that the claim is plausible only for rather special cases like "eclipse" and "thunder," which are names for connections of subject and attribute, not for attributes. "Screening by the earth" is a definition of neither moon nor a certain loss of light, just as "a loss of light from the moon due to screening by the earth," though a definition of an eclipse, is a definition neither of moon nor of loss of light. It remains that Aristotle describes nonimmediate connections as "something of the 'what it is.' "

58. For example, Ross, p. 519, and Barnes, p. 114.

59. See the section "Per Se 1 and Per Se 2" in Chapter 13.

60. This expression violates the distinction between per se relations and accidents implied by I.6 74b5–12, where per se 1 and per se 2 relations are necessary, accidents are not necessary, and everything belongs either per accidens or per se (i.e., per se 1 or per se 2), but not both. The least expensive solution is to distinguish between accidents and things that belong per accidens. This allows us to preserve the distinction between per se and per accidens relations, so that if *B* belongs to *A*, *B* belongs either per se or per accidens (but not both ways) to *A*. It follows that all things that belong per accidens are accidents and also some things that belong per se are accidents—all those that belong per se but are not in the essence (cf. *Met.* Δ.30 1025a30–32). The topic of per se accidents has been the subject of lively debate: see Barnes 1970, Wedin 1973, Graham 1975, and H. Granger 1981. See also Urbanas 1988. On this reading the only change needed in 74b5–12 is to understand "the accidents" (*tà sumbebēkóta* b12) as "the things that belong per accidens" (*tà katà sumbebēkòs hupárchonta*). (Similarly at I.4 73b4.) The per se accidents of a thing will be among the attributes that belong to it per se 1 and per se 2.

61. It is unfortunate that their role in science is not discussed more fully in I.22. See the section "Scientific Facts" in Chapter 8. At 83b26–27 Aristotle merely takes as given that a per se accident belongs to its subject, and then relates it to the species and subspecies of that subject. See Barnes, p. 171 for interpretation of the argument. See also Mansion 1976, pp. 160–62 and notes ad loc.

62. There is therefore nothing here either to support directly or to demolish Hintikka's thesis (Hintikka 1973, pp. 136f.) that necessity is conceived by Aristotle as an extensional, "in every case" relation.

63. I shall return to this point in Chapter 14 below.

CHAPTER VIII

1. Actually, only *some* of the premises are required to be of these sorts—a qualification we can ignore for the purposes of the present chapter. The roles of the different kinds of principles in proofs will be treated below in Chapters 12–15.

2. Recall that for Aristotle a definition is a statement of the essence; see e.g., II.10 93b29. See Bonitz 764b45–59.

3. *APo* I.6 74b5–12, discussed in Chapter 7 above.

4. The need for this qualification is discussed in the section "Individuals, Intelligibility, and Per Se Facts" later in the present chapter.

5. Throughout this chapter and the next I will presume (along with Aristotle) that

all facts can be expressed in subject-predicate form suitable to the syllogistic. For "*B* is *A*," understand throughout "All *B* is *A*," a universal affirmative predication that can be derived from a definition. If triangle is defined as a rectilinear plane figure with three sides, one of the associated univeral predications is "all triangles are figures." Another one is "all triangles have three sides," which in subject-predicate form might be put, "having three sides is predicated of triangle." This issue is discussed in the section "In Every Case" in Chapter 7 above.

6. See above, p. 100.

7. These claims will be discussed in the sections "Grounds of Truth or Grounds of Knowledge?" and "Priority of What is Better Known" in Chapter 17 below.

8. These lines speak of the premises of a syllogism, not a demonstration, but the context leaves no doubt that Aristotle is thinking of scientific demonstrations. Moreover, he is thinking of definitions, not the other kinds of scientific principles (I.8 75b30–32).

9. In Chapter 10 I take up the implications some of these passages have for the existence of the objects of scientific knowledge.

10. We are speaking here of the fully organized science, not of science in the making. The way in which our knowledge of scientific principles is based on acquaintance with particulars will be discussed below in Chapter 18.

11. An individual may fall under several universals treated by a single science. It may be an isosceles triangle, a triangle, and a figure. The most specific relevant universal will be most informative since the attributes of isosceles triangles include all the attributes of triangles and more besides.

12. Even its size is relevant (geometry studies the area of the triangle) and in some ways its position too. Geometry treats properties of polygons inscribed in a circle, but not the property of the circle's being in this room (cf. *Phys.* IV.1 208b22–25). We might specify the "geometrical position" of figures that geometry considers, in contrast to their "material position," which it does not.

13. See above, p. 60.

14. See the section "Per Se 3" in Chapter 7 above.

15. Recall that Greek can express this without a separate word for "thing."

16. Aristotle recognizes this point at *Met.* Z.4 1030a17–b13.

17. This assumes we have appropriate insurance policies to deal with the possibility that an individual may fall under several universals treated by the same science.

18. For an argument of the same form, see I.22 83b38–84a6.

19. Sorabji 1980, chaps. 12–13, esp. pp. 213–19.

20. Aristotle makes this point at *Met.* H.3 1043b36–1044a2. (I owe this reference to Frank Lewis.)

21. Cf. I.2 71b12, b15–16, I.4 73a21, b18–19, I.6 74b6.

22. The section "Existence and Necessity" in Chapter 10 below takes up the related issue of whether the theory requires that subjects of sciences exist eternally.

23. I will take up questions about the existence of universals and their dependence on particulars in Chapter 10 below.

CHAPTER IX

1. Axioms cannot do this work since they are common to more than one science, whereas essential facts have to do with subjects and attributes in a single subject genus. Neither can existence claims, which occur only for subjects, not for attributes (I.10 76b3–

10). Another interpretation stems from the division of principles into axioms, definitions, and hypotheses (I.2 72a14–24). If hypotheses include all immediate facts not found in definitions, our problem is solved. In fact, two possible translations of the key passage invite this interpretation: a hypothesis is "one [thesis] that assumes either of the members of a contradiction (e.g., I mean, that something exists or does not exist)" or "one [thesis] that assumes either of the members of a contradiction (i.e., I mean, that something is or is not the case)" (a19–20). Although these readings offer a blank check for scientific principles and so avoid the problem created by the stricter reading of hypotheses as solely existence claims, I prefer to remain with the stricter reading for the reasons given note f on I.2 72a14–24 in Chapter 3 above. There are reasons against these rival readings too, the strongest being that I.4 defines per se 1 and per se 2 relations in terms of *tí esti*, which has a direct connection with definitions (cf. also II.13 96b21–25).

2. How a derivative per se attribute like 2R is proved to hold of a subject like triangle is treated below in Chapter 12. Despite the difficulties of forcing mathematical proofs into the straightjacket of Aristotle's theory, it is still the case that the indemonstrable facts about a thing stated in the principles of the science (typically in the definition of the thing) are (together with the other principles) sufficient to prove all the derivative facts about it.

3. This account covers both per se 1 and per se 2 relations. Aristotle is concerned in I.22 to prove that *n* is finite.

4. The same method is found in a third passage, at II.5 91b28–30.

5. The other discussions of definitions in *Topics* (VI and VIII.3) are no more useful for our purposes.

6. I will use "prime*" to mark the second way of being prime identified in the passage.

7. "The final predicate" cannot be A_n, the final one of the A_i's, since *each* of the A_i's extends more widely than x.

8. Recall that attributes that belong "universally" are coextensive with their subject, but are sometimes if not always demonstrable and therefore not essential. See esp. I.4 73b25–74a3 and I.5, and discussion in the section "Universal" in Chapter 7 above.

9. It would be wrong to grant this. Accidental attributes may belong "more widely" than to the thing in question.

10. Esp. in Rolland-Gosselin 1912, Le Blond 1979.

11. Hence the distinction between definition and property (*Top.* I.5 102a18–19).

12. *Met.* I.8 also takes for granted that biological classification takes place by genus and differentiae. See G. Lloyd 1961, pp. 66f.

13. G. E. R. Lloyd notes some differences, but agrees that although Z.12 may be "somewhat later" than *APo* II.13, "Aristotle still believes that the infimae species will be revealed by a process of dividing and subdividing genera into opposites" (G. Lloyd 1961, pp. 67f.), and that *APr*, *APo*, and *Met.* retain division "as a useful method for definitions, and . . . formulate certain rules for its application to the problem of establishing the infimae species of each genus" (p. 70). The present discussion accords with Lloyd's view that *Met.* Z.12 is earlier than *PA* I.2–4 (ibid., p. 80).

14. In the Greek, the successive genera are mentioned in order, beginning with the first: animal two-footed wingless.

15. This is not to say that *PA* presents biology as a demonstrative science or that the methodology of *PA* agrees closely with the prescriptions of *APo*. I do not even require that the *APo* ideal of presenting sciences in demonstrative form was in Aristotle's mind when he wrote *PA*. My thesis is much more limited: that Aristotle continued to be inter-

ested in how to go about defining (a topic closely connected with his interests in biological classification) and that the view of definition he presents in *PA* marks a major change over his views in *APo* and solves a major difficulty for the theory of demonstration.

16. It requires only trivial adjustments to make sure that the conclusions of these demonstrations are per se predications of the subject. We shall see in the section "Existence and Necessity" in Chapter 10 below that the notion of hypothetical necessity that Aristotle invokes in *PA* I is entirely appropriate to demonstrative science.

17. There is one unimportant exception, at I.3 643a16–24.

18. Balme 1972, p. 101.

19. Since Aristotle is not committed to this definition of man (it also includes birds) there is no need to add *ḗ* between *dípoun* and *schizópoun* with Balme on the grounds that cloven-footed does not imply two-footed. The example merely shows what a definition resulting from dichotomous division *might* look like.

20. Aristotle need not accept all the popular classifications. He says it is right to follow their lead in using multiple criteria, not that they always choose the right criteria, combine them correctly, or reach correct results. In fact, he accepts genera of animals not found in popular classifications at all.

21. *PA* I.3 642b30–643a1, 643a4–5, a16–24.

22. This is the view of *Top*. VI.3 140a27–29.

23. "Per se accident" is equivalent to "derivative per se attribute." See the section "Scientific Facts" in Chapter 8 above.

CHAPTER X

1. See *APo* I.10 76b3–11 and note e in Chapter 3 above.

2. See the discussions of I.10 76a31–36 and I.1 71a11–17 in Chapter 3 above.

3. At II.2 90a2, the "is" with qualification is called "partial" (*epì mérous*).

4. Cf. pp. 62–63ff. above.

5. Cf. *Met*. Δ.13 1020a11–14.

6. The subject-genus term ("magnitude") may not appear in the definitions of all subjects, but we may suppose that there are ways of telling what subject genus we are in. One plausible solution is that since the definition of each term includes its genus, if we look at the definition of the genus and in turn of *its* genus, eventually we reach the subject genus term.

7. At 76a32–33 he says "what both the primary and the derivative things signify is assumed." This is surely more nearly correct to say than that what they *are* is assumed. Neither expression is quite correct, but given that Aristotle is going to use the same expression to apply to both, his choice is reasonable.

8. This passage is translated and discussed in the section "Six Requirements for Scientific Principles" in Chapter 2 above.

9. This passage is translated and discussed at the beginning of Chapter 7 above.

10. *Met*. Λ.7–8; *G&C* II.11 337b35–338a2.

11. "Aristotle often speaks as though the only principles are the definitions," Barnes, p. 109, with references.

12. *APo* I.6 74b5–12, discussed in Chapter 7 above.

13. *G&C* II.11 337b35–338a2.

14. Recall that "universal" refers to propositions or facts, not to the subject of

those propositions or facts. Similarly, "particular" means "particular occurrence or instance of a fact"—not the moon, but the moon undergoing eclipse tonight.

15. In I.31 particulars are locatable in place and time; in I.8 demonstrations apply to particulars at a particular time and in a way. Both discussions apply equally well to all three qualifications (time, place, and manner).

16. See the section "Individuals, Intelligibility, and Per Se Facts" in Chapter 8 above.

17. *Some* considerations of time and place may be scientifically relevant. Astronomy treats times of first visibility of stars and the synodic periods of planets, and Aristotelian physics treats of certain places, such as the center of the universe. In such cases, I would like to distinguish between universal and particular references to time and place, as between saying that all pieces of earth tend to move toward the center of the universe and saying that this particular piece of earth is in this particular place (even if this place happens to be the center of the universe), or between saying that any star's first visibility always occurs the same length of time from the vernal equinox and that this particular star's first visibility occurs on such and such a date. In each of these examples, the second statement treats a particular and mentions a particular time or place in its relation to that particular. Temporal and spatial considerations of this type help determine the particular star or piece of earth as a particular. Scientific knowledge of the corresponding universal facts will not include reference to times and places as the temporal and spatial locations of particulars, although the universal facts will be such that they can be applied to appropriate particulars, and when the application is made, individual times and places will be mentioned. The considerations presented in this note should be understood to hold throughout the discussion. I am indebted to Henry Mendell for drawing my attention to this issue.

18. It is given here that per se relations are "always." I take it that they are "always" for the reasons given in discussing the passages from I.8 and I.31.

19. This example is unsatisfactory since the timelessly true proposition refers to particulars. It is not difficult to restate the proposition generally, in terms of opaque objects interposed between objects and their sources of illumination.

20. The nature of lunar eclipses is perhaps too complicated for straightforward exposition, since it is a relation between three bodies and the behavior of light. The point can be seen more simply for triangle—2R.

CHAPTER XI

1. Hintikka 1974, p. 91. Hintikka's analysis begins with the conviction that Aristotelian sciences must be interesting and then identifies principles that make for "interesting" sciences (Hintikka 1972, 1974). But this approach is precisely backwards. See Frede 1974.

2. Notably Einarson 1936; H. Lee 1935; Ross 1949, pp. 56–59; Heath 1949, pp. 55–56; Heath 1921, 1:336–38; Heath 1925, 1:124; and Szabó 1978.

3. A total of 119 definitions are found in Heath's translation. However, see below, n. 55 for this chapter.

4. The manuscripts give five common notions. Proclus thinks that all five are genuine (*Comm.* 196.15–21). Heath translates all five, but argues for following Heron in keeping only 1–3 as genuine (Heath 1925, 1:225, 232). In what follows I accept Heath's recommendation.

5. The translation is from Heath 1925, 1:154f.; the labeling is mine.

6. H. Lee 1935, pp. 114–18.

7. Quotation from H. Lee 1935, p. 115. The view is expounded in Zeuthen 1896 and in Zeuthen 1902, pp. 98–103, and accepted by Heath 1925, 1:195, 221. It has recently been attacked by Knorr 1983.

8. H. Lee does not make anything of the fact that Aristotle says that geometry assumes the existence of points and lines (I.10 76b5) whereas Euclid's postulates allow for the construction of lines and circles.

9. Heath 1925, 1:191; also Heath 1949, pp. 27–30.

10. H. Lee 1935, p. 117.

11. This is the view of Heath 1925, 1:117; H. Lee 1935; and Szabó 1978.

12. This is the view of Solmsen 1975, pp. 126–28.

13. In Proclus, *Comm.* 66.7–8, 66.20–22, 67.12–15.

14. Pre-Euclidean concern for systematic organization is attested by Eudemus, who credits Leodamas, Archytas, and Theaetetus with making progress to a "more scientific arrangement" (Proclus, *Comm.* 66.17). Also, Theudius "arranged the elements well" (Proclus, *Comm.* 67.14). As Aristotle's pupil, Eudemus may have taken his ideas of scientific arrangement from his teacher.

15. Proclus, *Comm.* 68.7–10.

16. See esp. Proclus, *Comm.* 69.1–70.18.

17. Proclus, *Comm.* 75.5–77.6.

18. *APo* I.10 76b9; *Top.* VI.4 142a22–31, b7–10; *De An.* II.2 413a13–20; and see Heath 1925, 1:146–51, 155–92.

19. Heath 1925, 1:151; Szabó 1978, pp. 231–32.

20. Heath 1925, 1:202.

21. Fritz 1955, p. 392.

22. Esp. Tarán 1981, pp. 427f.

23. In an appendix of Mueller forthcoming.

24. Szabó 1978, pp. 274–76.

25. Proclus, *Comm.* 66.2.

26. Szabó follows Heath's conjecture (Heath 1921, 1:175f.) that Oenopides restricted the permissible constructions to those that can be carried out by means of ruler and compass, then asserts (falsely; cf. Zeuthen 1902, pp. 98–99) that P1–P3 are equivalent to postulating the use of the straightedge and compass in constructions. He then declares, "to say that Oenopides solved the problem 'by means of the ruler and compasses' is just to say that he solved it *by making conscious use of Euclid's first three postulates*" (Szabó 1978, p. 275, his emphasis). The illegitimacy of this reasoning is manifest.

27. So Hare 1965, pp. 22ff.

28. So Solmsen 1975, pp. 110ff.

29. Above, pp. 70–71.

30. Their role is discussed in the section "The Analogical Nature of Axioms" in Chapter 6 above.

31. Evidence for this claim is presented in the subsection "Definitions" just below.

32. Discussion of LNC goes back to Plato, *Republic* 436b. Aristotle's arguments in its favor are normally seen as directed against actual, not imaginary, opponents.

33. See references above in n. 13 for this chapter.

34. For Euclid's presupposing the existence of points, see Mueller 1981, pp. 27–29.

35. See Knorr 1983.

36. Zeuthen 1896, p. 223.

37. Such assumptions were not explicitly formulated. Cf. Heath 1925, 1:234ff.

38. See the treatment of Euclid, *Elements* I, prop. 1, at the beginning of Chapter 12 below.

39. For the role of constructions in geometrical proofs, see Hintikka and Remes 1974, chaps. 1, 5. They also call attention to the surprising failure of the ancients to recognize the function and importance of constructions in proofs.

40. Theorems are stated in declarative sentences and their conclusions are followed by "Q.E.D." (*hóper édei deîxai*). Construction problems are stated as infinitive clauses ("on a given line to construct an equilateral triangle") and the conclusions are followed by "Q.E.F." (*hóper édei poiêsai*).

41. See Heath 1925, 1:124–29.

42. See below, Chapter 15.

43. See Mueller 1981, pp. 14f., on the relation between P1–P3 and Hilbert's existence claims. Mueller (forthcoming) is less willing than I to associate Aristotle's existence claims with Euclid's construction postulates.

44. Heath 1921, 1:175.

45. See the list of terms in Mueller 1981, p. 39.

46. Heath 1925, 1:165.

47. Ibid., p. 171.

48. The notion of "fat definition" is defined at the beginning of Chapter 9 above.

49. Cf. *Top*. VI.3.

50. Proclus, *Comm*. 157.10–11.

51. Proclus, *Comm*. 157.17–158.2.

52. See Mueller's note in Morrow 1970, p. 125.

53. Heath 1925, 1:225. For a different view, see Knorr 1980a, pp. 159–62, and Mueller 1981, pp. 21–26.

54. In fact, the second part of def. 17 can be proved from *Elements* III, def. 11 and props. 31 and 24, though Euclid does not seem to have recognized it.

55. An examination of Heiberg's MS P in the Vatican Library and of MS B and other MSS of the *Elements* in the Bodleian Library revealed that (in those MSS) the definitions are not numbered (although B numbers the postulates and common notions). Usually definitions are separated by punctuation marks, or the initial word of each definition begins with a capital letter and/or is written in a different color of ink from the rest. There is great variation in the way the definitions are individuated. The material that Heiberg arranges as twenty-three definitions appears as sixteen in one MS and thirty in another.

56. P4 is an exception to this practice of grouping unproved propositions next to the definitions of relevant terms. According to this practice, P4 should be placed next to def. 10, which defines right angles. I discuss this issue at the end of the present chapter.

57. Cf. Hultsch 1885, index, s.v. ὅρος.

58. See Fritz 1955.

59. Book XI makes use of definitions that refer to the motion of figures—defs. 14, 15, 18–23—although motion is not a geometrical concept.

60. Proclus, *Comm*., pp. 76ff., 178ff.

61. Fritz 1955, pp. 370–74.

62. In his axiomatization of geometry, Hilbert (1971) has five groups of principles (plus definitions), each with a different subject matter (incidence, order, congruence, parallels, continuity).

63. So Zeuthen 1896, followed by Frajese 1951. See *contra*, Mueller 1981, p. 28, and above, n. 53 for this chapter.

CHAPTER XII

1. The translation is from Heath 1925, 1:241–42; the labeling is mine.

2. The traditional names for these parts are D1: *prótasis*, D2: *ékthesis*, D3: *diorismós*, D4: *kataskeuḗ*, D5: *apódeixis*, D6: *sumpérasma* (Proclus, *Comm.* CCIII.1–205.8).

3. In this case, the conclusion is stated in terms of the specimen case considered. Elsewhere Euclid frequently states the *sumpérasma* generally, in effect as a restatement of the *prótasis*.

4. For axioms, see above, pp. 70, 71–72. For definitions, see above, pp. 72, 124–25f. For existence claims, see above, pp. 124–25, 139.

5. Cf. also *Met.* N.2 1089a21–25, *APr* I.41 49b33–37.

6. Aristotle recognizes their value at *Met.* Θ.9 1051a21–29.

7. This point is made by Proclus, *Comm.* 207.4–25, and discussed by Mueller 1981, pp. 12–14.

8. This assertion as well as others in this section is based on the section "Differences between Aristotle and Euclid" in Chapter 11 above. See in particular the subsection "Hypotheses and Postulates."

9. By "guaranteed" I mean that it is either asserted or presupposed, ibid.

10. These reconstructions admittedly differ from Aristotle's assumptions that points and lines exist in that they postulate the existence of *certain* lines, ibid.

11. This must be understood also in Euclid's proof, which does not use a word for "radius."

12. This is not a trivial matter, but a significant omission in Euclid's geometry. See Heath 1925, 1:234–40.

13. We shall see this in greater detail in Chapter 13 below.

14. Solmsen 1975, 1941, 1951; Ross 1939; Ross 1949, pp. 6–23; Barnes 1981. Barnes 1981, n. 1, contains useful references to previous discussion.

15. This passage is discussed by Lear 1980, pp. 80–82.

16. See Lear 1980, pp. 95–97.

17. See Lear 1980, pp. 25–30, on the use of syllogistic in I.21.

18. See Lear 1980, pp. 10–14.

19. Lear 1980, p. 13.

20. Heath (1925, 1:320–21), following Heiberg, argues that Aristotle used the same construction as Euclid. Mendell (1984) argues that he used the "Pythagorean," not the "Euclidean," construction. Mendell's reading requires some changes in the proof, but nothing that affects the points I am making.

21. The translation is from Heath 1925, 1:316–17; the labeling is mine.

22. The book was called to my attention by Professor John Murdoch.

23. References to two nineteenth-century works that also pursue this peculiar exercise are to be found in Heath 1949, p. 26.

24. *APr* I.1 24a16–17; reaffirmed for the propositions of demonstrative sciences at *APo* I.22 83b17–18.

25. See the section "Subjects and Attributes" in Chapter 8 above.

26. See the section "Per Se 2 and Definitions of Attributes" in Chapter 7 above.

27. See also Mueller 1974, esp. pp. 40–42.

28. A demonstration is a fragment of a science and makes little sense outside the context of the whole science. Only there is it clear what propositions are principles and how each fact is dependent on the principles.

29. Cf. the examples of eclipse (II.8), nontwinkling planets (I.13), and the various kinds of "cause" in II.11.

30. This argument is discussed earlier in the present section.

31. The translation is from Heath 1949, p. 23.

32. So Heath 1949, p. 24.

33. Prôton, the word here translated "first," also means "primary." The significance of "primary" and its equivalence to "immediate," are discussed in the section "Principles Are Primary and Immediate" in Chapter 2 above.

34. Cf. I.14 79a29–31, also I.19 81b30–82a6, I.12 78a14–21.

35. Cf. *PA* I.4 644a23–b7.

36. *APo* I.10 76b10, also *Top.* I.1 100a27–29.

37. Cf. Ross's astute comment (1949, p. 36): "while he rightly (in *APo*) treats mathematical reasoning as the best example of strict scientific reasoning, he did not in fact pay close attention to the actual character of mathematical reasoning."

CHAPTER XIII

1. Presumably it *is* the only way, since we are never told of any other.

2. *APo* I.4 73a21–24, translated and discussed at the beginning of Chapter 7 above; also I.6 74b5–39.

3. *APo* I.6 75a29–31, 74b5–12, translated and discussed at the beginning of Chapter 7 above. This is also the point of statements that the conclusion belongs per se to a *genus* (I.7 75a40–41, a42–b2, I.10 76b13).

4. These types are per se 1 and per se 2, which are discussed at length in Chapter 7 above.

5. *APo* I.6 75a18–19.

6. For the notation, see above, p. 159.

7. These issues are discussed in detail below in Chapter 17.

8. *APo* I.7 75b10–11, I.9 76a8–9.

9. Problems arising from these claims are discussed in the section "Existence and Necessity" in Chapter 10 above.

10. See the section "Subjects and Attributes" in Chapter 8 above.

11. *APo* I.10 76b3–4, b6–10, b12–13, translated and discussed in Chapter 3 above; also I.7 75a42–b2.

12. For these terms, see the section "Scientific Facts" in Chapter 8 above. For the status of *B*—*C* as a subject-predicate proposition, see above, p. 154.

13. This view is defended in the section "Subjects and Attributes" in Chapter 8 above.

14. The notion of "fat definition" is defined at the beginning of Chapter 9 above.

15. Ackrill 1963, p. 82.

16. The different sorts of relations are discussed in the section "Per Se 1" in Chapter 7 above.

17. These adjustments were discussed above, pp. 160–63.

18. The proof is translated and discussed in the section "Proofs and Syllogisms" in Chapter 12 above.

19. This use of the term "essence" may appear somewhat unusual, but it is in line with the view developed above in Chapter 8 that the scientific essence of *x* is the basic

facts about x, which are stated in the definition of x. The essence of alternate angles is that they are interior and opposite angles made by a straight line falling on parallel lines.

20. The middle is identified as the grounds (II.2 90a6–7).

21. This word sometimes means "definition."

22. This is the simplest case. The more complex case where there are different explanations for a single attribute belonging to different subjects is treated at II.17 99a30–b7, discussed in the section "Commensurate Universals" later in the present chapter.

23. Cf. II.16 98a37–38, b4–16.

24. *Aitía*. Cf. *Gen. An.* V.3 783b10–23.

25. Actually, he says *lógos*, not *ho lógos*, but the article is omitted because it is a predicate noun. Cf. 99a21–22.

26. The move from "broad-leaved plants are deciduous" to "vines are deciduous" is the same as that from triangle—2R to isosceles triangle—2R. See the discussion of this kind of proof below in Chapter 14.

27. See the section "Are Conclusions of Demonstrations Per Se?" in Chapter 7 above.

28. *APo* I.4 73b16–24, I.6 75a28–31, I.7 75a40–41, b1–2, I.9 76a4–8, I.22 84a11–17.

29. Analogous claims hold for the kinds of demonstrations discussed below in Chapters 14–15. It is straightforward to apply the following arguments to those other kinds of proofs.

30. *APo* I.22 84a14–21, translated and discussed in Chapter 7 above.

31. Ross (p. 522) says "that every number is either odd or even, must be assumed." I disagree. This fact will be a provable proposition following from the definitions of number, even, and odd.

32. See I.4 73b23–74a3, I.5 74a16–17, a25–b4, I.24 85b5–13, 86a25–27.

33. Literally, "the chance one" (*tò tuchón*, I.4 73b33, b35, b36, b38, b39).

34. For the example, see I.24 85b38–86a3, II.17 99a18–21.

35. *APr* II.23 68b21.

36. *APo* II.16 98b38.

37. *APo* II.17 99b4, II.16 98b25.

38. *APo* II.16 98b26.

39. Aristotle makes provisions for cases where the attribute is used homonymously and analogically (II.17 99a1–16).

40. I shall treat the case where there is one primary subject. What I say applies *mutatis mutandis* where there is more than one.

41. This use of the word is discussed in the section "Universal" in Chapter 7 above.

42. These prepositions are used interchangeably. For *diá*, see I.5 74a13–16; for *katá*, see I.5 74a36, b2, I.24 85a24; for *hêi*, see I.5 74a35–36, I.24 85b6–14; for *hóti* and *dihóti*, see I.24 85b37–86a3.

43. For the notion of subtraction, see Cleary 1985.

44. Mere coextension is not enough to make a relation universal: See I.5 74a25–32, translated and discussed in Chapter 7 above.

45. "Right angles do not intersect" means "straight lines each of which is at right angles to one and the same straight line do not intersect." Heath 1949, pp. 40f.

46. Though the words for "know" are *eidénai* and *ginṓskein*, not *epístasthai* (cf. note a on I.2 71b9–12 in Chapter 2 above).

47. See II.17 99a30–b7, discussed earlier in the present section.

48. The example is given at II.17 99a23–29, translated and discussed in the section "Aristotelian Demonstration" earlier in the present chapter.

49. For the limitations imposed by requiring syllogistic arguments in proofs, see the section "Proofs and Syllogisms" in Chapter 12 above, and for a detailed discussion of the role of convertible premises in demonstrations, see Inwood 1979. I agree with much of that paper, but not that all *kath' hautà sumbebēkóta* are commensurately universal or that provable *kath' hautó* relations are immediate and self-explanatory.

CHAPTER XIV

1. *Prôton*. See my discussion of I.4 73b26–74a3 in Chapter 7 above.

2. *APo* I.6 74b5–12, discussed in Chapter 7 above.

3. My application arguments are what Lennox calls Type A demonstrations. My USA proofs are his Type B demonstrations. (Lennox 1987, pp. 92–97.)

4. This sort of arrangement cannot be expected to work for every subject genus. See the section "Interpretations of the Subject Genus" in Chapter 4 above.

5. This method fails for cases where there is no single subject to which an attribute belongs without extending more widely. (This sort of case is taken up in II.17.)

6. The question what is to count as "the" middle in those cases is taken up in II.18, which is translated and discussed just below.

7. This point is discussed in the sections "Individuals, Intelligibility, and Per Se Facts" and "Essence and Necessity" in Chapter 8 above.

8. Cf. pp. 155–56f. below.

9. This interpretation agrees with the suggestion proposed in the section "Existence and Necessity" in Chapter 10 above that universal scientific facts and demonstrations apply "always" in that they apply *whenever* there are appropriate universals.

10. I ignore the problems raised by quantifying over mass terms. That is not a problem raised specifically by the theory of demonstrative science.

11. This passage is translated and discussed in Chapter 10 above.

12. Nothing has been said yet about the crucial step of determining that thunder is noise caused by quenching of fire in clouds.

13. The difference between "occurs" and "is occurring" is invisible in Greek, which lacks distinct forms for the present progressive tense.

14. Van Fraassen (1980, pp. 23–25) makes this objection.

15. The passage is translated and discussed in Chapter 13 above.

16. The proof is translated and discussed in the section "Proofs and Syllogisms" in Chapter 12 above.

17. Barnes, p. 173. The discussion of compactness in Lear 1980 (Chap. 2, esp. pp. 30–32) is helpful on this passage.

CHAPTER XV

1. *APo* I.10 76a31–36 may refer to proofs of existence, but even if it does, the idea is not developed. See my discussion of I.10 76a31–36 in Chapter 3 above.

2. These questions are reminiscent of Meno's paradox, of which Aristotle was conscious in *APo* (I.1 71a29–30, II.19 99b25–26).

3. This is one kind of unscientific question: I.12 77b5–9.

4. Cf. I.12 77a36–b27.

5. To my knowledge, Aristotle knew no metalogical theorems sufficiently powerful to assure that a relation could be proved without making it clear what the middle of the proof was.

6. The relations between postulates and assumptions of existence and between constructions and existence are discussed in the sections "Differences between Aristotle and Euclid" in Chapter 11 above and "Existence vs. Constructibility" in Chapter 12 above.

7. The translation is from Heath 1925, 1:347–48; the labeling is mine. For the six parts of the proof, see the beginning of Chapter 12 above.

8. Allowing for the modifications discussed in the section "Proofs and Syllogisms" in Chapter 12 above.

9. The existence of *AB* as a legitimate subject is guaranteed by P1 and P2, subject to concerns expressed in the subsection "Hypotheses and Postulates" of the section "Differences between Aristotle and Euclid" in Chapter 11 above.

10. See above, Chapter 12, n. 12.

11. Cf. *Met.* Z.17 1041a10–20.

12. For this distinction, see note e on I.10 76b3–11 in Chapter 3 above and the section "The Role of Existence Claims" in Chapter 10 above.

13. These issues are also treated below on pp. 207–8f.

14. Any roles of primitive existence claims and definitions in subject-attribute proofs will also be found in existence proofs in the subject-attribute arguments that constitute step S2.

15. Cf. the use of CN1 and CN3 in *Elements* I, prop. 46.

16. This point is argued in the section "Existence and Necessity" in Chapter 10 above.

17. In typical Euclidean proofs, theorems and problems show no differences in these respects. In both cases steps D2 and D4, which involve or presuppose constructions, guarantee the existence of the entities in question. This function of Aristotelian existence claims is discussed in the section "The Role of Existence Claims" in Chapter 10 above.

18. *Met.* Γ.2 1003a33–b15 makes the point that all non*substances* are said to be by virtue of being related to substances. In the context of a demonstrative science, the same is true for attributes vis à vis subjects. Cf. Mansion 1976, p. 163.

19. Euclid, *Elements* I, def. 23.

20. So Mansion 1976, pp. 171f.

21. This point is made in the section "An Aristotelian Demonstration" in Chapter 13 above.

CHAPTER XVI

1. There is neither any textual reason nor, as far as I can see, any interpretive reason to follow Mansion 1976, p. 188 n. 119, who calls the preliminary answer a material definition and the final answer a formal definition, and supposes that the preliminary answer states the matter of the definiendum and the final answer its form. *Met.* Z.17 1041b2–9 makes this distinction, but I do not see any justification for reading it into the present context.

2. Cf. II.8 93a30, a37.

3. APo II.9 treats the case where it is the same.

4. With Barnes, p. 212.

5. On another interpretation II.9 93b23–24 says that we must posit or make known in another way both that there are such immediate "what it is" and what they are—i.e., which ones they are. On this reading, Aristotle is saying that we must accept that there are principles and specify which propositions *are* principles. But this requires taking "that they are" and "what they are" in ways so different from the use that dominates these chapters that it seems best not to adopt the reading, even though it gives a more straightforward sense.

6. Ross (p. 633) takes the passage differently: "By the things that have a cause other than themselves A. means, broadly speaking, properties and accidents; by those that have not, substances, the cause of whose being lies simply in their form." I disagree. Since II.2, relations and subjects have been assimilated. Most recently, II.8 93a22–24 lists relations and substances indiscriminately, and nothing in the text of II.9 indicates a reversal of this tendency. Ross sees in II.9 93b23–24 a reference ahead to II.13, which offers a method for arriving at definitions of substances (p. 633). But the reference is vague (Ross can only cite the words "in some other way") and is better taken as saying that since (for reasons given above) some essences cannot be reached by the method of II.8, we must obtain them in some other way.

7. Cf. I.5 74a8–9.

8. Cf. II.8 93a21–26.

9. Cf. II.8 93b4–5: investigating why *A* belongs to *C* is a matter of investigating what *B* is.

10. For this use of the word *lógos*, see Bonitz, 435a31–36.

11. Cf. I.15.

12. This is not quite right: at II.8 93b8 thunder is called quenching of fire in cloud. Aristotle realizes that it would be unacceptable to call thunder simply "quenching of fire," so he identifies where the quenching of fire occurs (cf. 93b11).

13. See Bonitz, 524b54–61.

14. We have seen that even the example of eclipse runs afoul of the fact that there are also solar eclipses.

15. I assume that "leg" is not being used homonymously here. Cf. *EN* V.1 1129a29–31.

16. Cf. II.13 97a28–30, *Met.* Z.12 1038a19–25.

CHAPTER XVII

1. This passage is translated and discussed in Chapter 2 above.

2. Neither do they catch the sense of *aítion*, which is apparently used as an equivalent in *APo*.

3. For defense of this translation, see the sections "Aristotelian Grounds and Humean Causes" and "Aristotelian Explanations and the Deductive-Nomological Theory" later in the present chapter.

4. This is stressed especially at 74b29–30 and 75a1–11, where in a proof that *A* belongs to *C*, *B* is the middle. Also 74b32–39, which says that if the middle can perish while the conclusion is preserved, knowledge through that middle is not scientific knowledge.

5. The passage is translated and discussed in Chapter 14 above.

6. Ibid.

7. In this I am in agreement with Patzig 1981.

8. *APo* II.2 90a9–11 and II.8 93a7–8, b4–5.

9. This point was made in my discussion of II.1 89b23–35 in Chapter 15 above.

10. *Eidénai*; cf. note a on I.2 71b9–12 in Chapter 2 above.

11. Ross finds four or five ways, pp. 552f.

12. The intervening lines 78b11–13, are discussed later in the present chapter.

13. Barnes objects to this (p. 150), but I doubt whether Aristotle would be moved by his comments, which apply to the example rather than the principle. If

A—B—C

is a scientific demonstration of *A—C*, then the principles *A—B* and *B—C* are the only grounds of *A—C*. Granted that *A—B* is true and immediate, then if *B—C* is true and immediate, *A—C* is shown to be true and is explained. If *B* belonged to no *C* (and if this fact were immediate, a principle), then the fact that *A* belongs to no *C* would similarly be shown to be true and be explained.

14. This claim of Aristotle's, together with his view that causes and effects are asymmetric—that is, if *A* is the cause of *B* then *B* is not the cause of *A*—constitutes a basic difference between Aristotle and the deductive-nomological view of scientific explanation. See Brody 1972, p. 23.

15. E.g., I.4 73b40–74a1 (translated and discussed in Chapter 7 above) I.5 74a11–13, II.17 99a34–35. See the section "Commensurate Universals" in Chapter 13 above; also see Inwood 1979.

16. In the case of proofs in the second figure, the primary subject of the predicate is the narrowest middle because that is the way to have the middle reciprocate with the attribute.

17. Cf. I.24 85b23–27.

18. Cf. I.8.

19. It fails to provide for cases such as that discussed in II.17 99a30–b7, where there is more than one primary subject of an attribute. In that case, being long-lived does not reciprocate with bilelessness or with dryness. The modification required is not difficult.

20. S1 is called only a *hóti*-syllogism, not a *dihóti*-syllogism; the opposite is true of S2. The genitive in *sullogismòs toû dihóti* is a genitive of quality, not an objective genitive (Patzig 1981, pp. 143–44).

21. I shall return to this point in the section "Priority of What Is Better Known" later in the present chapter.

22. The problem is identified just below, p. 217.

23. For this interpretation, see Ross, p. 552.

24. Nor do I concur with Hadgopoulos 1977, pp. 33–34, that it means "immediate as regards its conclusion," so that in

A—B—C—D,

A—C and C—D are immediate in this sense. I cannot believe that Aristotle would introduce such a different use of the word without warning.

25. Aristotle does not think the relation between nearness and not twinkling is immediate. See *De Caelo* II.8 290a14–24. (I owe this reference to Hadgopoulos 1977.)

26. The words quoted are the protasis of a conditional sentence, but are clearly understood to be true.

27. A more sophisticated account of deciduousness is that the sap congeals in the connection of the leaf stalk (II.17 99a23–29). This is the grounds of the deciduousness of broad-leaved plants. The passage is translated and discussed in Chapter 13 above.

28. I take this definition as equivalent to that found in II.2 and II.8: "eclipse is a certain loss of light from the moon due to screening of the earth."

29. See the discussion above on pp. 203–5.

30. Or is it only a factor in the grounds? It is not clear if the shift from *lógos* (II.8) to *en tôi lógōi* (II.16) bears any significance.

31. To be fair, II.16 is not concerned with our question and assumes that we *are* in a position to identify which of two reciprocating terms is the grounds of the other. This is not begging the question, but starting from a more advanced level.

32. "Account" easily can replace "what it is" since a definition is "an account of what it is" (*lógos toû tí esti*, II.10 93b29) and Aristotle shifts easily between talking in the material and in the formal mode.

33. Cf. II.10 93b38–94a7, esp. *lógos ho dēlôn dià tí estin* (93b39). The passage is translated and discussed in Chapter 16 above.

34. See the section "Per Se 1 and Per Se 2" in Chapter 13 above.

35. Pp. 86–87, 90–91f., 98–100ff. above.

36. See the section "An Aristotelian Demonstration" in Chapter 13 above for discussion of this essential point.

37. Ibid.

38. This passage is discussed in the section "Six Requirements for Scientific Principles" in Chapter 2 above.

39. This passage is discussed in the section "Principles Are better Known, Prior, and Grounds" in Chapter 2 above.

40. Cf. *Met.* Z.3 1029b3–12 and passages from *Top.* VI.4 translated and discussed in the section "Demonstration, Teaching, and Ideal Explanations" later in the present chapter.

41. Aristotle makes sensible points similar to this in *Top.* VI.4. See passages translated and discussed in the section "Demonstration, Teaching, and Ideal Explanations" later in the present chapter.

42. Discussed below in Chapter 18, esp. in the section "The Path to the Principles."

43. See the sections "The Path to the Principles" and "Grasping the Universals" in Chapter 18 below.

44. See the beginning of Chapter 14 above.

45. This is Aristotle's other example in I.13 (78b4–13).

46. Aristotle's theory will be compared with the covering-law model in the section "Aristotelian Explanations and the Deductive-Nomological Theory" later in the present chapter.

47. *APo* I.8 75b24–26, discussed in Chapter 14 above.

48. Although subordinate from the perspective of the theory and the overall structure of demonstrative science, some application arguments will be important nevertheless. See the section "Applications to Natural Phenomena" in Chapter 14 above.

49. Cf. II.17 99a23–29, discussed in Chapter 13 above.

50. E.g., Brody 1972; Patzig 1981, pp. 150–56.

51. Brody 1972 and Van Fraassen 1980 agree with Aristotle that scientific expla-

nations are asymmetric. They discuss implications of this point for the nature of scientific explanations.

52. For this conception of the structure of the subject genus, see the section "The Composition of the Subject Genus" in Chapter 4 above.

53. Burnyeat 1981, p. 126.

54. For discussion of the terms "primary" and "immediate," see the section "Principles Are Primary and Immediate" in Chapter 2 above.

55. See my discussion of I.10 76b23–34 in Chapter 3 above.

56. *APo* I.24 85b24–25, *hôi gàr kath' hautò hupárchei ti, toûto autò hautôi aítion*, may make the same point, but its context suggests a contrast between a "universal" and a "particular" demonstration, such as triangle—2R vs. isosceles triangle—2R. It would then be saying that 2R belongs to triangle because of itself (because of the nature of triangle) and belongs to isosceles triangle because of something else (because it belongs primarily to triangle and derivatively to isosceles triangle).

57. Barnes (p. 150) sees this as relevant to case 1, but it is not tied in to case 1's concern that the premises be immediate.

58. I offer this interpretation undogmatically. Quite possibly in these lines Aristotle attempts to loosen the convertibility requirement of case 2 and winds up with a situation where the deduction he has in mind just cannot be formed. Pacius's example "What is capable of laughing is an animal, Man is capable of laughing, Therefore, man is an animal (quoted, evidently with approval, by Ross, pp. 552f.) is not a happy one, since it is hardly better known that man is capable of laughing than that man is an animal. Besides, man's being an animal is not the grounds of his being capable of laughing, whereas Aristotle's point in specifying that the one that is not the grounds (of the other) is better known is presumably to imply that the one that *is* the grounds of the other is less well known. The same criticism applies to the example used by Barnes, p. 150.

59. On my interpretation, case 2A covers one situation where a premise is false.

60. Cf. I.12 77a36–37 on the identity of question and premise.

61. Cf. *paralogismós* (I.12 77b20, b27–33, b40–78a6).

62. Again, Aristotle is insufficiently aware of the role of previously proved propositions in proofs. See above, pp. 160–63.

63. We can read I.12 and I.13 as a section devoted to this topic—I.12 covering cases where we do not even reach KF, I.13 covering cases where we reach KF but not KRW.

64. See Patzig 1981, pp. 146–48, and McKirahan 1978, pp. 203–4.

65. The following discussion is perhaps too thorough, but the prominence of the "four causes" in Aristotle's philosophy and the frequency with which they are brought into discussions of *aítion* in *APo* (e.g., Van Fraassen 1980) make it important to set out the similarities and differences in detail.

66. The definition at *Met.* Δ.2 1013a24 is word-for-word identical with the *Physics* definition just quoted.

67. Ross 1924, 1:292.

68. Ibid.; cf. Ross 1936, p. 511.

69. Ross says that the premises are said to be material causes of the conclusion "by a metaphor" (1949, p. 639).

70. In the second case outlined above, both essences count as middle terms (II.17 99a25–6). See the section "An Aristotelian Demonstration" in Chapter 13 above.

71. See the section "Applications to Species" in Chapter 14 above.

72. *APo* II.2 says this repeatedly. The significance of the doctrine is worked out in Chapter 15 above.

73. For difficulties in finding the "material cause" in Aristotle's list, see Ross, pp. 638–39, and Barnes's reply, p. 216.

74. Cf. Moravcsik 1975.

75. This difference may not amount to much, since Aristotle is willing to construe subject-attribute relations so broadly. See above, p. 75.

76. E.g., Hempel 1965; Nagel 1979.

77. See Brody 1972; Patzig 1981; Burnyeat 1981, pp. 122–26; Van Fraassen 1980.

78. Nagel 1979, p. 33.

79. Ibid., pp. 33, 34, 42.

80. The generality criterion here seems to play a role analogous to that of Aristotle's order of natural intelligibility. I believe that it will be more profitable for future research to explore this point of contact than concentrating on the kind of scientific explanation corresponding to Aristotle's application arguments.

81. Cf. Evans 1977, esp. pp. 68–73.

82. In view of the uninterpreted nature of modern abstract mathematics, examples from applied mathematics and theoretical science are more appropriate here.

83. It is possible, as Aristotle says, to parrot proofs without understanding them. *EN* VII.3 1147a19–22.

CHAPTER XVIII

1. This topic is taken up in the section "Differences between Aristotle and Euclid" in Chapter 11 above, esp. in the subsection "Definitions."

2. *PA* II.1 646a8–12; *HA* I.6 491a10–12; *Phys.* I.1 184a10ff.

3. E.g., the introduction of *Meteor.*, esp. I.1 338b1–339a6.

4. Esp. *Top.*, *Met.*, and *PA* I.

5. Balme 1975, p. 183.

6. This is especially true of *APo* I. *APo* II, which has less unity than *APo* I, explores issues relevant to demonstrative science and offers a brief sketch (II.19) of how principles come to be known.

7. This is discussed below, p. 240.

8. The expression *tà katà méros* is used in two ways: to refer to particulars as opposed to universals, and to less general as opposed to more general universals. In the present context, the former use occurs to connect induction with perception. But some forms of induction proceed from the less general to the more general, especially the so-called "perfect" induction that commentators have generally found in *APr* I.24, and this type of induction has a place in demonstrative science too. See the section "Induction" later in the present chapter.

9. Cf. I.8 75b34, I.31 87b29–30, II.19 100a17–b1.

10. See below, p. 256, for a discussion of this statement.

11. Barnes (p. 185f.) has a different text at 88a17 and a different interpretation of *háma*.

12. For the meaning of *gnṓrimos* and *gnōrízein*, see the section "Grasping the Universals" later in the present chapter. The present translation is neutral among the possibilities.

13. For criticism of Aristotle's argument, see Barnes, p. 251.

14. *APo* II.19 99b36–100a3 identifies three levels of ability: (a) perception, (b) the level where the perception remains, and (c) the level where an account is developed out of the perception that remains. These levels are presented chiefly as distinctions among types of animals: some cannot get beyond level (a), and none but humans can attain level (c). Since humans must have perceptions, and these must remain before an account can be reached, the passage has developmental overtones. Level (a) is stage (1) in 100a3ff., and level (b) is stage (2). I take stages (3), (4), and (5) to be a more detailed analysis of level (c).

15. *Téchnē*: the disciplined knowledge of a craft. For the difference between science and art, see Evans 1977, pp. 75–77.

16. Ross 1924, 1:116.

17. I suppose this kind of teaching would not count as the "intellectual" teaching of *APo* I.1 71a1.

18. Esp. I.13, which analyzes cases of *hóti* and *dihóti* knowledge in the same science and cases where one science knows *hóti* and another knows *dihóti*. In the former kind of case, *dihóti* knowledge is proper scientific knowledge, and mere *hóti* knowledge is deficient: there is something wrong with the demonstration, so that the real explanation of the conclusion is not given.

19. *Met.* A.1 981b10–13; cf. *APo* I.31 87b39–88a8.

20. Mure; Ross; Barnes; Tricot 1947; and Le Blond 1970, p. 129 n. 1.

21. This view is held by St. Thomas, followed by Trépanier 1948.

22. There is no need to worry why *arché* is singular. At each stage one thing (memory, etc.), always mentioned in the singular, arises from another (which is sometimes singular, sometimes plural). These "things" are *héxeis*, cognitive states or dispositions, not the objects of those states. The question of the name of this stage of intellectual comprehension is reserved for the next paragraph (though recall that *noûs* is called the *archè epistḗmēs*, I.33 88b36). Aristotle does not name the stage where the universal is at rest in the soul presumably because there was no name in common use for that stage. *Met.* A.1 tends to confirm this. I shall argue below (in the section "*Noûs*") that he calls this state *nóēsis* although that word is not a technical term.

23. The *arché* is *ek pantòs ēremḗsantos toû kathólou*, just as memory is *ex aisthḗseōs* and experience is *ek mnḗmēs*. Zabarella 1966, p. 1270b, and Pacius 1966, p. 347, identify the universal with the "principium artis & sapientiae," but this interpretation takes *téchnēs archè kaì epistḗmēs* as parallel with *hò àn en hápasin hèn enêi ekeínois tò autó*, which leaves the sentence beginning with *ek d' empeirías* incomplete.

24. For this conception of principle, see the excellent discussion in Wieland 1970, chap. 1, esp. sec. 5. Also Kosman 1973 and Burnyeat 1981.

25. This is a more careful statement of what the *Met.* passage expresses simply in terms of explanation: *aitía, dihóti, dià tí*.

26. *Enhupárchein* does not *mean* "be innate," only "be in." But the kind of "being in" in question is innateness (Barnes, p. 250).

27. *Gnōstikōtérōn*: literally, "more cognitive." I take the word as equivalent to *gnōrimṓteros têi phúsei*. See I.2 71b33–72a5.

28. I follow Philoponus, Zabarella, and Ross in this interpretation of *héōs epì archèn êlthen*. For another view, see Barnes (p. 254), who points out that the same simile occurs at *Probl.* XVIII.7 917a28–32 and XXVI.8 941a9–13. But these passages do not contain the difficult final phrase "until . . . position." Perhaps the phrase in question continues the thought of the passage before the simile: these states . . . come from percep-

tion one after the next (like soldiers recovering one by one from a rout) until we reach knowledge of the principles.

29. Pacius calls it an "*eleganti exemplo*."

30. For another, see *De An.* III.4 429b10–22.

31. In another way the highest universals will have more parts than anything else, since they can be divided into the most subclasses. Cf. *Met.* Δ.25 1023b23–25.

32. Or some? Aristotle does not specify that there be "perfect" induction.

33. Aristotle does not indicate whether the most general universals are the categories substance, quality, etc.

34. See below, pp. 253–55.

35. I follow Ross and Barnes in thinking that *pálai* refers to what shortly precedes.

36. The subject term must be universal. Whether the predicate term must also be universal depends on what it is for an attribute to belong to a subject. *APo* does not go into this issue, but even if Owen's view (Owen 1986b) holds for this work, all terms in proofs constructed by concatenating first-figure syllogisms must be universal except the attribute of the conclusion, and this term may be universal too.

37. See the discussion of universal connections in I.4–5.

38. Cf. *De An.* II.2 413a11–12.

39. Cf. *Met.* Z.3 1029b3–12.

40. This comment also applies to Barnes's rendering, "familiar."

41. For this interpretation see Evans 1977, pp. 68–73.

42. Cf. *Top.* VI.4 142a2–4 with *APo* I.2 72a1–5.

43. The relations between the expert and what is intelligible without qualification are complicated. Each is used to determine the other: the intelligible is what experts agree on, but it is also a standard for deciding who are experts.

44. Le Blond 1970, pp. 131ff., gives this interpretation, in my opinion with little warrant.

45. For a defense of this translation see discussion of I.2 71b33–72a5 in the section "Principles Are Better Known, Prior, and Grounds" in Chapter 2 above.

46. Cf. I.18 81b3–4.

47. See Engberg-Pedersen 1979, p. 317, and, *contra*, Hamlyn 1976, pp. 180f.

48. This interpretation presupposes no decision on whether 100a15ff. amplifies on the whole process from sense perception to scientific principles, or on both the move from experience to knowledge of universals and the move from that to principles, or just on the former move. Translating *prôta* as "principles" instead of "primary things" would rule out the last of these, and translating it "primary universals" would favor the last.

49. E.g., I.2 71b21, 72a6–7, I.3 72b5, I.6 74b24–25, I.9 76a30.

50. Ross, p. 675.

51. So Barnes, p. 256.

52. Hamlyn 1976 holds this last view.

53. Bonitz gives examples only of technical uses of *epaktikós*.

54. Ross, pp. 481ff.

55. Engberg-Pedersen 1979, pp. 301–5.

56. Ibid., p. 305.

57. See Ross, p. 482, fifth group of passages.

58. *Top.* I.12 105a14–15 contains a paradigm example of an argument by *epagōgḗ*, which matches this description.

59. Cf. *APo* I.1 71a1ff.

60. See I.31 88a12ff. and II.2 90a24–30 for cases where a single observation can lead to knowledge of the relevant universal.

61. Engberg-Pedersen 1979, pp. 311–14, offers an alternative interpretation: "Aristotle is not talking . . . about one species of induction, viz. perfect induction. Rather he is talking about any kind of *epagōgḗ* (from one particular case or more) and what he is wishing to bring out is what we imply when on the basis of attending to particular cases we assert a universal proposition" (p. 313). Even on this analysis, the passage does not accurately represent the paradigm inductive argument of *Top.* I.12 105a14–15. In any case, the view of induction I am proposing embraces *APr* II.23 on Engberg-Pedersen's interpretation.

62. It is objected to *APr* II.23 that the conjuncts of the complex premise "men are long-lived, horses are long-lived, etc." presuppose uses of *epagōgḗ* that cannot be perfect inductions since it is impossible to examine the longevity of all humans, horses, etc., that have ever lived, are now alive, and ever will live. See Fritz 1964a, p. 631.

63. *Met.* M.4 1078b28.

64. Hamlyn 1976, pp. 168, 170–71.

65. *APr* II.23 68b13–14; cf. *APo* I.18 81a40.

66. E.g., *EN* VI.3 1139b28f.

67. Cf. *Top.* I.1 100a25ff.

68. The two are frequently opposed, e.g., at I.18 81a40.

69. Esp. I.18.

70. See McKirahan 1983.

71. In this passage, *hoútō* naturally refers to *epagōgḗ*.

72. I.e., both the *kaí* and the *gár* in 100b4 are accounted for.

73. See above, p. 250.

74. A similar assertion appears at I.31 87b28–30: "Even if perception is of what is such and such and not a particular individual, we must perceive a particular individual at a particular place and time."

75. Barnes, p. 184.

76. Ross, pp. 677f.

77. Owen 1986b.

78. Cf. *De An.* II.6 418a20–24.

79. Also *De An.* II.12 424a17–21.

80. *De An.* II.12 424a18, III.8 432a5; cf. *De An.* III.12 434a29–30.

81. Hicks 1907, p. 416.

82. *De An.* III.4 429b10–22, in distinguishing between x and what it is to be x, says that we discriminate heat, cold, and the things of which flesh is a proportion by means of our faculty of perception, but unhelpfully says only that we judge what it is to be flesh either with a different faculty or with the same faculty in a different state. This statement even overlooks the difference between perceiving hot, etc., and perceiving flesh.

83. This interpretation avoids the embarrassment of requiring two kinds of universal, "the one which is grasped in perception (however we are to understand that idea) and the other that is only grasped by animals that go beyond that of perception alone" (Engberg-Pedersen 1979, p. 317).

84. I take the distinction between "by means of seeing" (*tôi horân*) and "as a result of seeing" (*ek toû horân*) (88a13–14) and between "seeing" and "grasping" (*noêsai*, a16) as signifying at least this much.

85. E.g., Ross, p. 86; Le Blond 1970, pp. 131–40.

86. Barnes; Lesher 1973.

87. At 100a10 *héxeis* refers to these levels.

88. The same point is made in an aside at I.33 88b36.

89. I take it that this is the point of the aphoristic concluding sentence of the work: *hē mèn archè tês archês eíē án, hē dè pâsa prâgma homoíōs échei pròs tò pân prâgma* (100b15–17). That is to say, *noûs* (the *arché* of the state *epistḗmē*) has as its objects the *archaí* (the principles of science [*epistḗmē*]), and the whole of *epistḗmē* (the state) has as its objects the whole subject matter of the science. This interpretation agrees with Barnes.

90. So Barnes, pp. 257–59.

91. *EN* VI.6 1141a3–8 does the same.

92. Lesher 1973.

93. For discussion, see Lesher 1973, pp. 53f., 56, 57f.

94. Ibid., pp. 57f.

95. Lesher 1973, pp. 59ff., is wrong to identify *archè epistḗmēs* with *tò kathólou*. Cf. above, n. 23 of the present chapter, for a decisive objection.

96. See Plato, *Republic* 511d, 534b. Plato also calls this *nóēsis* (511d, though at 534a *nóēsis* covers the two highest stages), and he also calls it *epistḗmē* (533d, 534 a,c). Although Plato says he does not quibble about the terminology (533e), Aristotle distinguishes between *noûs* and *nóēsis* as between *noûs* and *epistḗmē* (esp. II.19 100b5ff.) and between *nóēsis* and *epistḗmē* (I.31 88a5ff.). For *noûs* and *noeîn* in the earlier philosophical tradition, see Fritz 1943, 1974, and 1964. Although Plato and Aristotle developed the concepts of *noûs* and *noeîn* in special ways, we can recognize antecedents of Aristotle's *noûs* as early as Homer, for whom "the *noos* has a deeper insight than our eyes and ears" (Fritz 1974, p. 71) and penetrates beyond surface appearances of things and situations to discover the real truth about the matter (ibid., p. 25). Also, in many of the early thinkers, *noûs* is not discursive, deductive thought, but a sudden and direct perception or intuition of the truth (ibid., pp. 26, 40, 65). Moreover, Democritus distinguished between *noûs* and *aísthēsis* as giving us access to distinct realms of reality (*noûs* to atoms and void, *aísthēsis* to sensible appearances).

97. For *akríbeia* as depending on proximity to principles, see I.24 86a13–21. For principles as being "truer" than what depends on them, see *Met.* α.1 993b26–29. Also see Lesher 1973, pp. 62–64.

98. A dialectical premise is "a premise which assumes either member of a contradiction" (I.2 72a9–10, trans. Barnes).

99. These points occur also at *SE* 11 172a12–21: dialectic treats *all* things, not just one kind, and since it proceeds by asking questions that demand "yes" or "no" answers (*Top.* VIII.2 158a14–17), not by asserting the truth of certain propositions, the dialectician can investigate any subject without being an expert. However, dialectic's role in demonstrative science is more than merely to produce non-expert questions on any subject.

100. These works may have been composed before Aristotle had worked out the theory of demonstrative science in detail, but there is no reason not to use passages concerning sciences and their principles. For present purposes all that is needed is the notion, clearly present in *Top.* and *SE*, of an organized body of knowledge based on principles and proving its conclusions.

101. *Top.* VIII.1 155b3ff.

102. Cf. the distinction between eristic and dialectic in *SE* 11.

103. Cf. *Top.* VIII.14 163b9–16 and *SE* 16 175a5–12 for claims that knowledge of dialectic is useful to philosophers (where "philosopher" means quite generally a person engaged in any positive intellectual inquiry). See Evans 1977, p. 34.

104. The richness of this concept has been sketched by Owen (1986d), and it is

necessary only to state some of his conclusions. For argument and evidence, see Owen's paper.

105. *Top.* I.1 100a27–b23.

106. *APo* offers an instructive parallel. Although every science has a subject genus, common principles, and attributes, some sciences need not explicitly set out the existence of their subjects or the nature of their attributes, or of the common principles if these are obvious, though we must be explicit about things that are problematic (I.10 76b16ff.). The problematic subjects and attributes more than others will require dialectical treatment. Cf. the treatment of place, void, and time in *Phys.* IV.

107. Thus, the claim holds for *éndoxa* both in the sense of common conceptions about established forms of language or philosophical theses claiming to account for observed facts and the way words are used, and also in the sense of *phainómena katà tēn aísthēsin*. Cf. Owen 1986d, pp. 242–44.

108. The final words do not refer just to *Top.* I.14. Cf. Evans 1977, pp. 32f.

109. Here *empeiría* is used differently from the way it is used in *APo* II.19 and *Met.* A.1.

110. *APr* I.30 46a22–27 seems to tell the simpler, less satisfactory story.

111. E.g., *Top.* I.2 101a36–b4.

112. *Peirastikḗ* (*SE* 8 169b25, 11 171b4).

113. *Ērōtētikḗ* (*SE* 11 172a18).

114. See *Top.* VIII.2 158a15ff. for the examples.

115. Thus the philosopher is concerned with selecting the correct *tópos* (*Top.* VIII.1 155b7f.).

116. *SE* 9, 11.

117. *Top.* I.1 100b21–23.

118. This doctrine is put forward in II.1–2. See translation and discussion in Chapter 15 above.

119. See above, pp. 203–5.

120. See *Met.* Z.3 1029b7–8, translated in Chapter 17 above.

121. This issue is discussed in the section "Priority of What Is Better Known" in Chapter 17 above.

122. Cf. I.5 74a18–25.

123. See discussion in the section "Axioms and Subordinate Sciences" in Chapter 6 above.

124. The distinction between definitional and nondefinitional per se facts is discussed in the section "Scientific Facts" in Chapter 8 above.

125. The notion of "fat definitions" is defined at the beginning of Chapter 9 above.

126. For this point see the section "An Aristotelian Demonstration" in Chapter 13 above.

127. *APo* II.14 98a1–12, translated and discussed in Chapter 14 above.

128. Cf. Ross 1924, 1:124, on *Met.* A.2 983a21–23.

129. E.g., *EN* X.7 1177a21, a28, X.8 1178b21–22, b25–32.

130. *EN* X.7 1177a32–34.

BIBLIOGRAPHY

Ackrill, J. L., trans. and comm. 1963. *Aristotle's "Categories" and "De Interpretatione."* Clarendon Aristotle Series. Oxford.

———. 1981. "Aristotle's Theory of Definition: Some Questions on *Posterior Analytics* II 8–10." In Berti 1981, pp. 359–84.

Allen, D. J. 1961. "Quasi-mathematical Method in the *Eudemian Ethics*." In S. Mansion 1961, pp. 303–18.

Anonymous. 1909. *In Analyticorum Posteriorum Alterum Commentarium*. Edited by M. Wallies. Commentaria in Aristotelem Graeca 13.3. Berlin.

Anonymous. 1953. *Analytica Posteriora*. Edited by L. Minio-Paluello. Aristoteles Latinus 4.2. Bruges and Paris.

Apostle, H. G., trans. and comm. 1981. *Aristotle's "Posterior Analytics."* Grinnell, Iowa.

Aquinas, St. Thomas. 1970. *Commentary on the "Posterior Analytics" of Aristotle*. Translated by F. R. Larcher. Albany, N.Y.

Arpe, C. 1938. *Das τί ἦν εἶναι bei Aristoteles*. Hamburg.

Aubenque, P. 1962. *Le problème de l'être chez Aristote*. Paris.

Bailey, C., trans. and comm. 1947. *Titi Lucreti Cari De Rerum Natura Libri Sex*. 3 vols. Oxford.

Balme, D. M., trans. and comm. 1972. *Aristotle's "De Partibus Animalium" I and "De Generatione Animalium" I (with passages from II.1–3)*. Clarendon Aristotle Series. Oxford.

———. 1975. "Aristotle's Use of Differentiae in Biology." In Barnes, Schofield, and Sorabji 1975, pp. 183–93. (Original version published in S. Mansion 1961, pp. 195–212.)

———. 1987. "Aristotle's Use of Division and Differentiae." In Gotthelf and Lennox 1987, pp. 69–89.

Bambrough, R., ed. 1965. *New Essays on Plato and Aristotle*. London.

Barnes, J. 1970. "Property in Aristotle's *Topics*." *Archiv für Geschichte der Philosophie* 52:136–55.

Barnes, J., trans. and comm. 1975a. *Aristotle's "Posterior Analytics."* Clarendon Aristotle Series. Oxford.

———. 1975b. "Aristotle's Theory of Demonstration." In Barnes, Schofield, and Sorabji 1975, pp. 65–87. (Original version published in *Phronesis* 14 [1969]: 123–52.)

———. 1981. "Proof and the Syllogism." In Berti 1981, pp. 1–59.

Barnes, J., M. Schofield, and R. Sorabji, eds. 1975. *Articles on Aristotle. I: Science.* London.

———, eds. 1979. *Articles on Aristotle. III: Metaphysics.* London.

Bassenge, F. 1960. "Das τὸ ἑνὶ εἶναι, τὸ ἀγαθῷ εἶναι, etc. etc. und das τὸ τί ἦν εἶναι bei Aristoteles." *Philologus* 104:14–47, 201–22.

Becker, O. 1959. "Die Archai in der griechischen Mathematik." *Archiv für Begriffsgeschichte* 4:210–26.

Berka, K. 1963. "Aristoteles und die axiomatische Methode." *Altertum* 9:200–205.

Berti, E., ed. 1981. *Aristotle on Science: The "Posterior Analytics."* Proceedings of the Eighth Symposium Aristotelicum. Padova.

Blanche, R. 1962. *Axiomatics.* Translated by G. B. Keene. London.

Bolton, R. 1976. "Essentialism and Semantic Theory in Aristotle: *Posterior Analytics* II,7–10." *Philosophical Review* 85:514–44.

———. 1987. "Definition and Scientific Method in Aristotle's *Posterior Analytics* and *Generation of Animals.*" In Gotthelf and Lennox 1987, pp. 120–66.

Bonitz, H. 1866. "Aristotelische Studien IV." *Sitzungsberichte der kaiserliche Akademie der Wissenschaften, phil.-hist. Kl.* 52:347–423.

———. 1870. *Index Aristotelicus.* Berlin.

Bourgey, L. 1955. *Observation et experience chez Aristote.* Paris.

Brody, B. A. 1972. "Towards an Aristotelian Theory of Scientific Explanation." *Philosophy of Science* 39:20–31.

Brunschwig, J. 1981. "L'objet et la structure des *Seconds Analytiques* d'après Aristote." In Berti 1981, pp. 61–96.

Buchanan, E. 1962. *Aristotle's Theory of Being.* Greek, Roman and Byzantine Monographs, No. 2. University, Miss., and Cambridge, Mass.

Burkert, W. 1972. *Lore and Science in Ancient Pythagoreanism.* Revised translation by E. L. Minar. Cambridge, Mass. (Originally published as *Weisheit und Wissenschaft.* Nuremberg, 1962.)

Burnyeat, M. 1981. "Aristotle on Understanding Knowledge." In Berti 1981, pp. 97–139.

Cassini, A. 1988. "La función de la teoría de la demostración scientifica en Aristoteles." *Revista Latinoamericana de Filosofía* 14:165–77.

Catan, J. 1970. "Recollection and *Posterior Analytics.*" *Apeiron* 4, no. 2, 34–57.

Charlton, W., trans. and comm. 1970. *Aristotle's "Physics," Books 1–2.* Clarendon Aristotle Series. Oxford.

Chavarri, E. 1971, 1972. "Naturaleza de la demostración 'propter quid' en los *Analiticos Posteriores.*" *Estudios Filosoficos* 20:39–90, 21:3–58, 283–338, 559–85.

Cherniss, H. F. 1944. *Aristotle's Criticism of Plato and the Academy.* Baltimore.

Cleary, J. J. 1985. "On the Terminology of 'Abstraction' in Aristotle." *Phronesis* 30:13–45.

———. 1987. "Science, Universals and Reality." *Ancient Philosophy* 7:95–130.

Corcoran, J., ed. 1974. *Ancient Logic and Its Modern Interpretations.* Dordrecht.

Couloubaritsis, L. 1978. "*Sophia* et *philosophia* chez Aristote." *Annales de l'Institut de Philosophie de l'Université Libre de Bruxelles*, pp. 7–38.
———. 1978–79. "Dialectique et philosophie chez Aristote." *Philosophia* (Athens) 8–9:229–56.
———. 1980. "Y-a-t-il une intuition des principes chez Aristote?" *Revue Internationale de Philosophie* 133–34:440–71.
Cresswell, M. J. 1971. "Essence and Existence in Plato and Aristotle." *Theoria* 37:97–113.
Croissant, J. 1986. "Sur la theorie de la définition dans les *Seconds Analytiques.*" In J. Croissant, *Etudes de philosophie ancienne*, pp. 146–60. Cahiers de Philosophie Ancienne, 4. Brussels. (A summary of this essay appears in *Actes du 3e Congrès des Sociétés de Philosophie de Lanque Française*, pp. 225–28. Louvain, Paris, 1947.)
Demoss, D., and D. Devereux. 1988. "Essence, Existence, and Nominal Definition in Aristotle's *Posterior Analytics* II 8–10." *Phronesis* 33:133–54.
Diels, H., and W. Kranz. 1956. *Die Fragmente der Vorsokratiker*. 5th ed. Berlin.
Dod, B. G. 1986. "Eclipses and Thunderstorms: The Argument of Aristotle's *Posterior Analytics* II.1–10." *Classica et Mediaevalia* 37:123–35.
Drees, L. 1968. *Olympia*. London.
Einarson, B. 1936. "On Certain Mathematical Terms in Aristotle's Logic." *American Journal of Philology* 57:33–54, 151–72.
Engberg-Pedersen, T. 1979. "More on Aristotelian Epagoge." *Phronesis* 24:301–19.
Engmann, J. 1973. "Aristotle's Distinction between Substance and Universal." *Phronesis* 18:139–55.
———. 1978. "Aristotelian Universals." *Classical Philology* 73:17–23.
Eucken, R. 1872. *Die Methode des aristotelischen Forschung in ihrem Zusammenhang mit des philosophischen Grundprincipien des Aristoteles*. Berlin.
Eustratius. 1907. *In Analyticorum Posteriorum Librum Secundum Commentarium*. Edited by M. Hayduck. Commentaria in Aristotelem Graeca 21.1. Berlin.
Evans, J. D. G. 1977. *Aristotle's Concept of Dialectic*. Cambridge.
Evans, M. G. 1958–59. "Causality and Explanation in the Logic of Aristotle." *Philosophy and Phenomenological Research* 19:466–85.
Ferejohn, M. T. 1982. "Definition and the Two Stages of Aristotelian Demonstration." *Review of Metaphysics* 36:375–95.
Fowler, H. W., and F. G. Fowler. 1905. *The Works of Lucian of Samosata*. 4 vols. Oxford.
Frajese, A. 1951. "Sur la signification des postulats euclidiens." *Archives Internationales D'histoire des Sciences* 30:383–92.
Franciosi, F. 1980. "Some Remarks on the Controversy between Prof. Knorr and Prof. Szabo." In Hintikka, Gruender, and Agazzi 1980, pp. 187–91.
Frede, D. 1974. "Comment on Hintikka's Paper, 'On the Ingredients of an Aristotelian Science.' " *Synthese* 28:78–89.
Friedlein, G. 1873. *Proclus. In Primum Euclidis Elementorum Librum Commentarii*. Leipzig.
Fritz, K. von. 1943. "*Noos* and *Noein* in the Homeric Poems." *Classical Philology* 38:79–93.
———. 1955. "Die APXAI in der griechischen Mathematik." *Archiv für Begriffsgeschichte* 1:13–103. (Reprinted in Fritz 1971, pp. 335–429.)
———. 1964a. "Die ΕΠΑΓΩΓΗ bei Aristoteles." In Fritz 1971, pp. 623–76. (Origi-

nally published in *Sitzungsberichte des bayerischen Akademie der Wissenschaften, phil.-hist. Kl.*, no. 3. Munich, 1964.)
———. 1964b. "Der NOYΣ des Anaxagoras." *Archiv für Begriffsgeschichte* 9:87–102.
———. 1971. *Grundprobleme der Geschichte der antiken Wissenschaft*. Berlin.
———. 1974. "*Nous, Noein*, and Their Derivatives in Pre-Socratic Philosophy (Excluding Anaxagoras)." In Mourelatos 1974, pp. 23–85. (Originally published in *Classical Philology* 40 [1945]: 223–42 and 41 [1946]: 12–34.)
Gohlke, P. 1936. *Die Entstehung der aristotelischen Logik*. Neue Deutsche Forschungen, Abt. Philosophie, Bd. 14. Berlin.
Gomez-Lobo, A. 1976–77. "Aristotle's Hypotheses and the Euclidean Postulates." *Review of Metaphysics* 30:430–39.
———. 1978. "Aristotle's First Philosophy and the Principles of Particular Disciplines." *Zeitschrift für philosophische Forschung* 32:183–94.
———. 1980. "The So-called Question of Existence in Aristotle, *An. Post.* 2.1–2." *Review of Metaphysics* 34:71–89.
———. 1981. "Definitions in Aristotle's Posterior Analytics." In D. O'Meara, ed., *Studies in Aristotle*, 9:25–46. Washington.
Goodwin, W. W. 1890. *Syntax of the Moods and Tenses of the Greek Verb*. Boston.
Gotthelf, A., ed. 1985a. *Aristotle on Nature and Living Things. Philosophical and Historical Studies Presented to David M. Balme on His Seventieth Birthday*. Pittsburgh and Bristol.
———. 1985b. "Notes towards a Study of Substance and Essence in Aristotle's *Parts of Animals* II–IV." In Gotthelf 1985a, pp. 27–54.
———. 1987. "First Principles in Aristotle's *Parts of Animals*." In Gotthelf and Lennox 1987, pp. 167–98.
Gotthelf, A., and J. G. Lennox, eds. 1987. *Philosophical Issues in Aristotle's Biology*. Cambridge.
Graham, W. 1975. "Counterpredicability and *Per Se* Accidents." *Archiv für Geschichte der Philosophie* 57:182–87.
Granger, G. G. 1976. *La théorie aristotélicienne de la science*. Paris.
Granger, H. 1980. "Aristotle and the Genus-Species Relation." *Southern Journal of Philosophy* 18:37–50.
———. 1981. "The Differentia and the Per Se Accident in Aristotle." *Archiv für Geschichte der Philosophie* 63:118–29.
Guariglia, O. N. 1982. "La definición y la explicación causal segun Aristoteles." *Revista Latinoamericana de Filosofía* 8:119–46.
Guerriere, D. 1975. "The Aristotelian Concept of Episteme." *Thomist* 39:341–48.
Guthrie, W. K. C. 1971–81. *A History of Greek Philosophy*. 6 vols. Cambridge.
Hadgopoulos, D. J. 1975a. "Demonstration and the Second Figure in Aristotle." *New Scholasticism* 49:62–75.
———. 1975b. "A Note on Aristotle's Notions of Universality and Necessity." *Logique et Analyse* 18:171–74.
———. 1976a. "The Definition of the Predicables in Aristotle." *Phronesis* 21:59–63.
———. 1976b. "Reduction to Immediate Premises." *Apeiron* 10:29–37.
———. 1977. "διὰ μέσων or δι' ἀμέσων: *Posterior Analytics* II, viii, 93a36." *Apeiron* 11, no. 1: 32–39.
Hamelin, O. 1931. *Le système d'Aristote*. Paris.
Hamlyn, D. W. 1961. "Aristotle on Predication." *Phronesis* 6:110–26.
———. 1976. "Aristotelian Epagoge." *Phronesis* 21:167–84.

Hare, R. M. 1965. "Plato and the Mathematicians." In Bambrough 1965, pp. 21–38.
Heath, T. L. 1921. *A History of Greek Mathematics*. 2 vols. Oxford.
———, trans. and comm. 1925. *The Thirteen Books of Euclid's "Elements."* 2d ed. 3 vols. Cambridge. (Reprinted in New York, 1956.)
———. 1949. *Mathematics in Aristotle*. Oxford.
Hempel, C. G. 1965. *Aspects of Scientific Explanation and Other Essays in the Philosophy of Science*. New York.
Herlinus, C., and C. Dasypodius. 1566. *Analyseis Geometricae sex librorum Euclidis*. Schola Argentinensi.
Hesz, W. 1970. "Erfahrung und Intuition bei Aristoteles." *Phronesis* 15:48–81.
Hicks, R. D., trans. and comm. 1907. *Aristotle's "De Anima."* Cambridge.
Hilbert, D. 1971. *Foundations of Geometry*. 10th ed. La Salle, Ill.
Hintikka, J. 1972. "On the Ingredients of an Aristotelian Science." *Nous* 6:55–69.
———. 1973. *Time and Necessity*. Oxford.
———. 1974. "Reply to Dorothea Frede." *Synthese* 28:91–96. (See Frede 1974.)
———. 1979. "Necessity, Universality and Time in Aristotle." In Barnes, Schofield, and Sorabji 1979, pp. 108–24. (Originally published in *Ajatus* 20 [1957]: 65–70.)
———. 1980a. "Aristotelian Axiomatics and Geometrical Axiomatics." In Hintikka, Gruender, and Agazzi 1980, pp. 133–44.
———. 1980b. "Aristotelian Induction." *Revue Internationale de Philosophie* 34:422–39.
Hintikka, J., D. Gruender, and E. Agazzi, eds. 1980. *Pisa Conference Proceedings*. Vol. 1. Dordrecht.
Hintikka, J., and U. Remes. 1974. *The Method of Analysis*. Dordrecht and Boston.
Hultsch, F. 1885. *Autolyci de Sphaera quae movetur liber. De Ortibus et occasibus libri duo*. Leipzig.
Inwood, B. 1979. "A Note on Commensurate Universals in the *Posterior Analytics*." *Phronesis* 24:321–29.
Irwin, T. H. 1988. *Aristotle's First Principles*. Oxford.
Jaeger, W. 1911. *Emendationum Aristotelearum Specimen*. Berlin.
Joachim, H. H. 1922. *Aristotle on Coming-to-be and Passing-away*. Oxford. (Reprinted in Hildesheim, 1970.)
Jope, J. 1972. "Subordinate Demonstrative Science in the Sixth Book of Aristotle's *Physics*." *Classical Quarterly* 22:279–92.
Kahn, C. H. 1966. *The Greek Verb "to Be" and the Concept of Being*. Foundations of Language, 2. Dordrecht.
———. 1981. "The Role of *Nous* in the Cognition of First Principles in *Posterior Analytics* II 19." In Berti 1981, pp. 385–414.
Kal, V. 1988. *On Intuition and Discursive Reasoning in Aristotle*. Leiden.
Kato, M. 1987. "Aristoteles über den Ursprung wissenschaftlicher Erkenntnis." *Phronesis* 32:188–205.
Kirchmann, J. H. von. 1878. *Erläuterungen zu den zweiten Analytiken des Aristoteles*. Leipzig.
Kirwan, C., trans. and comm. 1971. *Aristotle's "Metaphysics," Books Γ, Δ, E*. Clarendon Aristotle Series. Oxford.
Kneale, W., and M. Kneale. 1962. *The Development of Logic*. Oxford.
Knorr, W. 1975. *Evolution of the Euclidean Elements*. Dordrecht.
———. 1980a. "On the Early History of Axiomatics: The Interaction of Mathematics

and Philosophy in Greek Antiquity." In Hintikka, Gruender, and Agazzi 1980, pp. 145–86.

———. 1980b. "On the Early History of Axiomatics: A Reply to Some Criticisms." In Hintikka, Gruender, and Agazzi 1980, pp. 193–95.

———. 1983. "Construction as Existence Proof in Ancient Geometry." *Ancient Philosophy* 3:125–48.

———. 1986. *The Ancient Tradition of Geometric Problems*. Boston, Basel, Stuttgart.

Kosman, L. A. 1973. "Understanding, Explanation and Insight in Aristotle's *Posterior Analytics*." In Lee, Mourelatos, and Rorty 1973, pp. 374–92.

Koterski, J. W. 1980. "Aristotle on Signifying Definitions." *New Scholasticism* 54:75–86.

Kuhn, T. S. 1970. *The Structure of Scientific Revolutions*. 2d ed. Chicago. (First edition published in 1962.)

———. 1977. *The Essential Tension*. Chicago.

Kullmann, W. 1974. *Wissenschaft und Methode. Interpretationen zur aristotelischen Theorie der Naturwissenschaft*. Berlin.

———. 1975. "Zur wissenschaftlichen Methode des Aristoteles." In G. A. Seeck, ed., *Die Naturphilosophie des Aristoteles*, pp. 301–38. Darmstadt.

———. 1981. "Die Funktion der mathematischen Beispeile in Aristoteles' *Analytica Posteriora*." In Berti 1981, pp. 245–70.

Kung, J. 1977. "Aristotle on Essence and Explanation." *Philosophical Studies* 31:361–83.

Landor, B. 1981. "Definitions and Hypotheses in *Posterior Analytics* 72a19–25 and 76b35–77a4." *Phronesis* 26:308–18.

———. 1985. "Aristotle on Demonstrating Essence." *Apeiron* 19:116–32.

Lear, J. 1980. *Aristotle and Logical Theory*. Cambridge.

Leavitt, F. J. 1988. "Aristotle's Theory of Explanation" (in Hebrew). *Iyyun* 37:207–14. (An English summary of this essay appears in *Philosopher's Index*, 1989, p. 492.)

Le Blond, J. M. 1970. *Logique et méthode chez Aristote*. 2d ed. Paris. (First edition published in 1949.)

———. 1979. "Aristotle on Definition." In Barnes, Schofield, and Sorabji 1979, pp. 63–79. (Originally published as "La définition chez Aristote." *Gregorianum* 20 [1939]: 351–80.)

Lee, E. N., A. P. D. Mourelatos, and R. M. Rorty, eds. 1973. *Exegesis and Argument*. Assen.

Lee, H. D. P. 1935. "Geometrical Method and Aristotle's Account of First Principles." *Classical Quarterly* 29:113–24.

Lennox, J. G. 1987. "Divide and Explain. The *Posterior Analytics* in Practice." In Gotthelf and Lennox 1987, pp. 90–119.

Lesher, J. H. 1973. "The meaning of ΝΟΥΣ in the *Posterior Analytics*." *Phronesis* 18:44–68.

Leszl, W. 1970. *Logic and Metaphysics in Aristotle*. Studia Aristotelica, 5. Padua.

———. 1972–73. "Knowledge of the Universal and Knowledge of the Particular in Aristotle." *Review of Metaphysics* 26:278–313.

———. 1980. "Unity and Diversity of the Sciences: The Methodology of the Mathematical and of the Physical Sciences and the Role of the Nominal Definition." *Revue Internationale de Philosophie* 34:384–421.

———. 1981. "Mathematics, Axiomatization and the Hypotheses." In Berti 1981, pp. 271–328.

Liddell, H. G., R. Scott, and H. S. Jones. 1968. *A Greek-English Lexicon (with Supplement)*. Oxford.

Lloyd, A. C. 1981. "Necessity and Essence in the *Posterior Analytics*." In Berti 1981, pp. 157–71.

Lloyd, G. E. R. 1961. "The Development of Aristotle's Theory of the Classification of Animals." *Phronesis* 6:59–81.

———, ed. and trans. 1978. *Hippocratic Writings*. Harmondsworth, Eng.

———. 1979. *Magic, Reason, and Experience*. Cambridge.

———. 1987. *The Revolutions of Wisdom*. Berkeley.

Lumpe, A. 1955. "Der Terminus 'Prinzip' (ἀρχή) von den Vorsokratikern bis auf Aristoteles." *Archiv für Begriffsgeschichte* 1:104–16.

McKeon, R. 1947. "Aristotle's Conception of the Development and Nature of Scientific Method." *Journal of the History of Ideas* 8:3–44.

McKirahan, R. 1978. "Aristotle's Subordinate Sciences." *British Journal for the History of Science* 11:197–220.

———. 1983. "Aristotelian Epagoge in *Prior Analytics* 2.21 and *Posterior Analytics* 1.1." *Journal of the History of Philosophy* 21:1–13.

Mansion, A. 1961. "L'origine du syllogisme et la théorie de la science chez Aristote." In S. Mansion 1961, pp. 57–81.

Mansion, S., ed. 1961. *Aristote et les problèmes de méthode*. Louvain.

———. 1976. *Le jugement d'existence chez Aristote*. 2d ed. Louvain. (First edition published in 1946.)

———. 1981. "La signification de l'universel d'après *An. Post.* I 1." In Berti 1981, pp. 329–42. (Reprinted in S. Mansion 1984a.)

———. 1984a. *Études Aristotéliciennes*. Louvain-la-Neuve.

———. 1984b. " 'Plus Connu en Soi,' 'Plus Connu pour Nous.' Une distinction epistémologique importante chez Aristote." In S. Mansion 1984a, pp. 213–22. (Originally published in *Pensamiento* 35 [1979]: 161–70.)

Matthen, M., ed. 1987a. *Aristotle Today: Essays on Aristotle's Ideal of Science*. Edmonton.

———. 1987b. "Introduction. The Structure of Aristotelian Science." In Matthen 1987a, pp. 1–23.

Mendell, H. 1984. "Two Geometrical Examples from Aristotle's *Metaphysics*." *Classical Quarterly* 34:359–72.

Mignucci, M. 1965. *La teoria Aristotelica della scienza*. Firenze.

———. 1975. *L'argomentazione dimostrativa in Aristotele. Commenti agli "Analitici Secondi" I*. Padova.

———. 1981. "Ὡς ἐπὶ τὸ πολύ et nécessaire dans la conception aristotélicienne de la science." In Berti 1981, pp. 173–203.

Moravcsik, J. M. E., ed. 1967. *Aristotle: A Collection of Critical Essays*. Garden City.

———. 1974. "Aristotle on Adequate Explanations." *Synthese* 28:3–17.

———. 1975. "AITIA as Generative Factor in Aristotle's Philosophy." *Dialogue* 14:622–38.

Moreau, J. 1961. "Aristote et la vérité anteprédicative." In S. Mansion, 1961, pp. 21–33.

Morrow, G. R. 1970. *Proclus: A Commentary on the First Book of Euclid's "Elements."* Princeton.

Morsink, J. 1982. "The Mandate of *Topics* I,2." *Apeiron* 16:102–28.

Moser, S. 1935. *Zur Lehre von der Definition bei Aristoteles, I. Teil: Organon und Metaphysik*. Innsbruck.

Mourelatos, A. P. D., ed. 1974. *The Presocratics: A Collection of Critical Essays*. Garden City, N.Y.

Mueller, I. 1969. "Euclid's *Elements* and the Axiomatic Method." *British Journal for the Philosophy of Science* 20:289–309.

———. 1974. "Greek Mathematics and Greek Logic." In Corcoran 1974, pp. 35–70.

———. 1979. "Aristotle on Geometrical Objects." In Barnes, Schofield, and Sorabji 1979, pp. 96–107. (Originally published in *Archiv für Geschichte der Philosophie* 52 [1970]: 158–71.)

———. 1981. *Philosophy of Mathematics and Deductive Structure in Euclid's Elements*. Cambridge, Mass.

———. Forthcoming. "On the Notion of a Mathematical Starting Point in Plato, Aristotle, and Euclid."

Mulhern, J. J. 1958. "Aristotle on Universality and Necessity." *Logique et Analyse* 12:288–99.

Mure, G. R. G., trans. 1928. *Analytica Posteriora*. In J. A. Smith and W. D. Ross, eds., *The Works of Aristotle Translated into English*, vol. 1. Oxford.

Nagel, E. 1979. *The Structure of Science*. Indianapolis.

Novak, J. A. 1978. "A Geometrical Syllogism: *Posterior Analytics*, II, 11." *Apeiron* 12, no. 2: 26–33.

Owen, A. F., trans. 1853. *The Organon, or Logical Treatises of Aristotle, with the Introduction of Porphyry*. London.

Owen, G. E. L. 1986a. "Aristotle: Method, Physics, Cosmology." In G. Owen 1986c, pp. 151–64. (Originally published in C. C. Gillespie, ed., *Dictionary of Scientific Biography*, 1:250–58. New York, 1970.)

———. 1986b. "Inherence." In G. Owen 1986c, pp. 252–58. (Originally published in *Phronesis* 10 [1965]: 97–105.)

———. 1986c. *Logic, Science and Dialectic*. London.

———. 1986d. "Τιθέναι τὰ φαινόμενα." In G. Owen 1986c, pp. 239–51. (Originally published in S. Mansion 1961, pp. 83–103; reprinted in Moravcsik 1967, pp. 167–90, and in Barnes, Schofield, and Sorabji 1975, pp. 113–26.)

Pacius, I. 1966. *In Porphyrii Isogagen et Aristotelis Organum Commentarius Analyticus*. Reprint. Hildesheim. (Originally published in Frankfurt, 1597.)

Parvu, I., and A. Parvu 1980. "The Postulate Problem in the Works of Aristotle, Euclid and Proclus." *Philosophie et Logique* 4:333–44. (Originally published in Romanian in *Revista de Filozofie* (Bucharest) 25 [1978]: 721–29.)

Patzig, G. 1981. "Erkenntnisgründe, Realgründe und Erklärung (zu *An. Post.* A 13)." In Berti 1981, pp. 141–56.

Pellegrin, P. 1982. *La Classification des animaux chez Aristote. Statut de la biologie et unité de l'aristotélisme*. Paris.

———. 1986. *Aristotle's Classification of Animals*. Translated by A. Preus. Berkeley. (Revised ed. of Pellegrin 1982.)

Philoponus, J. 1909. *In Aristotelis Analytica Posteriora Commentaria*. Edited by M. Wallies. Commentaria in Aristotelem Graeca 13,3. Berlin.

Pickard-Cambridge, W. A., trans. 1928. *Topica*. In J. A. Smith and W. D. Ross, eds., *The Works of Aristotle Translated into English*. Vol. 1. Oxford.

Rescher, N., and Z. Parks, 1970–71. "A New Approach to Aristotle's Apodeictic Syllogisms." *Review of Metaphysics* 24:678–89.

Rijen, J. van. 1984. "The Principle of Plenitude, the De Omni / Per Se Distinction, and the Development of Modal Thinking." *Archiv für Geschichte der Philosophie* 66:61–88.

Robin, L. 1909–10. "Sur la conception aristotélicienne de la causalité." *Archiv für Geschichte der Philosophie* 23:1–28, 184–210.

Rodier, G. 1900. *Aristote: Traitéde l'âme*. 2 vols. Paris.

Rolfes, E., trans. 1975. *Aristoteles' Lehre vom Beweis oder Zweite Analytik (Organon IV)*. Reprint of Leipzig 1922 ed. with new introduction and bibliography by O. Hoeffe. Hamburg.

Rolland-Gosselin, M. S. 1912. "Les méthodes de définition d'Aristote." *Revue des Sciences Philosophiques et Théologiques* 6:236–52, 661–75.

Rose, V. 1886. *Aristotelis Qui Ferebantur Librorum Fragmenta*. Leipzig.

Ross, W. D. 1924. *Aristotle's "Metaphysics."* 2 vols. Oxford.

———, trans. 1925. *EthicaNicomachea*. In J. A. Smith and W. D. Ross, eds., *The Works of Aristotle Translated into English*. Vol. 9. Oxford.

———. 1936. *Aristotle's "Physics."* Oxford.

———. 1939. "The Discovery of the Syllogism." *Philosophical Review* 48:251–72.

———. 1949. *Aristotle's Prior and Posterior Analytics*. Oxford.

Rutten, C. 1964. "Note sur le syllogisme de l'essence chez Aristote." *Phronesis* 9:72–81.

Scholz, H. 1975. "The Ancient Axiomatic Theory." Translated by J. Barnes. In Barnes, Schofield, and Sorabji 1975, pp. 50–64. (Originally published as "Die Axiomatik der Alten." *Blätter für deutsche Philosophie* 4 [1930]: 259–78.)

Schroeder, J. 1984. "Eine aristotelische Argument zur Unbeweisbarkeit von Definition." *Archiv für Geschichte der Philosophie* 66:225–42.

Seidl, H. 1984. *Beiträge zu Aristoteles' Erkenntnislehre und Metaphysik*. Elementa. Schriften zur Philosophie und ihrer Problemgeschichte, 35. Wurzburg and Amsterdam.

Simard, E. 1946. "La *Solertia* et la découverte des hypothèses." *Laval Théologique et Philosophique* 2:220–25.

Simon, Y. R., and K. Menger. 1947–48. "Aristotelian Demonstration and Postulational Method." *Modern Schoolman* 25:183–92.

Simplicius. 1882–95. *In Aristotelis Physicorum Libris Commentaria*. Edited by H. Diels. Commentaria in Aristotelem Graeca, vols. 9, 10. Berlin.

Smith, R. 1982a. "The Relationship of Aristotle's Two Analytics." *Classical Quarterly* 32:327–35.

———. 1982b. "The Syllogism in *Posterior Analytics* I." *Archiv für Geschichte der Philosophie* 64:113–35.

———. 1986. "Immediate Propositions and Aristotle's Proof Theory." *Ancient Philosophy* 6:47–68.

Solmsen, F. 1941. "The Discovery of the Syllogism." *Philosophical Review* 50:410–21.

———. 1951. "Aristotle's Syllogism and Its Platonic Background." *Philosophical Review* 60:563–71.

———. 1975. *Die Entwicklung der aristotelischen Logik und Rhetorik*. Neue Philologische Untersuchungen, 4. Reprint. Zurich. (Originally published in Berlin, 1929.)

Sorabji, R. 1969. "Aristotle and Oxford Philosophy." *American Philosophical Quarterly* 6:127–35.

———. 1980. *Necessity, Cause and Blame: Perspectives on Aristotle's Theory*. London.

Sorabji, R. 1981. "Definitions: Why Necessary and in What Way?" In Berti 1981, pp. 205–44.

Sparshott, F. 1987. "Aristotle's World and Mine." In Matthen 1987a, pp. 25–50.

Stewart, J. A. 1892. *Notes on the "Nicomachean Ethics" of Aristotle*. 2 vols. Oxford.

Stocks, J. L. 1933. "The Composition of Aristotle's Logical Works." *Classical Quarterly* 27:115–24.

Strycker, E. de. 1968. "Prédicats univoques et prédicats analogiques dans le 'Protreptique' d'Aristote." *Revue Philosophique de Louvain* 66:597–618.

———. 1969. "Sur un emploi technique de μᾶλλον chez Aristote." *Mnemosyne* 22:303–4.

Szabó, A. 1956. "Wie ist die Mathematik zu einer deduktiven Wissenschaft geworden?" *Acta Antiqua Academiae Scientiarum Hungaricae* 4:109–52.

———. 1967. "Greek Dialectic and Euclid's Axiomatics." In I. Lakatos, ed., *Problems of the Philosophy of Mathematics*, pp. 1–8 and discussion pp. 9–27. Amsterdam.

———. 1969. *Anfänge der griechischen Mathematik*. Munich and Vienna.

———. 1978. *The Beginnings of Greek Mathematics*. Translated by A. M. Ungar, Budapest and Dordrecht. (English translation of Szabó 1969.)

Tannery, P. 1887. *La géométrie grecque*. Paris.

Tarán, L. 1981. *Speusippus of Athens*. Leiden.

Thayer, H. S. 1979. "Aristotle on the Meaning of Science." *Philosophical Inquiry* 1:87–104.

Themistius. 1900. *In Posteriora Analytica Paraphrasis*. Edited by M. Wallies. Commentaria in Aristotelem Graeca 5.1. Berlin.

Tiles, J. E. 1983. "Why the Triangle Has Two Right Angles Kath' Hauto." *Phronesis* 28:1–16.

Tredennick, H., trans. 1960. *Aristotle. "Posterior Analytics."* Loeb Classical Library. London and Cambridge, Mass.

Trépanier, E. 1948. "La connaissance des premiers principes." *Laval Théologique et Philosophique* 4:289–310.

Tricot, J., trans. 1947. *Organon IV: Les "Seconds Analytiques."* Paris.

Upton, T. V. 1980. "Imperishable Being and the Role of Technical Hypotheses in Aristotelian Demonstration." *Nature and System* 2:91–99.

———. 1981. "A Note on Aristotelian Epagoge." *Phronesis* 26:172–76.

———. 1983. "Aristotle on Hypothesizing the Genus and Scientific Explanation." *Nature and System* 5:161–68.

———. 1984. "The Role of Dialectic and Objections in Aristotelian Science." *Southern Journal of Philosophy* 22:241–56.

———. 1985. "Aristotle on Hypothesis and the Unhypothesized First Principle." *Review of Metaphysics* 39:283–301.

Urbanas, A. 1988. *La notion d'accident chez Aristote*. Montreal and Paris.

VanderWeel, R. L. 1969. "The *Posterior Analytics* and the *Topics*." *Laval Théologique et Philosophique* 25:130–41.

Van Fraassen, B. C. 1980. "A Re-examination of Aristotle's Philosophy of Science." *Dialogue* 19:20–45.

Verdenius, W. J. 1981. "Notes on Some Passages from [*Posterior Analytics*] Book 1." In Berti 1981, pp. 343–57.

Waerden, B. L. van der. 1988. *Science Awakening*. 5th ed. Dordrecht and Princeton Junction, N.J.

Wedin, V. E., Jr. 1973. "A Remark on *Per Se* Accidents and Properties." *Archiv für Geschichte der Philosophie* 55:30–35.

Weil, E. 1975. "The Place of Logic in Aristotle's Thought." Translated by J. and J. Barnes. In Barnes, Schofield, and Sorabji 1975, pp. 88–112. (Originally published as "La place de la logique dans la pensée aristotélicienne." *Revue de Metaphysique et de Morale* 66 [1951]: 283–315.)

Wieland, W. 1970. *Die aristotelische Physik*. 2d ed. Göttingen. (First edition published in 1962.)

Wilkins, B. T. 1970. "Aristotle on Scientific Explanation." *Dialogue* 9:337–55.

Zabarella, I. 1582. *In duos Aristotelis libros Posteriorum Analyticorum Commentarius*. Venice. (Also in Zabarella 1966.)

———. 1966. *Opera Logica*. Reprint. Hildesheim. (Originally published in Venice, 1586.)

Zeuthen, H. G. 1896. "Die geometrische Construction als 'Existenzbeweis' in der antiken Geometrie." *Mathematische Annalen* 47:222–28.

———. 1902. *Histoire des Mathématiques dans l'antiquité et le moyen age*. Translated by J. Muscart. Paris.

Ziegelmeyer, E. H. 1945. "The Discovery of the First Principles According to Aristotle." *Modern Schoolman* 22:132–43.

INDEX LOCORUM

Pages containing translations or principal discussions of passages from Aristotle and Euclid are in boldface type.

GENERAL INDEX

GPSR Authorized Representative: Easy Access System Europe - Mustamäe tee 50, 10621 Tallinn, Estonia, gpsr.requests@easproject.com

www.ingramcontent.com/pod-product-compliance
Ingram Content Group UK Ltd.
Pitfield, Milton Keynes, MK11 3LW, UK
UKHW052215060725
460447UK00006B/444

* 9 7 8 0 6 9 1 6 0 3 2 5 4 *